AF249023

Rajiv Dutta

Fundamentals of Biochemical Engineering

Rajiv Dutta

Fundamentals of
Biochemical Engineering

Ane Books India

 Springer

Author:
Professor Rajiv Dutta
Deputy Director & Head
Amity Institute of Biotechnology
Lucknow, India
E-mail: rajivd@lko.amity.edu

ISBN 978-81-8052-202-4 Ane Books India

ISBN 978-3-540-77900-1 Springer Berlin Heidelberg New York

Library of Congress Control Number: 2008920288

Springer is a part of Springer Science + Business Media
springer.com

Ane Books India
4821 Parwana Bhawan, 1st Floor
24 Ansari Road, Darya Ganj, New Delhi -110 002, India
Tel: +91 (011) 2327 6843-44, 2324 6385
Fax: +91 (011) 2327 6863
e-mail: kapoor@anebooks.com
Website: www.anebooks.com

Printed in India

Cover design :KünkelLopka GmbH, Heidelberg, Germany

Printed on 80 gsm, TA Maplitho.

In affectionate reverance
this book is dedicated
to my
Lord and Master

Ganesha

Preface

The biotechnology and chemical engineering students at various Indian Universities and engineering institutions are required to take up the 'Biochemical Engineering' course either as an elective or a compulsory subject. This book is written keeping in mind the need for a text book on the aforesaid subject for students from both engineering and biology backgrounds. The main feature of this book is that it contains solved problems, which help the students to understand the subject better.

The book is divided into three sections: Enzyme mediated bioprocess, whole cell mediated bioprocess and the engineering principle in bioprocess

In the first section of this book, the brief introduction about biochemical engineering is given in chapter 1. The second chapter deals with basics of enzyme reaction kinetics. The third chapter deals with an important aspect in enzyme bioprocess i.e. immobilization of enzyme and its kinetics. Chapter 4 is concerned about the industrial bioprocess involving starch and cellulose.

The second section of this book is related to whole cell mediated bioprocess. Chapter 5 introduces students with basics of microbiology and cell culture techniques for both plant and animal cells. The area emerged as most important because of latest development in pharmaceutical industry. Chapter 6 deals with cell kinetics and fermenter design. The kinetics analysis based on structure and unstructured model is also included. In chapter 7, genetic engineering aspects were explained, and genetic stability problems and important engineering aspects of genetically modified cells were also addressed.

The third section deals with engineering principles of bioprocess. The chapter 8 deals with sterilization process and its engineering considerations (up stream process). The ninth chapter includes agitation and aeration in cellular growth and its impact on designing the bioreactors. The last chapter deals with brief introduction to down stream processing.

I wish to thank to Neeta, my wife and Ayuushi and Anukritii, my daughters for their untiring support in realizing this mission. This book would not have existed without the untiring effort of Miss Pragati Sahay, Lecturer, AIB and my Ph.D. students, who did all the proof reading jobs. I will fail in my duties if I don't acknowledge professional studies (M.E. Biotechnology) students of BITS, Pilani who helped my in preparing internet notes while teaching this course which later took shape of this book. I wish to extend my sincere thanks to Mr. Aseem K. Chauhan, Additional President, RBEF, mother organization of Amity University Uttar Pradesh (AUUP) and Maj Gen KK Ohri, AVSM (Retd), Director General of AUUP-Lucknow Campus for this constant encouragements for the mission to complete. I am also thankful to all my colleagues at Amity Institute of Biotechnology (AIB), Lucknow helping me at different levels for the success of this mission.

Rajiv Dutta

Contents

1

Introduction

Biochemical engineering is concerned with conducting biological processes on an industrial scale. This area links biological sciences with chemical engineering. The role of biochemical engineers has become more important in recent years due to the dramatic developments of biotechnology.

1.1 BIOTECHNOLOGY

Biotechnology can be broadly defined as "Commercial techniques that use living organisms, or substances from those organisms, to make or modify a product, including techniques used for the improvement of the characteristics of economically important plants and animals and for the development of microorganisms to act on the environment ..." (Congress of the United States, 1984). If biotechnology is defined in this general sense, the area cannot be considered new. Since ancient days, people knew how to utilize microorganisms to ferment beverage and food, though they did not know what was responsible for those biological changes. People also knew how to crossbreed plants and animals for better yields. In recent years, the term *biotechnology* is being used to refer to novel techniques such as *recombinant DNA* and *cell fusion.*

Recombinant DNA allows the direct manipulation of genetic material of individual cells, which may be used to develop microorganisms that produce new products as well as useful organisms. The laboratory technology for the genetic manipulation within living cells is also known as *genetic engineering.* A major objective of this technique is to splice a foreign gene for a desired product into circular forms of DNA (plasmids), and then to insert them into an organism, so that the foreign gene can be expressed to produce the product from the organism.

Cell fusion is a process to form a single hybrid cell with nuclei and cytoplasm from two different types of cells in order to combine the desirable characteristics of the two. As an example, specialized cells of the immune system can produce useful antibodies. However, it is difficult to cultivate those cells because their growth rate is very slow.

On the other hand, certain tumor cells have the traits for immortality and rapid proliferation. By combining the two cells by fusion, a hybridoma can be created that has both traits. The monoclonal antibodies (MAbs) produced from the hybridoma cells can be used for diagnosis, disease treatment, and protein purification.

The applications of this new biotechnology are numerous, as listed in Table 1.1. Previously expensive and rare pharmaceuticals such as insulin for diabetics, human growth hormone to treat children with dwarfism, interferon to fight infection, vaccines to prevent diseases, and monoclonal antibody for diagnostics can be produced from genetically modified cells or hybridoma cells inexpensively and also in large quantities. Disease-free seed stocks or healthier, higher-yielding food animals can be developed. Important crop species can be modified to have traits that can resist stress, herbicide, and pest. Furthermore, recombinant DNA technology can be applied to develop genetically modified microorganisms so that they can produce various chemical compounds with higher yields than unmodified microorganisms can.

1.2 BIOCHEMICAL ENGINEERING

The recombinant DNA or cell fusion technologies have been initiated and developed by pure scientists, whose end results can be the development of a new breed of cells in minute quantities that can produce a product. Successful commercialization of this process requires the development of a large-scale process that is technologically viable and economically efficient. To scale up a laboratory-scale operation into a large industrial process, we cannot just make the vessel bigger. For example, in a laboratory scale of 100 mL, a small Erlenmeyer flask on a shaker can be an excellent way to cultivate cells, but for a large-scale operation of 2,000 L, we cannot make the vessel bigger and shake it. We need to design an effective bioreactor to cultivate the cells in the most optimum conditions. Therefore, biochemical engineering is one of the major areas in biotechnology important to its commercialization.

To illustrate the role of a biochemical engineer, let's look at a typical biological process (bioprocess) involving microbial cells as shown in Figure 1.1. Raw materials, usually biomass, are treated and mixed with other ingredients that are required for cells to grow well. The liquid mixture, the medium, is sterilized to eliminate all other living microorganisms and introduced to a large cylindrical vessel, bioreactor or fermenter, typically equipped with agitators, baffles, air spargers, and various sensing devices for the control of fermentation conditions. A pure strain of microorganisms is introduced into the vessel. The number of cells will start to multiply exponentially after a certain period of lag time and reach a maximum cell concentration as the medium is depleted. The fermentation will be stopped and the contents will be pumped out for the product recovery and purification. This process can be operated either by batch or continuously.

Table 1.1 Applications of Biotechnology

Area	*Products or Applications*
Pharmaceuticals	Antibiotics, antigens (stimulate antibody response), endorphin (neurotransmitter), gamma globulin (prevent infections), human growth hormone (treat children with dwarfism), human serum albumin (treat physical trauma), immune regulators, insulin, interferon (treat infection), interleukins (treat infectious decease or cancer), lymphokines (modulate immune reaction), monoclonal antibody (diagnostics or drug delivery), neuroactive peptides (mimic the body's pain-controlling peptides), tissue plasminogen activator (dissolve blood clots), vaccines
Animal Agriculture	Development of disease-free seed stocks healthier, higher-yielding food animals.
Plant Agriculture	transfer of stress-, herbicide-, or pest-resistance traits to crop species, development of plants with the increased abilities of photosynthesis or nitrogen fixation, development of biological insecticides and non-ice nucleating bacterium.
Specialty Chemicals	amino acids, enzymes, vitamins, lipids, hydroxylated aromatics, biopolymers.
Environmental Applications	mineral leaching, metal concentration, pollution control, toxic waste degradation, and enhanced oil recovery.
Commodity Chemicals	acetic acid, acetone, butanol, ethanol, many other products from biomass conversion processes.
Bioelectronics	Biosensors, biochips.

To carry out a bioprocess on a large scale, biochemical engineers need to work together with biological scientists:

1. to obtain the best biological catalyst (microorganism, animal cell, plant cell, or enzyme) for a desired process
2. to create the best possible environment for the catalyst to perform by designing the bioreactor and operating it in the most efficient way

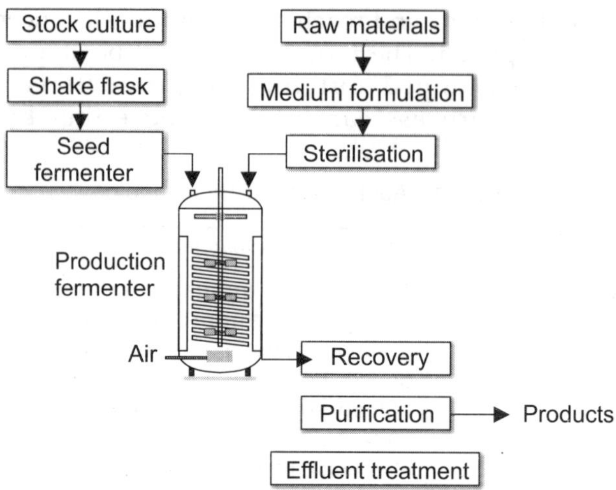

Fig. 1.1 Typical biological process

3. to separate the desired products from the reaction mixture in the most economical way

The preceding tasks involve process design and development, which are familiar to chemical engineers for the chemical processes. Similar techniques which have been working successfully in chemical processes can be employed with modifications. The basic questions which need to be asked for the process development and design are as follows:

1. *What change can be expected to occur?*
   To answer this question, one must have an understanding of the basic sciences for the process involved. These are microbiology, biochemistry, molecular biology, genetics, and so on. Biochemical engineers need to study these areas to a certain extent. It is also true that the contribution of biochemical engineers in selecting and developing the best biological catalyst is quite limited unless the engineer receives specialized training. However, it is important for biochemical engineers to get involved in this stage, so that the biological catalyst may be selected or genetically modified with a consideration of the large-scale operation.

2. *How fast will the process take place?*
   If a certain process can produce a product, it is important to know how fast the process can take place. Kinetics deals with rate of a reaction and how it is affected by various chemical and physical conditions. This is where the expertise of chemical engineers familiar with chemical kinetics and reactor design plays a major role. Similar techniques can be employed to deal with enzyme or cell kinetics. To design an effective bioreactor

for the biological catalyst to perform, it is also important to know how the rate of the reaction is influenced by various operating conditions. This involves the study of thermodynamics, transport phenomena, biological interactions, clonal stability, and so on.

3. *How can the system be operated and controlled for the maximum yield?*

   For the optimum operation and control, reliable on-line sensing devices need to be developed. On-line optimization algorithms need to be developed and used to enhance the operability of bioprocess and to ensure that these processes are operated at the most economical points.

4. *How can the products be separated with maximum purity and minimum costs?*

   For this step, the downstream processing (or bioseparation), a biochemical engineer can utilize various separation techniques developed in chemical processes such as distillation, absorption, extraction, adsorption, drying, filtration, precipitation, and leaching. In addition to these standard separation techniques, the biochemical engineer needs to develop novel techniques which are suitable to separate the biological materials. Many techniques have been developed to separate or to analyze biological materials on a small laboratory scale, such as chromatography, electrophoresis, and dialysis. These techniques need to be further developed so that they may be operated on a large industrial scale.

1.3 BIOLOGICAL PROCESS

Industrial applications of biological processes are to use living cells or their components to effect desired physical or chemical changes. Biological processes have advantages and disadvantages over traditional chemical processes. The major advantages are as follows:

1. *Mild reaction condition:* The reaction conditions for bioprocesses are mild. The typical condition is at room temperature, atmospheric pressure, and fairly neutral medium pH. As a result, the operation is less hazardous, and the manufacturing facilities are less complex compared to typical chemical processes.

2. *Specificity:* An enzyme catalyst is highly specific and catalyzes only one or a small number of chemical reactions. A great variety of enzymes exist that can catalyze a very wide range of reactions.

3. *Effectiveness:* The rate of an enzyme-catalyzed reaction is usually much faster than that of the same reaction when directed by nonbiological catalysts. A small amount of enzyme is required to produce the desired effect.
4. *Renewable resources:* The major raw material for bioprocesses is biomass which provides both the carbon skeletons and the energy required for synthesis for organic chemical manufacture.
5. *Recombinant DNA technology:* The development of the recombinant DNA technology promises enormous possibilities to improve biological processes.

However, biological processes have the following disadvantages:

1. *Complex product mixtures:* In cases of cell cultivation (microbial, animal, or plant), multiple enzyme reactions are occurring in sequence or in parallel, the final product mixture contains cell mass, many metabolic by-products, and a remnant of the original nutrients. The cell mass also contains various cell components.
2. *Dilute aqueous environments:* The components of commercial interests are only produced in small amounts in an aqueous medium. Therefore, separation is very expensive. Since products of bioprocesses are frequently heat sensitive, traditional separation techniques cannot be employed. Therefore, novel separation techniques that have been developed for analytical purposes, need to be scaled up.
3. *Contamination:* The fermenter system can be easily contaminated, since many environmental bacteria and molds grow well in most media. The problem becomes more difficult with the cultivation of plant or animal cells because their growth rates are much slower than those of environmental bacteria or molds.
4. *Variability:* Cells tend to mutate due to the changing environment and may lose some characteristics vital for the success of process. Enzymes are comparatively sensitive or unstable molecules and require care in their use.

1.4 DEFINITION OF FERMENTATION

Traditionally, fermentation was defined as the process for the production of alcohol or lactic acid from glucose ($C_6H_{12}O_6$).

$$C_6H_{12}O_6 \xrightarrow{\text{Yeast}} 2C_2H_5OH + 2CO_2$$

$$C_6H_{12}O_6 \xrightarrow{\text{enzymes}} 2CH_3CHOHCOOH$$

A broader definition of fermentation is "an enzymatically controlled transformation of an organic compound" according to *Webster's New College Dictionary* (A Merriam-Webster, 1977) that we adopt in this text.

1.5 PROBLEMS

1.1 Read any one article as a general introduction to biotechnology. Bring a copy of the article and be ready to discuss or explain it during class.

1.6 REFERENCES

Congress of the United States, *Commercial Biotechnology: An International Analysis*, p. 589. Washington, DC: Office of Technology Assessment, 1984.

Journals covering general areas of biotechnology and bioprocesses:

Applied & Environmental Microbiology

Applied Microbiology and Biotechnology

Biotechnology and Bioengineering

CRC Critical Review in Biotechnology

Developments in Industrial Microbiology

Enzyme andMicrobial Technology

Journal of Applied Chemistry & Biotechnology

Journal of Chemical Technology & Biotechnology

Nature

Nature Biotechnology

Science

Scientific American

2

Enzyme Kinetics

2.1 INTRODUCTION

Enzymes are biological catalysts that are protein molecules in nature. They are produced by living cells (animal, plant, and microorganism) and are absolutely essential as catalysts in biochemical reactions. Almost every reaction in a cell requires the presence of a specific enzyme. A major function of enzymes in a living system is to catalyze the making and breaking of chemical bonds. Therefore, like any other catalysts, they increase the rate of reaction without themselves undergoing permanent chemical changes.

The catalytic ability of enzymes is due to its particular protein structure. A specific chemical reaction is catalyzed at a small portion of the surface of an enzyme, which is known as the active site. Some physical and chemical interactions occur at this site to catalyze a certain chemical reaction for a certain enzyme.

Enzyme reactions are different from chemical reactions, as follows:

1. An enzyme catalyst is highly specific, and catalyzes only one or a small number of chemical reactions. A great variety of enzymes exist, which can catalyze a very wide range of reactions.

2. The rate of an enzyme-catalyzed reaction is usually much faster than that of the same reaction when directed by nonbiological catalysts. Only a small amount of enzyme is required to produce a desired effect.

3. The reaction conditions (temperature, pressure, pH, and so on) for the enzyme reactions are very mild.

4. Enzymes are comparatively sensitive or unstable molecules and require care in their use.

2.1.1 Nomenclature of Enzymes

Originally enzymes were given nondescriptive names such as:

rennin   curding of milk to start cheese-making processcr
pepsin   hydrolyzes proteins at acidic pH
trypsin hydrolyzes proteins at mild alkaline pH

The nomenclature was later improved by adding the suffix *-ase* to the name of the substrate with which the enzyme functions, or to the reaction that is catalyzed.myfootnote[1] For example:

Name of substrate + *ase*

α-amylase	starch	$\longrightarrow$	glucose + maltose + oligosaccharides
lactase	lactose	$\longrightarrow$	glucose + galactose
lipase	fat	$\longrightarrow$	fatty acids + glycerol
maltase	maltose	$\longrightarrow$	glucose
urease	urea + H_2O $\longrightarrow$		$2NH_3 + CO_2$
cellobiase	cellobiose	$\longrightarrow$	glucose

Reaction which is catalyzed + *ase*

alcohol dehydrogenase ethanol + NAD^+ $\rightleftharpoons$ acetaldehyde + $NADH^2$

glucose isomerase   glucose $\rightleftharpoons$ fructose

glucose oxidase   D-glucose + O_2 + H_2O $\longrightarrow$ gluconic acid

lactic acid dehydrogenase lactic acid $\longrightarrow$ pyruvic acid

As more enzymes were discovered, this system generated confusion and resulted in the formation of a new systematic scheme by the International Enzyme Commission in 1964. The new system categorizes all enzymes into six major classes depending on the general type of chemical reaction which they catalyze. Each main class contains subclasses, subsubclasses, and subsubsubclasses. Therefore, each enzyme can be designated by a numerical code system. As an example, alcohol dehydrogenase is assigned as 1.1.1.1, as illustrated in Table 2.1. (Bohinski, 1970).

2.1.2 Commercial Applications of Enzymes

Enzymes have been used since early human history without knowledge of what they were or how they worked. They were used for such things as making sweets from starch, clotting milk to make cheese, and brewing soy sauce. Enzymes have been utilized commercially since the 1890s, when fungal cell extracts were first added to brewing vats to facilitate the breakdown of starch into sugars (Eveleigh, 1981). The fungal amylase takadiastase was employed as a digestive aid in the United States as early as 1894.

---

[1] The term *substrate* in biological reaction is equivalent to the term *reactant* in chemical reaction.

Table 2.1 Partial Outline of the Systematic Classification of Enzymes

---

1. Oxidoreductases

    1.1. Acting on = CH−OH group of substrates

    1.1.1. Requires NAD+ or NADP+ as hydrogen acceptor

    1.1.1.1. Specific substrate is ethyl alcohol

2. Transferases

    2.1. Transfer of methyl groups 2.2 . Transfer of glycosyl groups

3. Hydrolases

4. Lyases

5. Isomerases

6. Ligases

Example

Reaction: $CH_2CH_2OH + NAD \longrightarrow CH_3CHO + NADH + H^+$

Systematic Name: alcohol NAD oxidoreductase (1.1.1.1.)

Trivial Name: alcohol dehydrogenase

Because an enzyme is a protein whose function depends on the precise sequence of amino acids and the protein's complicated tertiary structure, large-scale chemical synthesis of enzymes is impractical if not impossible. Enzymes are usually made by microorganisms grown in a pure culture or obtained directly from plants and animals. The enzymes produced commercially can be classified into three major categories (Crueger and Crueger, 1984):

1. *Industrial enzymes,* such as amylases, proteases, glucose isomerase, lipase, catalases, and penicillin acylases

2. *Analytical enzymes,* such as glucose oxidase, galactose oxidase, alcohol dehydrogenase, hexokinase, muramidase, and cholesterol oxidase

3. *Medical enzymes,* such as asparaginase, proteases, lipases, and streptokinase

α-amylase, glucoamylase, and glucose isomerase serve mainly to convert starch into high-fructose corn syrup (HFCS), as follows:

$$\text{Corn starch} \xrightarrow{\text{α-amylase}} \text{Thinned starch} \xrightarrow{\text{Gluco-amylase}} \text{Glucose} \xrightarrow{\text{Glucose isomerase}} \text{HFCS}$$

HFCS is sweeter than glucose and can be used in place of table sugar (sucrose) in soft drinks.

Alkaline protease is added to laundry detergents as a cleaning aid, and widely used in Western Europe. Proteins often precipitate on soiled clothes or make dirt adhere to the textile fibers. Such stains can be dissolved easily by addition of protease to the detergent. Protease is also used for meat tenderizer and cheese making.

The scale of application of analytical and medical enzymes is in the range of milligrams to grams while that of industrial enzymes is in tons. Analytical and medical enzymes are usually required to be in their pure forms; therefore, their production costs are high.

2.2 SIMPLE ENZYME KINETICS

Enzyme kinetics deals with the rate of enzyme reaction and how it is affected by various chemical and physical conditions. Kinetic studies of enzymatic reactions provide information about the basic mechanism of the enzyme reaction and other parameters that characterize the properties of the enzyme. The rate equations developed from the kinetic studies can be applied in calculating reaction time, yields, and optimum economic condition, which are important in the design of an effective bioreactor.

Assume that a substrate (S) is converted to a product (P) with the help of an enzyme (E) in a reactor as

$$S \xrightarrow{\text{E}} P \tag{2.1}$$

If you measure the concentrations of substrate and product with respect to time, the product concentration will increase and reach a maximum value, whereas the substrate concentration will decrease as shown in Figure 2.1

The rate of reaction can be expressed in terms of either the change of the substrate C_S or the product concentrations C_P as follows:

$$r_S = -\frac{dC_S}{dt} \tag{2.2}$$

$$r_S = \frac{dC_P}{dt} \tag{2.3}$$

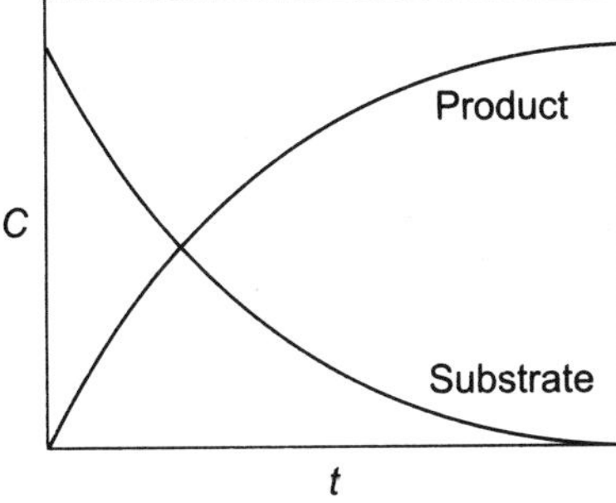

Fig. 2.1   The change of product and substrate
concentrations with respect to time.

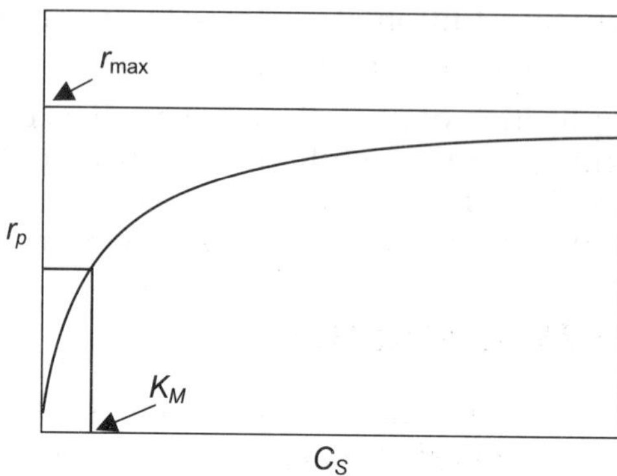

Fig. 2.2 The effect of substrate concentration on the initial reaction rate.

In order to understand the effectiveness and characteristics of an enzyme reaction, it is important to know how the reaction rate is influenced by reaction conditions such as substrate, product, and enzyme concentrations. If we measure the initial reaction rate at different levels of substrate and enzyme concentrations, we obtain a series of curves like the one shown in Figure 2.2. From these curves we can conclude the following:

1. The reaction rate is proportional to the substrate concentration (that is, first-order reaction) when the substrate concentration is in the low range.

2. The reaction rate does not depend on the substrate concentration when the substrate concentration is high, since the reaction rate changes gradually from first order to zero order as the substrate concentration is increased.

3. The maximum reaction rate r_{max} is proportional to the enzyme concentration within the range of the enzyme tested.

Henri observed this behavior in 1902 (Bailey and Ollis, p. 100, 1986) and proposed the rate equation

$$r = \frac{r_{max}\, C_S}{K_M + C_S} \tag{2.4}$$

where r_{max} and K_M are kinetic parameters which need to be experimentally determined. Eq. (2.4) expresses the three preceding observations fairly well. The rate is proportional to C_S (first order) for low values of C_S, but with higher values of C_S, the rate becomes constant (zero order) and equal to r_{max}. Since Eq. (2.4) describes the experimental results well, we need to find the kinetic mechanisms which support this equation.

Brown (1902) proposed that an enzyme forms a complex with its substrate. The complex then breaks down to the products and regenerates the free enzyme. The mechanism of one substrate-enzyme reaction can be expressed as

$$S + E \underset{k_2}{\overset{k_1}{\rightleftharpoons}} ES \qquad (2.5)$$

$$ES \xrightarrow{k_3} P + E \qquad (2.6)$$

Brown's kinetic inference of the existence of the enzyme-substrate complex was made long before the chemical nature of enzymes was known, 40 years before the spectrophotometric detection of such complexes.

One of the original theories to account for the formation of the enzyme-substrate complex is the "lock and key" theory. The main concept of this hypothesis is that there is a topographical, structural compatibility between an enzyme and a substrate which optimally favors the recognition of the substrate as shown in Figure 2.3.

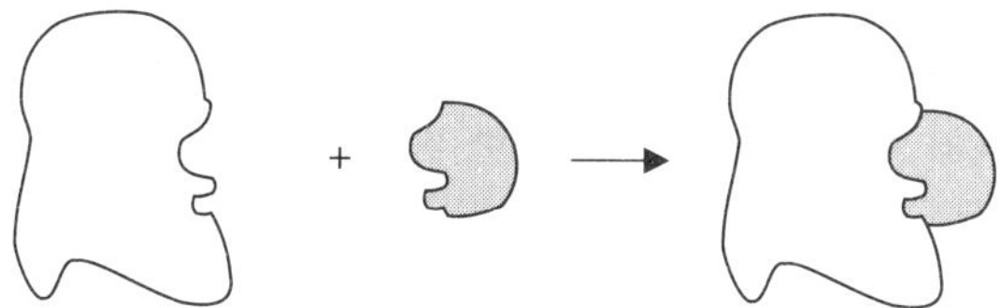

Fig. 2.3 Lock and key theory for the enzyme-substrate complex.

The reaction rate equation can be derived from the preceding mechanism based on the following assumptions:

1. The total enzyme concentration stays constant during the reaction, that is, $C_{E_0} = C_{ES} + C_E$
2. The amount of an enzyme is very small compared to the amount of substrate.[2] Therefore, the formation of the enzyme-substrate complex does not significantly deplete the substrate.
3. The product concentration is so low that product inhibition may be considered negligible.

In addition to the preceding assumptions, there are three different approaches to derive the rate equation:

1. **Michaelis-Menten approach** (Michaelis and Menten, 1913): It is assumed that the product-releasing step, Eq. (2.6), is much slower than the reversible reaction, Eq. (2.5), and the slow step determines the rate, while the other is at equilibrium. This is an assumption which is often employed in heterogeneous catalytic reactions in chemical kinetics.[3] Even though the enzyme is

---

[2] This is a reasonable assumption because enzymes are very efficient. Practically, it is also our best interests to use as little enzymes as possible because of their costs.

[3] For heterogeneous catalytic reactions, the first step is the adsorption of reactants on the surface of a catalyst and the second step is the chemical reaction between the reactants to produce products. Since the first step involves only weak physical or chemical interaction, its speed is much quicker than that of the second step, which requires complicated chemical interaction. This phenomena is fairly analogous to enzyme reactions.

soluble in water, the enzyme molecules have large and complicated three-dimensional structures. Therefore, enzymes can be analogous to solid catalysts in chemical reactions. Furthermore, the first step for an enzyme reaction also involves the formation of an enzyme-substrate complex, which is based on a very weak interaction. Therefore, it is reasonable to assume that the enzyme-substrate complex formation step is much faster than the product releasing step which involves chemical changes.

2. **Briggs-Haldane approach** (Briggs and Haldane, 1925): The change of the intermediate concentration with respect to time is assumed to be negligible, that is, $d(C_{ES})/dt = 0$. This is also known as the *pseudo-steady-state* (or *quasi-steady-state)* assumption in chemical kinetics and is often used in developing rate expressions in homogeneous catalytic reactions.

3. **Numerical solution:** Solution of the simultaneous differential equations developed from Eqs. (2.5) and (2.6) without simplification.

2.2.1   Michaelis-Menten Approach

If the slower reaction, Eq. (2.6), determines the overall rate of reaction, the rate of product formation and substrate consumption is proportional to the concentration of the enzyme-substrate complex as:[4]

$$r = \frac{dC_P}{dt} = \frac{dC_S}{dt} = k_3 C_{ES} \tag{2.7}$$

Unless otherwise specified, the concentration is expressed as molar unit, such as $kmol/m^3$ or mol/L. The concentration of the enzyme-substrate complex C_{ES} in Eq. (2.7), can be related to the substrate concentration C_S and the free-enzyme concentration C_E from the assumption that the first reversible reaction Eq. (2.5) is in

---

[4] It seems that the rate of substrate consumption should be expressed as

$$-\frac{dC_S}{dt} = -k_1 C_S C_E - k_2 C_{ES}$$

Then, it gives a contradictory result that the substrate concentration stays constant $(dC_S/dt = 0)$ because the first reversible reaction, Eq. (2.5), is assumed to be in equilibrium $(k_1 C_S C_E - k_2 C_E S) = 0)$. Since the rate of reaction is determined by the second slower reaction, Eq. (2.6), the preceding expression is wrong. Instead, the rate of substrate consumption must also be written by the second reaction as

$$-\frac{dC_S}{dt} = k_3 C_{ES}$$

For the Briggs-Haldane approach, the rate expression for substrate can be expressed by the first reversible reaction as explained in the next section.

equilibrium. Then, the forward reaction is equal to the reverse reaction so that

$$k_1 C_S C_E = k_2 C_{ES} \tag{2.8}$$

By substituting Eq. (2.8) into Eq. (2.7), the rate of reaction can be expressed as a function of C_S and C_E, of which C_E cannot be easily determined. If we assume that the total enzyme contents are conserved, the free-enzyme concentration C_E can be related to the initial enzyme concentration C_{E_0}

$$C_{E_0} = C_E + C_{ES} \tag{2.9}$$

So, now we have three equations from which we can eliminate C_E and C_{ES} to express the rate expression as the function of substrate concentration and the initial enzyme concentration. By substituting Eq. (2.8) into Eq. (2.9) for C_E and rearranging for C_{ES}, we obtain

$$C_{ES} = \frac{C_{E_0} C_S}{\dfrac{k_2}{k_1} + C_S} \tag{2.10}$$

Substitution of Eq. (2.10) into Eq. (2.7) results in the final rate equation

$$r = \frac{dC_P}{dt} = -\frac{dC_S}{dt} = \frac{k_3 C_{E_0} C_S}{\dfrac{k_2}{k_1} + C_S} = \frac{r_{max} C_S}{K_M + C_S} \tag{2.11}$$

which is known as Michaelis-Menten equation and is identical to the empirical expression Eq. (2.4). K_M in Eq. (2.11) is known as the Michaelis constant. In the Michaelis-Menten approach, K_M is equal to the dissociation constant K_1 or the reciprocal of equilibrium constant K_{eq} as

$$K_M = \frac{k_2}{k_1} = k_1 = \frac{C_S C_E}{C_{ES}} = \frac{1}{K_{eq}} \tag{2.12}$$

The unit of K_M is the same as C_S. When K_M is equal to C_S, r is equal to one half of r_{max} according to Eq. (2.11). Therefore, the value of K_M is equal to the substrate concentration when the reaction rate is half of the maximum rate r_{max} (see Figure 2.2). K_M is an important kinetic parameter because it characterizes the interaction of an enzyme with a given substrate.

Another kinetic parameter in Eq. (2.11) is the maximum reaction rate r_{max}, which is proportional to the initial enzyme concentration. The main reason for combining two constants k_3 and C_{E_0} into one lumped parameter r_{max} is due to the difficulty of expressing the enzyme concentration in molar unit. To express the enzyme concentration in molar unit, we need to know the molecular weight of enzyme and the exact amount of pure enzyme added, both of which are very difficult to determine. Since we often use enzymes which are not in pure form, the actual amount of enzyme is not known.

Enzyme concentration may be expressed in mass unit instead of molar unit. However, the amount of enzyme is not well quantified in mass unit because actual contents of an enzyme can differ widely depending on its purity. Therefore, it is common to express enzyme concentration as an arbitrarily defined unit based on its catalytic ability. For example, one unit of an enzyme, cellobiose, can be defined as the amount of enzyme required to hydrolyze cellobiose to produce 1 μmol of glucose per minute. Whatever unit is adopted for C_{E_0}, the unit for $k_3 C_{E_0}$ should be the same as r, that is, kmole/m^{3}s. Care should be taken for the consistency of unit when enzyme concentration is not expressed in molar unit.

The Michaelis-Menten equation is analogous to the Langmuir isotherm equation

$$\theta = \frac{C_A}{K + C_A} \tag{2.13}$$

where θ is the fraction of the solid surface covered by gas molecules and K is the reciprocal of the adsorption equilibrium constant.

2.2.2   Briggs-Haldane Approach

Again, from the mechanism described by Eqs. (2.5) Eq. (2.6), the rates of product formation and of substrate consumption are

$$\frac{dC_P}{dt} = k_3 C_{ES} \tag{214}$$

$$-\frac{dC_S}{dt} = k_1 C_S C_E - k_2 C_{ES} \tag{2.15}$$

Assume that the change of C_{ES} with time, dC_{ES}/dt, is negligible compared to that of C_P or C_S.

$$\frac{dC_{ES}}{dt} = k_1 C_S C_E - k_2 C_{ES} - k_3 C_{ES} = 0 \tag{2.16}$$

Substitution of Eq. (2.16) into Eq. (2.15) confirms that the rate of product formation and that of the substrate consumption are the same, that is,

$$r = \frac{dC_P}{dt} = -\frac{dC_S}{dt} = k_3 C_{ES} \tag{2.7}$$

Again, if we assume that the total enzyme contents are conserved,

$$C_{E_0} = C_E + C_{ES} \tag{2.9}$$

Substituting Eq. (2.9) into Eq. (2.16) for C_E, and rearranging for C_{ES}

$$C_{ES} = \frac{C_{E_0} C_S}{\dfrac{k_2 + k_3}{k_1} + C_S} \tag{2.17}$$

Substitution of Eq. (2.17) into Eq. (2.14) results

$$r = \frac{dC_P}{dt} = -\frac{dC_S}{dt} = \frac{k_3 C_{E_0} C_S}{\dfrac{k_2 + k_3}{k_1} + C_S} = \frac{r_{max} C_S}{K_M + C_S} \tag{2.18}$$

which is the same as the Michaelis-Menten equation, Eq. (2.11), except that the meaning of K_M is different. In the Michaelis-Menten approach, K_M is equal to the dissociation constant k_2/k_1, while in the Bnggs-Haldane approach, it is equal to $(k_2 + k_3)/k_1$. Eq. (2.18) can be simplified to Eq. (2.11) if $k_2 \gg k_3$ which means that the product-releasing step is much slower than the enzyme-substrate complex dissociation step. This is true with many enzyme reactions. Since the formation of the complex involves only weak interactions, it is likely that the rate of dissociation of the complex will be rapid. The breakdown of the complex to yield products will involve the making and breaking of chemical bonds, which is much slower than the enzyme-substrate complex dissociation step.

Example 2.1

When glucose is converted to fructose by glucose isomerase, the slow product formation step is also reversible as:

$$S + E \underset{k_2}{\overset{k_1}{\rightleftharpoons}} ES$$

$$ES \underset{k_4}{\overset{k_3}{\rightleftharpoons}} P + E$$

Derive the rate equation by employing (a) the Michaelis-Menten and (b) the Briggs-Haldane approach. Explain when the rate equation derived by the Briggs-Haldane approach can be simplified to that derived by the Michaelis-Menten approach.

Solution:

(a) Michaelis-Menten approach: The rate of product formation is

$$r_P = k_3 C_{ES} - k_4 C_P C_E \tag{2.19}$$

Since enzyme is preserved,

$$C_{E_0} = C_E + C_{ES} \tag{2.20}$$

Substitution of Eq. (2.20) into Eq. (2.19) for C_E yields

$$R_p = (k_3 + k_4 C_P)C_{ES} - k_4 C_P C_{E0} \tag{221}$$

Assuming the first reversible reaction is in equilibrium gives

$$C_{ES} = \frac{k_1}{k_2} C_E C_S \tag{2.22}$$

Substituting Eq. (2.22) into Eq. (2.20) for C_E and rearranging for C_{ES} yields

$$C_{ES} = \frac{C_{E_0} C_S}{\dfrac{k_2}{k_1} + C_S} \tag{223}$$

Substituting Eq. (2.23) into Eq. (2.8) gives

$$r_P = \frac{k_3 C_{E_0} \left(C_S - \dfrac{k_4}{k_3} \dfrac{k_2}{k_1} C_P \right)}{\dfrac{k_2}{k_1} + C_S} \tag{2.24}$$

(b) Briggs-Haldane approach: Assume that the change of the complex concentration with time, dC_{ES}/dt, is negligible. Then,

$$\frac{dC_{ES}}{dt} = k_1 C_S C_E - k_2 C_{ES} - k_3 C_{ES} + k_4 C_P C_E \cong 0 \tag{2.25}$$

Substituting Eq. (2.20) into Eq. (2.25) for C_E and rearranging gives

$$C_{ES} = \frac{C_{E_0}(k_1 C_S + k_4 C_P}{(k_2 + k_3) + k_1 C_S + k_4 C_P} \tag{2.26}$$

Inserting Eq. (2.26) into Eq. (2.19) for C_{ES} gives

$$r_P = \frac{k_3 C_{E_0} \left(C_S - \dfrac{k_4}{k_3} \dfrac{k_2}{k_1} C_P \right)}{\dfrac{k_2 + k_3}{k_1} + C_S + \dfrac{k_4}{k_1} C_P} \tag{2.27}$$

If the first step of the reaction, the complex formation step, is much faster than the second, the product formation step, k_1 and k_2 will be much larger than k_3 and k_4. Therefore, in Eq. (2.27),

$$\frac{k_2 + k_3}{k_1} \cong \frac{k_2}{k_1} \qquad (2.28)$$

and
$$\frac{k_4}{k_1} \cong 0 \qquad (2.29)$$

which simplifies Eq. (2.27) into Eq. (2.24).

2.2.3 Numerical Solution

From the mechanism described by Eqs. (2.5) and (2.6), three rate equations can be written for C_P C_{ES}, and C_S as

$$\frac{dC_P}{dt} = k_3 C_{ES} \qquad (2.14)$$

$$\frac{dC_{ES}}{dt} = k_1 C_S C_E - k_2 C_{ES} - k_3 C_{ES} \qquad (2.30)$$

$$\frac{dC_S}{dt} = -k_1 C_S C_E + k_2 C_{ES} \qquad (2.31)$$

Eqs. (2.14), (2.30), and (2.31) with Eq. (2.9) can be solved simultaneously without simplification. Since the analytical solution of the preceding simultaneous differential equations are not possible, we need to solve them numerically by using a computer. Among many software packages that solve simultaneous differential equations, Advanced Continuous Simulation Language (ACSL, 1975) is very powerful and easy to use.

The heart of ACSL is the integration operator, INTEG, that is,

$$R = \text{INTEG } (X, R0)$$

implies

$$R = R_0 + \int_0^t X dt$$

Original set of differential equations are converted to a set of first-order equations, and solved directly by integrating. For example, Eq. (2.14) can be solved by integrating as

$$C_P = C_{P_0} + \int_0^t k_3 C_{ES} dt$$

which can be written in ACSL as

$$C_P - \text{INTEG } (K3{*}CES, CP0)$$

For more details of this simulation language, please refer to the *ACSL User Guide (ACSL, 1975)*.

You can also use Mathematica (Wolfram Research, Inc., Champaign, IL) or MathCad (MathSoft, Inc., Cambridge, MA), to solve the above problem, though they are not as powerful as ACSL.

It should be noted that this solution procedure requires the knowledge of elementary rate constants, k_1, k_2, and k_3. The elementary rate constants can be measured by the experimental techniques such as pre-steady-state kinetics and relaxation methods (Bailey and Ollis, pp. 111 – 113, 1986), which are much more complicated compared to the methods to determine K_M and r_{max}. Furthermore, the initial molar concentration of an enzyme should be known, which is also difficult to measure as explained earlier. However, a numerical solution with the elementary rate constants can provide a more precise picture of what is occurring during the enzyme reaction, as illustrated in the following example problem.

Example 2.2

By employing the computer method, show how the concentrations of substrate, product, and enzyme-substrate complex change with respect to time in a batch reactor for the enzyme reactions described by Eqs. (2.5) and (2.6). The initial substrate and enzyme concentrations are 0.1 and 0.01 mol/L, respectively. The values of the reaction constants are: $k_1 = 40$ L/mols, $k_2 = 5$ s^{-1}, and $k_3 = 0.5$ s^{-1}.

Table 2.2   ACSL Program for Example 2.2

```
PROGRAM ENZY-EX2 ACSL
   INITIAL
      ALGORITHM IALG=5 $ 'RUNGE-KUTTA FOURTH ORDER'
      CONSTANT K1=40., K2=5., K3=0.5, CEO=0.01, ...
           CSO=0.1, CPO=0.0, TSTOP=130
      CINTERVAL CINT=0.2 $ 'COMMUNICAITON INTERVAL'
      NSTEPS NSTP=10
      VARIABLE TIME=0.0
   END           $ 'END OF INITIAL'
   DYNAMIC
      DERIVATIVE
         DCSDT=-K1*CS*CE+K2*CES
         CS=INTEG(DCSDT,CSO)
         DCESDT=K1*CS*CE-K2*CES-K3*CES
         CES=INTEG(DCESDT,0.0)
         CE=CEO-CES
         DCPDT=K3*CES
         CP=INTEG(DCPDT,CPO)
      END       $ 'END OF DERIVATIVE SECTION'
      TERMT(TIME.GE.TSTOP)
   END           $ 'END OF DYNAMIC SECTION'
   END           $ 'END OF PROGRAM'
```

Table 2.3 Executive Command Program for Example 2.2

```
SET TITLE = 'SOLUTION OF EXAMPLE 2.2'
SET PRN=9
OUTPUT TIME,CS,CP,CES,'NCIOUT'=50 $'DEFINE LIST TO BE PRINTED'
PREPAR TIME,CS,CP,CES $'DEFINE LIST TO BE SAVED'
START
SET NPXPPL=50, NPYPPL=60
PLOT 'XAXIS'=TIME, 'XLO'=0, 'XHI'=130, CS, CP, CES, 'LO'=0, 'HI'=0.1
STOP
ORIGIN:=1          Default origin is 0.
```

$kl := 40 \qquad k2 := 5 \qquad k3 := 0.5$

$$D(t,C) := \begin{bmatrix} -kl \cdot C_1 \cdot (0.01 - C_2) + k2 \cdot C_2 \\ kl \cdot C_r (0.01 - C_2) - k2 \cdot C_2 - k3 \cdot C_2 \\ k3 \cdot C_2 \end{bmatrix} \qquad \text{Note:} \quad \begin{aligned} C1 &= Cs \\ C2 &= Ces \\ C3 &= Cp \end{aligned}$$

$t0 := 0 \quad t1 := 130$ Initial and terminal values of independent variable

$$C0 := \begin{pmatrix} 0.1 \\ 0 \\ 0 \end{pmatrix}$$

Vector of initial function values

$N := 13$ Number of solution values on $[t0, t1]$

$S := Rkadapt(C0,t0,t1,N,D)$ Solve using adaptive Runge-Kutta method

$t := S^{\langle 1 \rangle} \qquad Cs := S^{\langle 2 \rangle} \qquad Ces := S^{\langle 3 \rangle} \qquad Cp := S^{\langle 4 \rangle}$

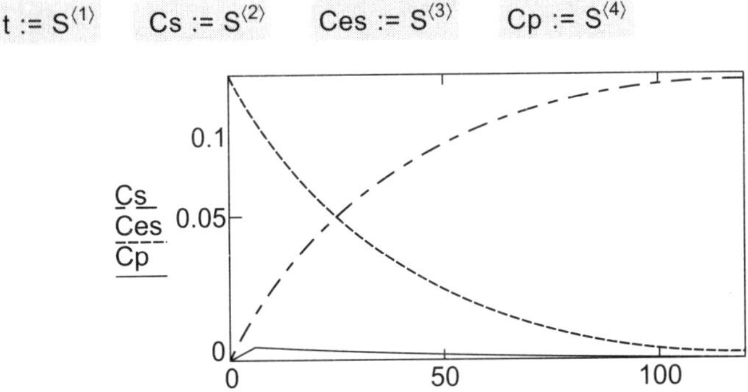

Fig. 2.4 Solution of Example 2.2 by using MathCad

Solution:

To determine how the concentrations of the substrate, product, and enzyme-substrate complex are changing with time, we can solve Eqs. (2.14), (2.30), and (2.31) with the enzyme conservation equation Eq. (2.9) by using ACSL.

The ACSL program to solve this problem is shown in Table 2.2, which is composed of four blocks: PROGRAM, INITIAL, DYNAMIC, and DERIVATIVE. Each block when present must be terminated with an END statement. For the integration algorithm (IALG), Runge-Kutta fourth order (IALG = 5) was selected, which is default if not specified.[5] The calculation interval (integration step size) is equal to the comunication interval (CINT) divided by the number of steps (NSTP). The run-time control program is shown in Table 2.3. Figure 2.4 shows the solution by MathCad.

2.3 EVALUATION OF KINETIC PARAMETERS

In order to estimate the values of the kinetic parameters, we need to make a series of batch runs with different levels of substrate concentration. Then the initial reaction rate can be calculated as a function of initial substrate concentrations. The results can be plotted graphically so that the validity of the kinetic model can be tested and the values of the kinetic parameters can be estimated.

The most straightforward way is to plot r against C_S as shown in Figure 2.2. The asymptote for r will be r_{max} and K_M is equal to C_S when r = 0.5 r_{max}. However, this is an unsatisfactory plot in estimating r_{max} and K_M because it is difficult to estimate asymptotes accurately and also difficult to test the validity of the kinetic model. Therefore, the Michaelis-Menten equation is usually rearranged so that the results can be plotted as a straight line. Some of the better known methods are presented here. The Michaelis-Menten equation, Eq. (2.11), can be rearranged to be expressed in linear form. This can be achieved in three ways:

$$\frac{C_S}{r} = \frac{K_M}{r_{max}} \frac{C_S}{r_{max}} \tag{2.32}$$

$$\frac{1}{r} = \frac{1}{r_{max}} + \frac{K_M}{r_{max}} \frac{1}{C_s} \tag{2.33}$$

$$r = r_{max} - KM \frac{r}{C_S} \tag{2.34}$$

---

[5] Other algorithms are also available for the selection. They are Adams-Moulton (IALG = 1), Gears Stiff (IALG = 2), Runge-Kutta first order or Euler (IALG = 3), and Runge-Kutta second order (IALG = 4). The Adams-Moulton and Gear's Stiff are both variable-step, variable-order integration routines. For the detailed description of these algorithms, please refer to numerical analysis textbooks, such as Gerald and Wheatley (1989), Chappra and Canale (1988), Carnahan et al. (1969), and Burden and Faires (1989). }

An equation of the form of Eq. (2.32) was given by Langmuir (Carberry, 1976) for the treatment of data from the adsorption of gas on a solid surface. If the Michaelis-Menten equation is applicable, the *Langmuir plot* will result in a straight line, and the slope will be equal to $1/r_{max}$. The intercept will be K_M/r_{max}, as shown in Figure 2.5.

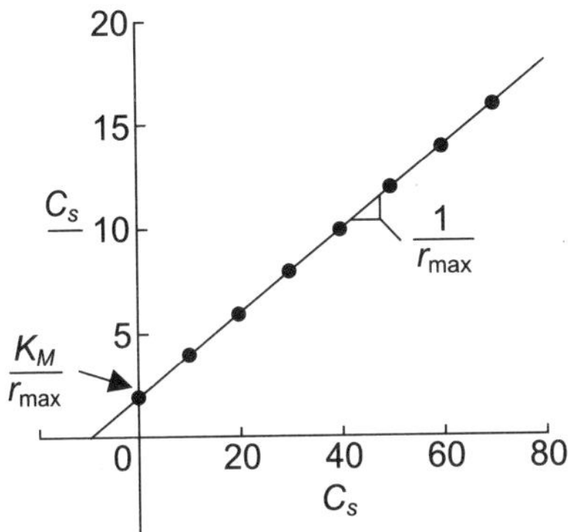

Fig. 2.5   The Langmuir plot (K_M = 10, r_{max} = 5).

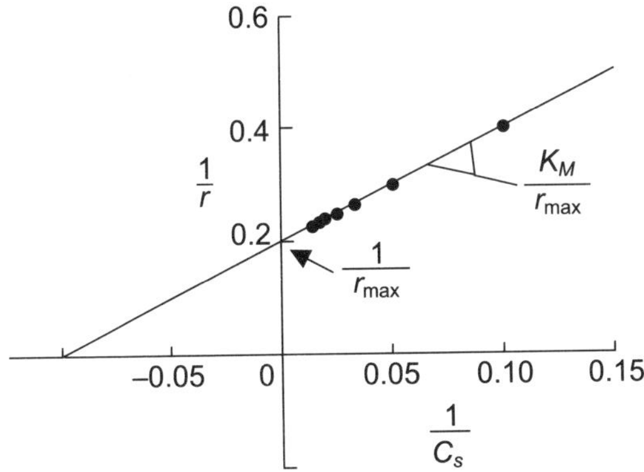

Fig. 2.6   The Lineweaver-Burk plot (K_M = 10, r_{max} = 5).

Similarly, the plot of $1/r$ versus $1/C_S$ will result in a straight line according to Eq. (2.33), and the slope will be equal to K_M/r_{max}. The intercept will be $1/r_{max}$, as shown in Figure 2.6. This plot is known as *Lineweaver-Burk* plot (Lineweaver and Burk, 1934).

The plot of r versus r/C_S will result in a straight line with a slope of $-K_M$ and an intercept of r_{max}, as shown in Figure 2.7. This plot is known as the *Eadie-Hofstee* plot (Eadie, 1942; Hofstee, 1952).

The Lineweaver-Burk plot is more often employed than the other two plots because it shows the relationship between the independent variable C_S and the dependent variable r. However, $1/r$ approaches infinity as C_S decreases, which gives undue weight to inaccurate measurements made at low substrate concentrations, and insufficient weight to the more accurate measurements at high substrate concentrations. This is illustrated in Figure 2.6. The points on the line in the figure represent seven equally spaced substrate concentrations. The space between the points in Figure 2.6 increases with the decrease of C_s.

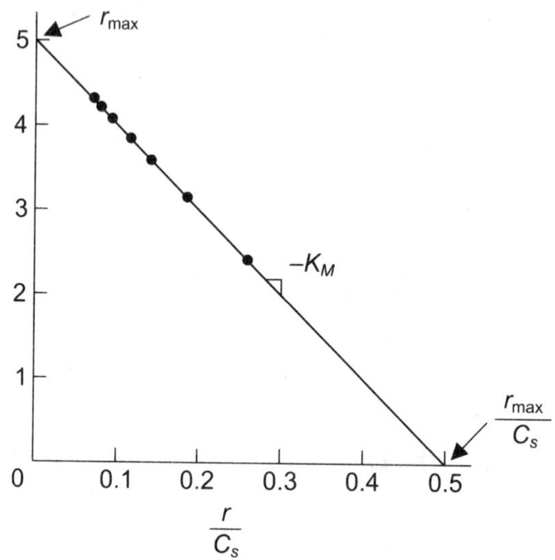

Fig. 2.7   The Eadie-Hofstee plot (K_M = 10, r_{max} = 5).

On the other hand, the Eadie-Hofstee plot gives slightly better weighting of the data than the Lineweaver-Burk plot (see Figure 2.7). A disadvantage of this plot is that the rate of reaction r appears in both coordinates while it is usually regarded as a dependent variable. Based on the data distribution, the Langmuir plot (C_S/r versus C_S) is the most satisfactory of the three, since the points are equally spaced (see Figure 2.5).

The values of kinetic parameters can be estimated by drawing a least-squares line roughly after plotting the data in a suitable format. The linear regression can be also carried out accurately by using a calculator with statistical functions, spreadsheet programs such as Excel (Microsoft., Redmond, WA) or other software packages such as MathCad (MathSoft, Inc., Cambridge, MA). However, it is important to examine the plot visually to ensure the validity of the parameters values obtained when these numerical techniques are used.

Another approach for the determination of the kinetic parameters is to use the SAS NLIN (NonLINear regression) procedure (SAS, 1985) which produces weighted least-squares estimates of the parameters of nonlinear models. The advantages of this technique are that: (1) it does not require linearization of the Michaelis-Menten equation, (2) it can be used for complicated multiparameter models, and (3) the estimated parameter values are reliable because it produces weighted least-squares estimates.

In conclusion, the values of the Michaelis-Menten kinetic parameters, r_{max} and K_M, can be estimated, as follows:

1. Make a series of batch runs with different levels of substrate concentration at a constant initial enzyme concentration and measure the change of product or substrate concentration with respect to time.

2. Estimate the initial rate of reaction from the C_S or C_P versus time curves for different initial substrate concentrations.

3. Estimate the kinetic parameters by plotting one of the three plots explained in this section or a nonlinear regression technique. It is important to examine the data points so that you may not include the points which deviate systematically from the kinetic model as illustrated in the following problem.

Example 2.3

From a series of batch runs with a constant enzyme concentration, the following initial rate data were obtained as a function of initial substrate concentration.

Substrate Concentration mmol/L	Initial Reaction Rate mmol/L min
1	0.20
2	0.22
3	0.30
5	0.45
7	0.41
10	0.50
15	0.40
20	0.33

a. Evaluate the Michaelis-Menten kinetic parameters by employing the Langmuir plot, the Lineweaver-Burk plot, the Eadie-Hofstee plot, and nonlinear regression technique. In evaluating the kinetic parameters, do not include data points which deviate systematically from the Michaelis-Menten model and explain the reason for the deviation.

b. Compare the predictions from each method by plotting r versus C_S curves with the data points, and discuss the strengths and weaknesses of each method.

c. Repeat part (a) by using all data.

Solution:

a. Examination of the data reveals that as the substrate concentration increased up to 10mM, the rate increased.

However, the further increases in the substrate concentration to 15mM decreased the initial reaction rate. This behavior may be due to substrate or product inhibition. Since the Michaelis-Menten equation does not incorporate the inhibition effects, we can drop the last two data points and limit the model developed for the low substrate concentration range only ($C_S \leq 10$mM). Figure 2.8 shows the three plots prepared from the given data. The two data points which were not included for the linear regression were noted as closed circles.

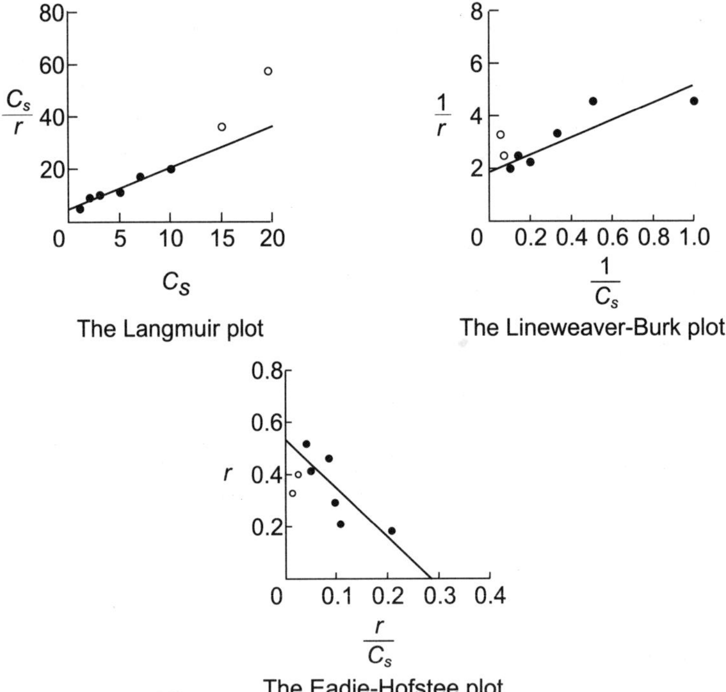

Fig. 2.8 Solution to Example 2.3.

Table 2.4 shows the SAS NLIN specifications and the computer output. You can choose one of the four iterative methods: modified Gauss-Newton, Marquardt, gradient or steepest-descent, and multivariate secant or false position method (SAS, 1985). The Gauss-Newton iterative methods regress the residuals onto the partial derivatives of the model with respect to the parameters until the iterations converge. You also have to specify the model and starting values of the parameters to be estimated. It is optional to provide the partial derivatives of the model with respect to each parameter.

b. Figure 2.9 shows the reaction rate versus substrate concentration curves predicted from the Michaelis-Menten equation with parameter values obtained by four different methods. All four methods predicted the rate reasonably well within the range of concentration *(Cs ≤ 10mmol/L)* from which the parameter values were estimated. However, the rate predicted from the Lineweaver-Burk plot fits the data accurately when the substrate concentration is the lowest and deviates as the concentration increases. This is because the Lineweaver-Burk plot gives undue weight for the low substrate concentration as shown in Figure 2.6. The rate predicted from the Eadie-Hofstee plot shows the similar tendency as that from the Lineweaver-Burk plot, but in a lesser degree. The rates predicted from the Langmuir plot and nonlinear regression technique are almost the same which give the best line fit because of the even weighting of the data.

Table 2.4 SAS NLIN Specifications and Computer Output for Example 2.3

Computer Input:

```
DATA Example3;
 INPUT CS R @@;
 CARDS;
 1 0.20 2 0.22 3 0.30 5 0.45 7 0.41 10 0.50;
 PROC NLIN METHOD=GAUSS; /*Gaussian method is default. */
 PARAMETERS RMAX=0.5 KM=1; /*Starting estimates of parameters.*/
 MODEL R=RMAX*CS/(KM+CS); /* dependent=expression */
 DER.RMAX=CS/(KM+CS); /* Partial derivatives of the model */
 DER.KM=-RMAX*CS/((KM+CS)*(KM+CS));/* with respect to each parameter. */
```

Computer Output:

PARAMETER	ESTIMATE	ASYMPTOTIC STD. ERROR	95 % CONFIDENCE INTERVAL	
			LOWER	UPPER
RMAX	0.6344375	0.08593107	0.395857892	0.87301726
KM	2.9529776	1.05041396	0.036600429	5.86935493

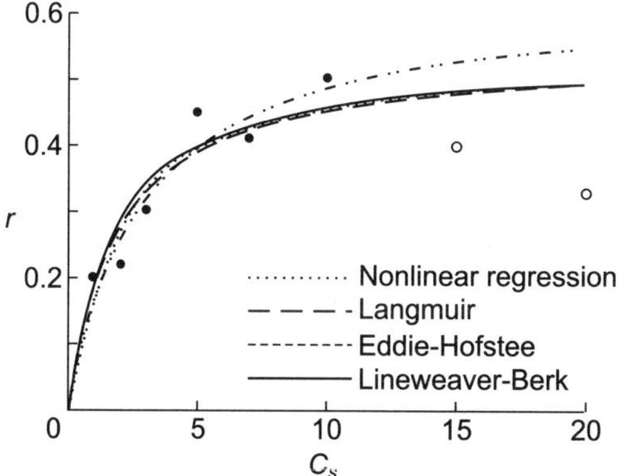

Fig. 2.9 The reaction rate versus substrate concentration curves predicted from the Michaelis-Menten equation with parameter values obtained by four different methods.

c. The decrease of the rate when the substrate concentration is larger than 10mmol/L may be interpreted as data scatter due to experimental variations. In that case, all data points have to be included in the parameter estimation. Table 2.5 also summarizes the estimated values of r_{max} (mmol/L min) and K_M(mmol/L) from the four different methods when all data were used. By adding two data points for the high substrate concentration, the parameter values changes significantly.

Table 2.5   Estimated Values of Michaelis-Menten Kinetic Parameters for Example 2.3

Method	For C_S < 10 mmol/L		For All Data	
	r_{max}	K_M	r_{max}	K_M
Langmuir	0.63	2.92	0.37	-0.04
Lineweaver-Burk	0.54	1.78	0.45	1.37
Eadie-Hofstee	0.54	1.89	0.45	1.21
Nonlinear Regression	0.63	2.95	0.46	1.30

However, in the case of the Lineweaver-Burk plot and the Eadie-Hofstee plot, the changes of the parameter values are not as large as the case of the Langmuir plot, which is because both plots have undue weight on low substrate concentration and the deviation of data at the high substrate concentration level are partially ignored. In the case of the Langmuir plot, the change of the parameter estimation was so large that the K_M value is even negative, indicating that the Michaelis-Menten model cannot be employed. The nonlinear regression techniques are the best way to estimate the parameter values in both cases.

2.4   ENZYME REACTOR WITH SIMPLE KINETICS

A *bioreactor* is a device within which biochemical transformations are caused by the action of enzymes or living cells. The bioreactor is frequently called a fermenter[6] whether the transformation is carried out by living cells or *in vivo*[7] cellular components (that is, enzymes). However, in this text, we call the bioreactor employing enzymes an *enzyme reactor* to distinguish it from the bioreactor which employs living cells, *the fermenter.*

---

[6] Fermentation originally referred to the metabolism of an organic compound under anaerobic conditions. Therefore, the fermenter was also limited to a vessel in which anaerobic fermentations are being carried out. However, modern industrial fermentation has a different meaning, which includes both aerobic and anaerobic large-scale culture of organisms, so the meaning of fermenter was changed accordingly.

[7] Literally, in life; pertaining to a biological reaction taking place in a living cell or organism.

2.4.1   Batch or Steady-State Plug-Flow Reactor

The simplest reactor configuration for any enzyme reaction is the batch mode. A batch enzyme reactor is normally equipped with an agitator to mix the reactant, and the pH of the reactant is maintained by employing either a buffer solution or a pH controller. An ideal batch reactor is assumed to be well mixed so that the contents are uniform in composition at all times.

Assume that an enzyme reaction is initiated at $t = 0$ by adding enzyme and the reaction mechanism can be represented by the Michaelis-Menten equation

$$-\frac{dC_S}{dt} = \frac{r_{max}C_S}{K_M + C_S} \tag{2.35}$$

An equation expressing the change of the substrate concentration with respect to time can be obtained by integrating Eq. (2.35), as follows:

$$\int_{C_{S_0}}^{C_S} -\left(\frac{K_M + C_S}{C_S}\right) dC_S = \int_0^t r_{max}\,dt \tag{2.36}$$

and
$$K_M \ln \frac{C_{S_0}}{C_S} + (C_{S_0} - C_S) = r_{max}t \tag{2.37}$$

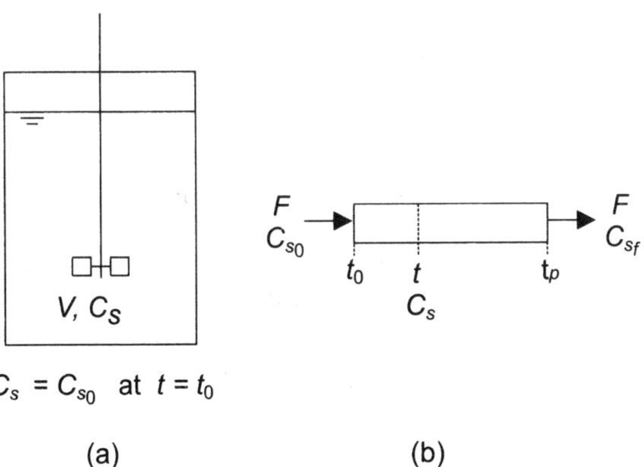

Fig. 2.10   Schematic diagram of (a) a batch stirred-tank reactor and (b) a plug-flow reactor.

This equation shows how C_S is changing with respect to time. With known values of r_{max} and K_M, the change of C_S with time in a batch reactor can be predicted from this equation.

In a *plug-flow enzyme reactor* (or tubular-flow enzyme reactor), the substrate enters one end of a cylindrical tube which is packed with immobilized enzyme and the product stream leaves at the other end. The long tube and lack of stirring device prevents complete mixing of the fluid in the tube. Therefore, the properties of the flowing stream will vary in both longitudinal and radial directions. Since the variation in the radial direction is small compared to that in the longitudinal direction, it is called a plug-flow reactor. If a plug-flow reactor is operated at steady state, the properties will be constant with respect to time. The ideal plug-flow enzyme reactor can approximate the long tube, packed-bed, and hollow fiber, or multistaged reactor.

Eq. (2.37) can also be applied to an ideal steady-state plug-flow reactor, even though the plug-flow reactor is operated in continuous mode. However, the time t in Eq. (2.37) should be replaced with the residence time t in the plug-flow reactor, as illustrated in Figure 2.10.

Rearranging Eq. (2.37) results in the following useful linear equation which can be plotted (Levenspiel, 1984):

$$\frac{C_{S0} - C_S}{\ln(C_{S0}/C_S)} = -K_M + \frac{r_{max}t}{\ln(C_{S0}/C_S)} \tag{2.38}$$

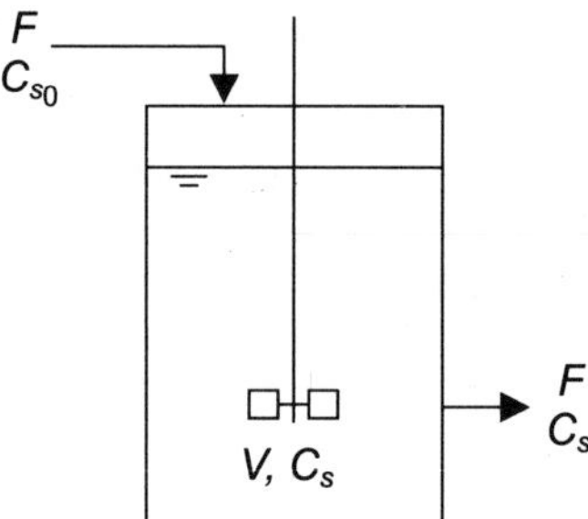

Fig. 2.11 Schematic diagram of a continuous stirred-tank reactor (CSTR).

The plot of $(C_{S_0} - C_S)/\ln(C_{S_0}/C_S)$ versus $t/\ln(C_{S_0}/C_S)$ may yield a straight line with a slope of r_{max} and an intercept of $-K_M$.

2.4.2 Continuous Stirred-Tank Reactor

A continuous stirred-tank reactor (CSTR) is an ideal reactor which is based on the assumption that the reactor contents are well mixed. Therefore, the concentrations of the various components of the outlet stream are assumed to be the same as the concentrations of these components in the reactor. Continuous operation of the enzyme reactor can increase the productivity of the reactor significantly by eliminating the downtime. It is also easy to automate in order to reduce labor costs.

The substrate balance of a CSTR (see Figure 2.11) can be set up, as follows:

Input − Output + Generation = Accumulation

$$FC_{S_0} \quad - \quad FC_S \quad + \quad r_S V = V\frac{dC_S}{dt} \qquad (2.39)$$

where F is the flow rate and V is the volume of the reactor contents. It should be noted that r_S is the rate of substrate consumption for the enzymatic reaction, while dC_S/dt is the change of the substrate concentration in the reactor. As can be seen in Eq. (2.39), r_S is equal to dC_S/dt when F is zero, which is the case in batch operation.

For the steady-state CSTR, the substrate concentration of the reactor should be constant. Therefore, dC_S/dt is equal to zero. If the Michaelis-Menten equation can be used for the rate of substrate consumption (r_S), Eq. (2.39) can be rearranged as:

$$\frac{F}{V} = D = \frac{1}{\tau} = \frac{r_{max}C_S}{(C_{S_0} - C_S)(K_M + C_S)} \qquad (2.40)$$

where D is known as *dilution rate,* and is equal to the reciprocal of the residence time (τ) [8]

Eq. (2.40) can be rearranged to give the linear relationship:

$$C_S = -K_M + \frac{r_{max}C_S\tau}{C_{S_0} - C_S} \qquad (2.41)$$

Michaelis-Menten kinetic parameters can also be estimated by running a series of steady-state CSTR runs with various flow rates and plotting C_S versus $(C_S\tau)/(C_{S_0} - C_S)$. Another approach is to use the Langmuir plot $(C_S r$ vs $C_S)$ after calculating the reaction rate at different flow rates. The reaction rate can be calculated from the relationship: $r = F\,(C_{S_0} - C_S)/V$. However, the initial rate approach described in Section 2.2.4 is a better way to estimate the kinetic parameters than this method because steady-state CSTR runs are much more difficult to make than batch runs.

2.5   INHIBITION OF ENZYME REACTIONS

A *modulator* (or effector) is a substance which can combine with enzymes to alter their catalytic activities. An *inhibitor* is a modulator which decreases enzyme activity. It can decrease the rate of reaction either competitively, noncompetitively, or partially competitively.

---

[8] It is common in biochemical engineering to use the term *dilution rate* rather than the term *residence time,* which chemical engineers are more familiar. In this book, both terminologies are used.

2.5.1 Competitive Inhibition

Since a competitive inhibitor has a strong structural resemblance to the substrate, both the inhibitor and substrate compete for the active site of an enzyme. The formation of an enzyme-inhibitor complex reduces the amount of enzyme available for interaction with the substrate and, as a result, the rate of reaction decreases. A competitive inhibitor normally combines reversibly with enzyme. Therefore, the effect of the inhibitor can be minimized by increasing the substrate concentration, unless the substrate concentration is greater than the concentration at which the substrate itself inhibits the reaction. The mechanism of competitive inhibition can be expressed as follows:

$$E + S \underset{k_2}{\overset{k_1}{\rightleftharpoons}} ES$$

$$E + I \underset{k_4}{\overset{k_3}{\rightleftharpoons}} EI \tag{2.42}$$

$$ES \xrightarrow{k_5} E + P$$

If the slower reaction, the product formation step, determines the rate of reaction according to the Michaelis-Menten assumption, the rate can be expressed as:

$$r_p = k_5 C_{ES} \tag{2.43}$$

The enzyme balance gives

$$C_{E0} = C_E + C_{ES} + C_{EI} \tag{2.44}$$

From the two equilibrium reactions,

$$\frac{C_E C_S}{C_{ES}} = \frac{k_2}{k_1} = K_S \tag{2.45}$$

$$\frac{C_E C_I}{C_{EI}} = \frac{k_4}{k_3} = K_I \tag{2.46}$$

where K_S and K_I are dissociation constants which are the reciprocal of the equilibrium constants. Combining the preceding four equations to eliminate C_E, C_{ES}, and C_{EI} yields

$$r_p = \frac{r_{max} C_S}{C_S - K_{MI}} \tag{2.47}$$

where

$$K_{MI} = K_S \left(1 + \frac{C_I}{K_I} \right) \tag{2.48}$$

Therefore, since K_{MI} is larger than K_S, the reaction rate decreases due to the presence of inhibitor according to Eqn Eq. (2.48). It is interesting to note that the maximum reaction rate is not affected by the presence of a competitive inhibitor. However, a larger amount of substrate is required to reach the maximum rate. The graphical consequences of competitive inhibition are shown in Figure 2.12.

2.5.2 Noncompetitive Inhibition

Noncompetitive inhibitors interact with enzymes in many different ways. They can bind to the enzymes reversibly or irreversibly at the active site or at some other region. In any case the resultant complex is inactive. The mechanism of noncompetitive inhibition can be expressed as follows:

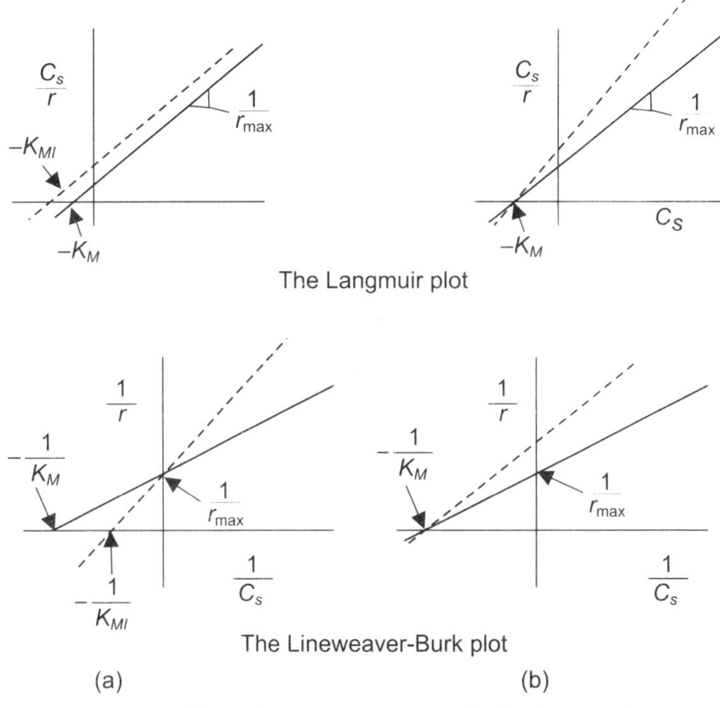

Fig. 2.12 The effect of inhibitors as seen in the Langmuir and Lineweaver-Burk plots: (a) competitive, (b) noncompetitive.

$$E + S \underset{k_2}{\overset{k_1}{\rightleftharpoons}} ES$$

$$E + I \underset{k_4}{\overset{k_3}{\rightleftharpoons}} EI$$

$$EI + S \underset{k_6}{\overset{k_5}{\rightleftharpoons}} EIS \tag{2.49}$$

$$ES + I \xrightleftharpoons[k_8]{k_7} ESI$$

$$ES \xrightarrow{k_9} E + P$$

Since substrate and inhibitor do not compete for a same site for the formation of enzyme-substrate or enzyme-inhibitor complex, we can assume that the dissociation constant for the first equilibrium reaction is the same as that of the third equilibrium reaction, as

$$\frac{k_2}{k_1} = K_S = \frac{k_6}{k_5} = K_{IS} \tag{2.50}$$

Similarly,
$$\frac{k_4}{k_3} = K_I = \frac{k_8}{k_7} = K_{SI} \tag{2.51}$$

As shown in the previous section, the rate equation can be derived by employing the Michaelis-Menten approach as follows:where

$$r_p = \frac{r_{I,max} C_S}{C_S + K_S} \tag{2.52}$$

where

$$r_{I,max} = \frac{r_{max}}{1 + C_I / K_I}$$

Therefore, the maximum reaction rate will be decreased by the presence of a noncompetitive inhibitor, while the Michaelis constant K_S will not be affected by the inhibitor. The graphical consequences of noncompetitive inhibition are shown in Figure 2.12. Note that making these plots enables us to distinguish between competitive and noncompetitive inhibition.

Several variations of the mechanism for noncompetitive inhibition are possible. One case is when the enzyme-inhibitor-substrate complex can be decomposed to produce a product and the enzyme-inhibitor complex. This mechanism can be described by adding the following slow reaction to Eq. (2.49)

$$EIS \xrightarrow{K_{10}} EI + P \tag{2.54}$$

This case is known as partially competitive inhibition. The derivation of the rate equation is left as an exercise problem.

2.6 OTHER INFLUENCES ON ENZYME ACTIVITY

The rate of an enzyme reaction is influenced by various chemical and physical conditions. Some of the important factors are the concentration of various components (substrate, product, enzyme, cofactor, and so on), pH, temperature, and shear. The effect of the various concentrations has been discussed earlier. In this section, the effect of pH, temperature, and shear are discussed.

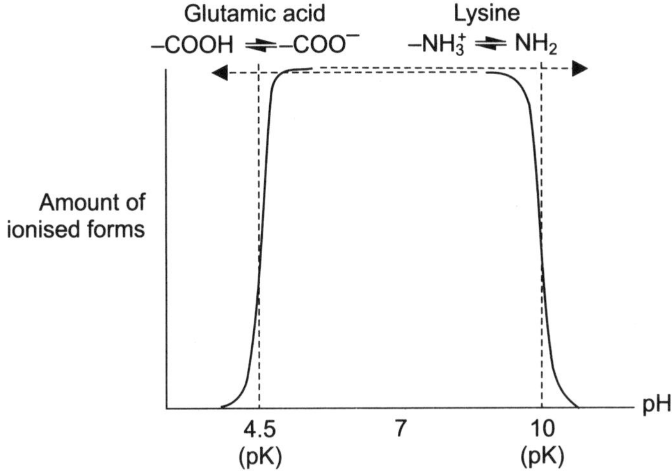

Fig. 2.13 Control of pH optimum by ionizable groups of amino acid residues (Wiseman and Gould, 1970)

2.6.1 Effect of pH

The rate of an enzyme reaction is strongly influenced by the pH of the }reaction solution both *in vivo* and *in vitro*[9] The typical relationship between the reaction velocity and pH shows a bell-shaped curve Figure 2.13. The optimum pH is different for each enzyme. For example, pepsin from the stomach has an optimum pH between 2 and 3.3, while the optimum pH of amylase, from saliva, is 6.8. Chymotrypsin, from the pancreas, has an optimum pH in the mildly alkaline region between 7 and 8.

The reason that the rate of enzyme reaction is influenced by pH can be explained as follows:

1. Enzyme is a protein which consists of ammo acid residues (that is, amino acids minus water).

$$H_2N-\underset{\underset{H}{|}}{\overset{\overset{R}{|}}{C}}-COOH \qquad -HN-\underset{\underset{H}{|}}{\overset{\overset{R}{|}}{C}}-CO-$$

Amino acid Amino acid residue

---

9 Literally, in glass; pertaining to a biological reaction taking place in an artificial apparatus.

2. The amino acid residues possess basic, neutral, or acid side groups which can be positively or negatively charged at any given pH. As an example (Wiseman and Gould, 1970), let's consider one acidic amino acid, glutamic acid, which is acidic in the lower pH range. As the pH is

$$
\begin{array}{c}
\text{COOH} \\
| \\
(\text{CH}_2)_2 \\
| \\
-\text{HN}-\text{C}-\text{CO}- \\
| \\
\text{H}
\end{array}
\quad
\underset{+\text{H}^+}{\overset{-\text{H}^+}{\rightleftharpoons}}
\quad
\begin{array}{c}
\text{COO}^- \\
| \\
(\text{CH}_2)_2 \\
| \\
-\text{HN}-\text{C}-\text{CO}- \\
| \\
\text{H}
\end{array}
$$

increased, glutamic acid is ionized as
which can be expressed as

$$
A \underset{k_2}{\overset{k_1}{\rightleftharpoons}} A^- + H^+
\tag{2.55}
$$

In equilibrium,
$$
\frac{C_{A^-} C_{H^+}}{C_A} = \frac{k_1}{k_2} = K
\tag{2.56}
$$

When $C_{A^-} = C_A$, pH is equal to pK. For glutamic acid, pK = 4.5.

On the other hand, an amino acid, lysine, is basic in the range of higher pH value. As the pH is decreased, lysine is ionized as

$$
\begin{array}{c}
\text{NH}_3^+ \\
| \\
(\text{CH}_2)_4 \\
| \\
-\text{HN}-\text{C}-\text{CO}- \\
| \\
\text{H}
\end{array}
\quad
\underset{+\text{H}^+}{\overset{-\text{H}^+}{\rightleftharpoons}}
\quad
\begin{array}{c}
\text{NH}_2 \\
| \\
(\text{CH}_2)_4 \\
| \\
-\text{HN}-\text{C}-\text{CO}- \\
| \\
\text{H}
\end{array}
$$

Similarly, the pK value of lysine is 10.0, at which half of the residues are ionized.

3. An enzyme is catalytically active when the amino acid residues at the active site each possess a particular charge. Therefore, the fraction of the catalytically active enzyme depends on the pH.

Let's suppose that one residue of each of these two ammo acids, glutamic acid and lysine, is present at the active site of an enzyme molecule and that, for example, the charged form of both amino acids must be present if that enzyme molecule is to function. Since glutamic acid is charged when its pH $\geq$ 4.5 and lysine is charged when its pH $\leq$ 10.0, the enzyme will be most active when 4.5 $\leq$ pH $\leq$ 10.0, as shown in Figure 2.13.

2.6.2   Effect of Temperature

The rate of a chemical reaction depends on the temperature according to Arrhenius equation as

$$k = A_0 e^{-E/RT} \tag{2.57}$$

Consequently, when In k is plotted versus $1/T$, a straight line with slope $-E/R$ is obtained.

The temperature dependence of many enzyme-catalyzed reactions can be described by the Arrhenius equation. An increase in the temperature increases the rate of reaction, since the atoms in the enzyme molecule have greater energies and a greater tendency to move. However, the temperature is limited to the usual biological range. As the temperature rises, denaturation processes progressively destroy the activity of enzyme molecules. This is due to the unfolding of the protein chain after the breakage of weak (for example, hydrogen) bonds, so that the overall reaction velocity drops. For many proteins, denaturation begins to occur at 45 to 50°C. Some enzymes are very resistant to denaturation by high temperature, especially the enzymes isolated from thermophilic organisms found in certain hot environments.

2.6.3   Effect of Shear

Enzymes had been believed to be susceptible to mechanical force, which disturbs the elaborate shape of an enzyme molecule to such a degree that denaturation occurs. The mechanical force that an enzyme solution normally encounters is fluid shear, generated either by flowing fluid, the shaking of a vessel, or stirring with an agitator. The effect of shear on the stability of an enzyme is important for the consideration of enzyme reactor design, because the contents of the reactor need to be agitated or shook in order to minimize mass-transfer resistance.

However, conflicting results have been reported concerning the effect of shear on the activity of enzymes. Charm and Wong (1970) showed that the enzymes catalase, rennet, and carboxypeptidase were partially inactivated when subjected to shear in a coaxial cylinder viscometer. The remaining activity could be correlated with a dimensionless group gammatheta, where gamma and theta are the shear rate and the time of exposure to shear, respectively.[10] In the case

---

[10] The shear rate γ is defined as:

$$\gamma = \frac{du}{dy} = \frac{\tau}{\mu}$$

where τ is the shear stress [N/m^2] and μ is the viscosity of the fluid.

of catalase, about 50 percent of the activity was lost when $\gamma\theta$ was 0.5×10^7.

However, Thomas and Dimnill (1979) studied the effect of shear on catalase and urease activities by using a coaxial cylindrical viscometer that was sealed to prevent any air-liquid contact. They found that there was no significant loss of enzyme activity due to shear force alone at shear rates up to 10^6 sec^{-1}. They reasoned that the deactivation observed by Charm and Wong (1970) was the result of a combination of shear, air-liquid interface, and some other effects which are not fully understood. Charm and Wong did not seal their shear apparatus. This was further confirmed, as cellulase deactivation due to the interfacial effect combined with the shear effect was found to be far more severe and extensive than that due to the shear effect alone (Jones and Lee, 1988).

2.7 EXPERIMENT: ENZYME KINETICS

Objectives:

The objectives of this experiment are:

1. To give students an experience with enzyme reactions and assay procedures
2. To determine the Michaelis-Menten kinetic parameters based on initial-rate reactions in a series of batch runs
3. To simulate batch and continuous runs based on the kinetic parameters obtained

Materials:

1. Spectrophotometer
2. l0g/L glucose standard solution
3. Glucose assay kit (No. 16-UV, Sigma Chemical Co., St. Louis, MO)
4. Cellobiose
5. Cellobiase enzyme (Novozym 188, Novo Nordisk Bioindustrials Inc., Danbury, CT) or other cellulase enzyme
6. 0.05M (mol/L) sodium acetate buffer (pH 5)
7. 600 mL glass tempering beaker (jacketed) (Cole-Parmer Instrument Co., Chicago, IL) with a magnetic stirrer
8. Water bath to control the temperature of the jacketed vessel

Calibration Curve for Glucose Assay:

1. Prepare glucose solutions of 0, 0.5, 1.0, 3.0, 5.0 and 7.0g/L by diluting l0g/L glucose standard solution.

2. Using these standards as samples, follow the assay procedure described in the brochure provided by Sigma Chemical Co.

3. Plot the resulting absorbances versus their corresponding glucose concentrations and draw a smooth curve through the points.

Experiment Procedures:

1. Prepare a 0.02M cellobiose solution by dissolving 3.42 g in 500 mL of 0.05M NaAc buffer (pH 5).

2. Dilute the cellobiase-enzyme solution so that it contains approximately 20 units of enzyme per mL of solution. One unit of cellobiase is defined as the amount of enzyme needed to produce 1 μ-mol of glucose per min.

3. Turn on the bath circulator, making sure that the temperature is set at 50°C.

4. Pour 100 mL of cellobiose solution with a certain concentration (20, 10, 5, 2, or ImM) into the reactor, turn on the stirrer, and wait until the solution reaches 50°C. Initiate the enzyme reaction by adding 1 mL of cellobiase solution to the reaction mixture and start to time.

5. Take a 1 mL sample from the reactor after 5- and 10-minutes and measure the glucose concentration in the sample.

Data Analysis:

1. Calculate the initial rate of reaction based on the 5 and 10 minute data.

2. Determine the Michaelis-Menten kinetic parameters as described in this chapter.

3. Simulate the change of the substrate and product concentrations for batch and continuous reactors based on the kinetic parameters obtained. Compare one batch run with the simulated results. For this run, take samples every 5 to 10 minutes for 1 to 2 hours.

2.8 NOMENCLATURE

A_0 constant called frequency factor in Arrhenius equation, dimensionless

C   concentration, $kmol/m^3$

D   dilution rate, s^{-1}

E   activation energy (kcal/kmol)

F   flow rate, m^3/s

k	rate constant
K_M	Michaelis constant, $kmol/m^3$
K	dissociation constant
K_{eq}	equilibrium constant
R	gas constant (kcal/kmol °K)
r	rate of reaction per unit volume, $kmol/m^3s$
r_{max}	maximum rate of reaction per unit volume, $kmol/m^3s$
T	temperature °K
t	time, s
V	working volume of reactor, m^3
τ	residence time, s

Subscript

E	enzyme
El	enzyme-inhibitor complex
ES	enzyme-substrate complex
I	inhibitor
P	product
S	substrate

2.9 PROBLEMS

2.1 In order to measure the enzyme activity and the initial rate of reaction, 5 mL of cellobiose (100mumol/mL) and 44 mL of buffer solution were placed in a stirred vessel. The reaction was initiated by adding 1 mL of enzyme (beta-glucosidase) solution which contained 0.1 mg of protein per mL. At 1, 5, 10, 15, and 30 minutes, 0.1mL of sample was removed from the reaction mixture and its glucose content was measured. The results were as follows:

Time Min	Glucose Concentration μmol/mL
1	0.05
5	0.23
10	0.38
15	0.52
30	1.03

   a. What is the activity of the β-glucosidase in units/mL of enzyme solution and in umts/mg protein? A unit is defined as the enzyme activity which can produce Imumol of product per minute.

   b. What is the initial rate of reaction?

2.2 Suppose that the following sequence describes the reactions of two different substrates catalyzed by one enzyme:

$$E + S_1 \underset{k_2}{\overset{k_1}{\rightleftharpoons}} ES_1$$

$$ES_1 + S_2 \underset{k_4}{\overset{k_3}{\rightleftharpoons}} ES_1S_2$$

$$ES_1S_2 \xrightarrow{k_5} E + P$$

   a. Derive the rate equation by making the Michaelis-Menten assumption.

   b. If the concentration of S_1 is much higher than that of S_2, how can the rate equation be simplified?

2.3 In some enzyme-catalyzed reactions, multiple complexes are involved as follows:

$$E + S \underset{k_2}{\overset{k_1}{\rightleftharpoons}} ES$$

$$(ES)_1 \underset{k_4}{\overset{k_3}{\rightleftharpoons}} (ES)_2$$

$$(ES)_2 \xrightarrow{k_5} E + P$$

Develop a rate expression using (a) the Michaelis-Menten approach and (b) the Bnggs-Haldane approach.

2.4 Suppose that two substrates are required for an enzymatic conversion according to the following mechanism:

$$E + S_1 \underset{k_2}{\overset{k_1}{\rightleftharpoons}} ES_1$$

$$E + S_2 \underset{k_4}{\overset{k_3}{\rightleftharpoons}} ES_2$$

$$ES_1 + S_2 \underset{k_4}{\overset{k_3}{\rightleftharpoons}} ES_1S_2$$

$$ES_1S_2 \xrightarrow{k_5} E + P$$

Develop suitable rate expression using (a) the Michaelis-Menten approach and (b) Bnggs-Haldane approach. Please note that the rate constants of the second equilibrium reaction are the same as those of the third reaction.

2.5 Eadie (1942) measured the initial reaction rate of hydrolysis of acetylcholme (substrate) by dog serum (source of enzyme) and obtained the following data:

Substrate Concentration mol/L	Initial Reaction Rate mol/L min
0.0032	0.111
0.0049	0.148
0.0062	0.143
0.0080	0.166
0.0095	0.200

Evaluate the Michaelis-Menten kinetic parameters by employing (a) the Langmuir plot, (b) the Lineweaver-Burk plot, (c) the Eadie-Hofstee plot, and (d) non-linear regression procedure.

2.6 The Michaelis-Menten approach assumes that the product releasing step is much slower than the first complex forming step of the simple enzyme-reaction mechanism:

$$E + S \underset{k_2}{\overset{k_1}{\rightleftharpoons}} ES$$

$$ES \xrightarrow{k_5} E + P$$

Derive a rate equation for the case in which the enzyme-substrate formation step is much slower than the product releasing step, that is, $k_1 \ll k_3$, $k_2 \ll k_3$. State your assumptions.

2.7 A carbohydrate (S) decomposes in the presence of an enzyme (E). The Michaelis-Menten kinetic parameters were found to be as follows:

$$K_M = 200 \ \text{mol/m}^3$$

$$r_{max} = 100 \ \text{mol/m}^3\text{min}$$

a. Calculate the change of substrate concentration with time in a batch reactor the initial substrate concentration is 300 mol/m^3.

b. Assume that you obtained the C_S versus t curve you calculated in part (a) experimentally. Estimate K_M and r_{max} by plotting the $(C_{S_0} - C_S)/\ln(C_{S_0}/C_S)$ versus $t/\ln(C_{S_0}/C_S)$ curve according to Eq. (2.38). Is this approach reliable?

c. Chemostat (continuously stirred-tank reactor) runs with various flow rates were carried out. If the inlet substrate concentration is 300 mol/m^3 and the flow rate is 100 cm^3/min, what is the steady-state substrate concentration of the outlet? The reactor volume is 300 cm^3. Assume that the enzyme concentration in the reactor is constant so that the same kinetic parameters can be used.

2.8 The K_M value of an enzyme is known to be 0.01 mol/L. To measure the maximum reaction rate catalyzed by the enzyme, you measured the initial rate of the reaction and found that 10 percent of the initial substrate was consumed in 5 minutes. The initial substrate concentration is 3.4 times 10^{-4} mol/L. Assume that the reaction can be expressed by the Michaelis-Menten kinetics.

    a. What is the maximum reaction rate?

    b. What is the concentration of the substrate after 15 minutes?

2.9 A substrate is converted to a product by the catalytic action of an enzyme. Assume that the Michaelis-Menten kinetic parameters for this enzyme reaction are:

$$K_M = 0.03 \text{ mol/L}$$

$$r_{max} = 13 \text{ mol/L min}$$

    a. What should be the size of a steady-state CSTR to convert 95 percent of incoming substrate ($C_{S_0} = 10$ mol/L) with a flow rate of 10L/hr?

    b. What should be the size of the reactor if you employ a plug-flow reactor instead of the CSTR in part (a)?

2.10 A substrate is decomposed in the presence of an enzyme according to the Michaelis-Menten equation with the following kinetic parameters:

$$K_M = 10 \text{ g/L}$$

$$r_{max} = 7 \text{ g/L min}$$

If we operate two one-liter CSTRs in series at steady state, what will be the concentration of substrate leaving the second reactor? The flow rate is 0.5 L/min. The inlet substrate concentration is 50 g/L and the enzyme concentration in the two reactors is maintained at the same value all of the time. Is the two-reactor system more efficient than one reactor whose volume is equal to the sum of the two reactors?

2.11 Suppose that the following sequence describes an enzyme-substrate reaction with product inhibition:

$$E + S \underset{k_2}{\overset{k_1}{\rightleftharpoons}} ES$$

$$E + P \underset{k_4}{\overset{k_3}{\rightleftharpoons}} EP$$

$$ES + P \underset{k_6}{\overset{k_5}{\rightleftharpoons}} ESP$$

$$EP + S \underset{k_8}{\overset{k_7}{\rightleftharpoons}} EPS$$

$$ES \xrightarrow{k_9} E + P$$

   a. Derive the rate equation by making the Michaelis-Menten assumption.

   b. How can the rate equation derived in part (a) be simplified if ESP and EPS are the same?

   c. If $k_2/k_1 = k_8/k_7$ and $k_4/k_3 = k_6/k_5$ of the rate equation developed in part (b), what is the apparent maximum rate r_{max}^{app} and Michaelis constant?

   d. Explain how you can estimate the parameters of the rate equation in part (c) experimentally.

2.12 The kinetic model of lactose hydrolysis by *Aspergillus niger* lactase can be described as follows (Scott et al., 1985):

$$E + P \underset{k_2}{\overset{k_1}{\rightleftharpoons}} ES \xrightarrow{k_3} E + P + Q$$

$$E + S \underset{k_4}{\overset{k_3}{\rightleftharpoons}} ES$$

where S, P, Q, and E are lactose, galactose, glucose, and free enzyme.

   a. Derive the rate equation for the production of galactose by using Bnggs-Haldane approach.

   b. Does galactose inhibit the reaction competitively or noncompetitively?

2.13 Yang and Okos (1989) proposed an alternative kinetic model of lactose hydrolysis by *Aspergillus niger* lactase:

$$E + P \underset{k_2}{\overset{k_1}{\rightleftharpoons}} ES \xrightarrow{k_3} EP + Q$$

$$EP \underset{k_5}{\overset{k_4}{\rightleftharpoons}} E + P$$

where S, P, Q, and E are lactose, galactose, glucose, and free enzyme. In this mechanism, glucose is released from the enzyme-substrate complex first, leaving the enzyme-galactose complex, which subsequently releases the galactose molecules.

   a. Derive the rate equation for the production of galactose by using Bnggs-Haldane approach.

   b. How is the rate equation developed by this model simplified to that by Scott et al. (1985) in the previous problem?

2.14 Derive a rate equation for the following partially competitive inhibition using the Michaelis-Menten approach.

$$E + S \underset{k_2}{\overset{k_1}{\rightleftharpoons}} ES \qquad\qquad E + I \underset{k_4}{\overset{k_3}{\rightleftharpoons}} EI$$

$$ES + I \underset{k_6}{\overset{k_5}{\rightleftharpoons}} EIS \qquad\qquad EI + S \underset{k_8}{\overset{k_7}{\rightleftharpoons}} EIS$$

$$ES \xrightarrow{k_9} E + P \qquad\qquad EIS \xrightarrow{k_{10}} E + P$$

2.15 Eadie (1942) measured the initial reaction rate of hydrolysis of acetylcholme (substrate) by dog serum (source of enzyme) in the absence and presence of prostigmine (inhibitor), 1.5×10^{-7} mol/L and obtained the following data:

| | Initial Reaction Rate (mol/L min) | |
Substrate Cone. (mol/L)	Absence of Prostigmine	Presence of Prostigmine
0.0032	0.111	0.059
0.0049	0.148	0.071
0.0062	0.143	0.091
0.0080	0.166	0.111
0.0095	0.200	0.125

   a. Is prostigmine competitive or noncompetitive inhibitor?

   b. Evaluate the Michaelis-Menten kinetic parameters in the presence of inhibitor by employing the Langmuir plot.

2.16 Invertase hydrolyzes cane sugar into glucose and fructose. The following table shows the amount of sugar inverted in the first 10 minutes of reaction for various initial substrate concentrations. The amount of invertase was set constant.

Substrate Sugar Con. g/L	Sugar Inverted in 10 min g/L
48.9	1.9
67.0	2.1
98.5	2.4
199.1	2.7
299.6	2.5
400.2	2.3

A Lineweaver-Burk plot of the preceding data did not result in a straight line when the substrate concentration was high. To take into account the substrate inhibition effect, the following reaction mechanism was suggested:

$$E + S \underset{k_2}{\overset{k_1}{\rightleftharpoons}} ES$$

$$ES + S \underset{k_4}{\overset{k_3}{\rightleftharpoons}} ESS$$

$$ES \overset{k_5}{\longrightarrow} E + P$$

   a. Derive the rate equation using the Michaelis-Menten approach.

   b. Determine the three kinetic parameters of the equation derived in part (a) using the preceding experimental data.

2.17 You carried out an enzymatic reaction in a 5-L CSTR. The inlet substrate concentration was 100 mmol/L and the flow rate was set 1 L/hr. After a steady-state was reached, the outlet substrate concentration was 10 mmol/L.

   a. What is the reaction rate in the reactor?

   b. You measured the steady-state outlet substrate concentration as a function of the inlet flow rate and found the following results. Estimate Michaelis kinetic parameters by using the best plotting technique for the equal weight of all data points.

Flow rate L/hr	Outlet Substrate Cone. m mol/L
0.70	4.0
0.80	7.0
1.00	10.0
1.16	14.0

2.18 Suppose that the following sequence describes the reactions of two different substrates catalyzed by one enzyme:Enzyme Kinetics 2-43

$$E + S_1 \underset{k_2}{\overset{k_1}{\rightleftharpoons}} ES_1 \qquad\qquad k_1 = 70 \text{ L/mol s}$$

$$k_2 = 5 \text{ s}^{-1}$$

$$E + S_2 \underset{k_4}{\overset{k_3}{\rightleftharpoons}} ES_2 \qquad\qquad k_3 = 43.5 \text{ L/mol s}$$

$$k_4 = 9.6 \text{ s}^{-1}$$

$$ES_1 + I \xrightarrow{k_5} E + P_1 \qquad\qquad k_5 = 3.5 \text{ s}^{-1}$$

$$ES_2 \xrightarrow{k_6} E + P_2 \qquad\qquad k_6 = 2.8 \text{ s}^{-1}$$

The initial concentration of substrates, products, and enzyme are: $C_{S_1} = 0.1\text{mol/L}$, $C_{S_2} = 0.3\text{mol/L}$, $C_{P_1} = C_{P_2} = 0$, $C_{ES_1} = C_{ES_2} = 0$, and $C_E = 0.05\text{mol/L}$. Show the changes of C_{S_1}, C_{S_2}, C_{ES_1}, C_{ES_2}, C_{P_1} and C_{P_2} with respect to time by solving the simultaneous differential equations.

2.19 The initial rate of reaction for the enzymatic cleavage of deoxyguanosine triphosphate was measured as a function of initial substrate concentration as follows (Kornberg et al., *J.Biol.Chem.*, **233**, 159, 1958):

Substrate Concentration $\mu mol/L$	Initial Reaction Rate $\mu mol/L$ min
6.7	0.30
3.5	0.25
1.7	0.16

a. Calculate the Michaelis-Menten constants of the above reaction.

b. When the inhibitor was added, the initial reaction rate was decreased as follows:

Substrate $\mu mol/L$	Inhibitor $\mu mol/L$	Initial Reaction Rate $\mu mol/L$ min
6.7	146	0.11
3.5	146	0.08
1.7	146	0.06

Is this competitive inhibition or noncompetitive inhibition? Justify your answer by showing the effect of the inhibitor graphically. *[Contributed by Professor Gary F. Bennett, The University of Toledo, Toledo, OH]*

2.20 The effect of an inhibitor on an enzyme reaction was studied by measuring the initial rates at three different initial inhibitor concentrations. The obtained Michelis-Menten kinetic parameters are as follows:

Inhibitor $\mu mol/L$	r_{max} $\mu mol/L$ min	K_M $\mu mol/L$
0	0.70	5
2	0.20	5
4	0.11	5
6	0.08	5

a. Write the kinetic model for this enzyme reaction.

b. Derive the rate equation. State your assumptions for any simplification of the rate equation.

c. Estimate the value of inhibition kinetic parameter.

2.21 An enzyme (cathepsin) hydrolyzes L-glutamyl-L-tyrosine to carbobenzoxy-L-glutamic acid and L-tyrosine. It has been found (Frantz and Stephenson, *J. Biol. Chem.*, **169,** 359, 1947) that the glutamic acid formed in the hydrolysis, inhibits (competitively) the progress of the reaction by forming a complex with cathepsin. The course of the reaction is followed by adding tyrosme decarboxylase which evolves CO_2.

Substrate μmol/mL	Inhibitor μmol/mL	Initial Reaction Rate μmol/mL min
4.7	0	0.0434
4.7	7.57	0.0285
4.7	30.30	0.0133
10.8	0	0.0713
10.8	7.57	0.0512
10.8	30.30	0.0266
30.3	0	0.1111
30.3	7.57	0.0909
30.3	30.30	0.0581

Calculate (a) the value of Michaelis-Menten constants of the enzyme, K_S, and (b) the dissociation constant of enzyme-inhibitor complex, K_1 [*Contributed by Professor Gary F. Bennett, The University of Toledo, Toledo, OH*]

2.10 REFERENCES

Advanced Continuous Simulation Language: User Guide Reference Manual. Concord, MA: Mitchell and Gauthier, Assoc., Inc.,1975.

Bailey, J. E. and D. F. Ollis, *Biochemical Engineering Fundamentals.* New York, NY: McGraw-Hill Book Co., 1986.

Bohinski, R. C, *Modern Concepts in Biochemistry,* p. 105. Boston, MA: Allyn and Bacon, Inc., 1970.

Briggs, G. E. and J. B. S. Haldane, "A Note on the Kinetics of Enzyme Action," *Biochem. J.* **19** (1925): 338 – 339.

Brown, A. J., "Enzyme Action," *J. Chem. Soc.* **81** (1902): 373 – 388.

Burden, R. L. and J. D. Faires, *Numerical Analysis.* Boston, MA: PWS-KENT Publishing Co., 1989.

Carberry, J. J., *Chemical and Catalytic Reaction Engineering,* pp. 364 – 366. New York, NY: McGraw-Hill Book Co., 1976.

Carnahan, B., H. A. Luther, and J. O. Wilkes, *Applied Numerical Methods.* New York, NY: John Wiley & Sons, Inc., 1969.

Chapra, S. C. and R. P. Canale, *Numerical Methods for Engineers.* New York, NY: McGraw-Hill Book Co., 1988.

Charm, S. E. and B. L. Wong, "Enzyme Inactivation with Shearing," *Biotechnol. Bioeng.* **12** (1970): 1103 – 1109.

Crueger, W. and A. Crueger, *Biotechnology: A Textbook of Industrial Microbiology,* pp. 161 – 162. Madison, WI: Science Tech, Inc., 1984.

Eadie, G. S., "The Inhibition of Cholinesterase by Physostigmine and Prostigmme," *J. Biol. Chem.* **146** (1942): 85 – 93.

Eveleigh, D. E., "The Microbiological Production of Industrial Chemicals," *Scientific American* **245** (1981): 155 – 178.

Hofstee, B. H. J., "Specificity of Esterases: I. Identification of Two Pancreatic Aliesterases," *J. Biol. Chem.* **199** (1952): 357 – 364.

Gerald, C. F. and P. O. Wheatley, *Applied Numerical Analysis.* Reading, MA: Addison-Wesley Pub. Co., 1989.

Jones, E. O. and J. M. Lee, "Kinetic Analysis of Bioconversion of Cellulose in Attrition Bioreactor," *Biotechnol. Bioeng.* **31** (1988): 35 – 40.

Levenspiel, O., *The Chemical Reactor Omnibook*[+]. Corvalis, OR: Oregon State University, 1984.

Lineweaver, H. and D. Burk, "The Determination of Enzyme Dissociation Constants," *J. Amer. Chem. Soc.* **56** (1934): 658 – 666.

Michaelis, L. and M. L. Menten, "Die Kinetik der Invertinwirkung," *Bio-chem. Zeitschr.* 49 (1913): 333 – 369.

SAS User's Guide: Statistic (5th ed.), pp. 576 – 606. Cary, NC: SAS Institute, Inc., 1985.

Scott, T. C., C. G. Hill, and C. H. Amundson, "Determination of the Steady-state Behavior of Immobilized beta-Galactosidase Utilizing an Integral Reactor Scheme," *Biotechnol. Bioeng. Symp.* **15** (1985): 431 – 445.

Thomas, C. R. and P. Dunnill, "Action of Shear on Enzymes: Studies with Catalase and Urease," *Biotechnol Bioeng.* **21** (1979): 2279 – 2302.

Wiseman, A. and B. J. Gould, *Enzymes, Their Nature and Role,* pp. 70 – 75. London, UK: Hutchinson Educational Ltd., 1970.

Yang, S. T. and M. R. Okos, "A New Graphical Method for Determining Parameters in Michaelis-Menten Type Kinetics for Enzymatic Lactose Hydrolysis," *Biotechnol. Bioeng.* **34** (1989): 763 – 773.

3

Immobilized Enzyme

Since most enzymes are globular protein, they are soluble in water. Therefore, it is very difficult or impractical to separate the enzyme for reuse in a batch process. Enzymes can be immobilized on the surface of or inside of an insoluble matrix either by chemical or physical methods. They can be also immobilized in their soluble forms by retaining them with a semipermeable membrane.

A main advantage of immobilized enzyme is that it can be reused since it can be easily separated from the reaction solution and can be easily retained in a continuous-flow reactor. Furthermore, immobilized enzyme may show selectively altered chemical or physical properties and it may simulate the realistic natural environment where the enzyme came from, the cell.

3.1 IMMOBILIZATION TECHNIQUES

Immobilization techniques can be classified by basically two methods, the chemical and the physical method. The former is covalent bond formation dependent and the latter is noncovalent bond formation dependent.[1]

3.1.1 Chemical Method

Covalent Attachment: The covalent attachment of enzyme molecules via nonessential *amino acid residues* (that is, amino acids minus water) to water-insoluble, functionalized supports are the most widely used method for immobilizing enzymes. *Functional groups* of the nonessential amino acid residues that are suitable for the immobilization process are free α-, β-, or γ-carboxyl groups, α- or β-amino groups, and phenyl, hydroxyl, sulfhydryl, or imidazole groups.[2]

---

[1] Covalent bonds are the result of two atoms sharing electrons in order to fill their energy level.

[2] In any molecule, if an atom or a group of atoms is attached to the adjacent atom (usually carbon) by an easily ruptured bond, then this group reacts readily with other reagents. It is called a functional group. For example, OH is a functional group of C_2H_5OH.

Anothervariation of immobilization by covalent attachment is the *copolymerization* of the enzyme with a reactive monomer (M) such as

$$nM + E \longrightarrow M_nE$$

where $M_n E$ may have the following structure:

$$
\begin{array}{c}
M \\
| \\
M\text{—}M\text{—}E\text{—}M\text{—}M\text{—}M \\
| \\
M \\
| \\
M
\end{array}
$$

Commonly employed water-insoluble supports for the covalent attachment of enzymes include: *synthetic supports* such as acrylamide-based polymers, maleic anhydride-based polymers, methacrylic acid-based polymers, styrene-based polymers, and polypeptides, and *natural supports* such as agarose (Sepharose), cellulose, dextran (Sephadex), glass, and starch (Zaborsky, 1973).

Already active polymers such as maleic anhydride copolymers will be simply mixed with enzymes to produce immobilized enzymes. Normally, natural or synthetic polymers need to be activated by treating them with reagents before adding the enzyme. The activation involves the chemical conversion of a functional group of the polymer. The enzyme's active site should not be involved in the attachment, in which case the enzyme would lose its activity upon immobilization.

The following experiment illustrates the immobilization of *glucose oxidase* in agarose (Gutcho, 1974).

1. *Activation of agarose:* Eliminate excess water from the water-swollen, ball-shaped agarose particles (Sepharose 2B) by subjecting them to suction on a glass filter. Add 3.9 grams of the particles (corresponding to 100 mg of shrunken and dried agarose) to 4 mL of a cyanogen bromide solution containing 25 mg cyanogen bromide per mL of water. Allow the particles to be activated for 6 minutes at pH 11 by adding 2 M sodium hydroxide solution with an automatic titrator at 23°C. Wash the activated product on a glass filter with 1 L of ice water. Finally wash the particles rapidly with 0.1 M phosphate buffer (pH of 7.4).
2. *Binding with glucose oxidase:* React the activated polymer for 10 hours (with mild stirring) with 24 mg of glucose oxidase dissolved in 1 mL of 0.1 M phosphate buffer at pH 7.4.

Cross-linking Using Multifunctional Reagents: Water-insoluble enzymes can be prepared by using multifunctional agents that are all bifunctional in nature and have low molecular weight, such as glutaraldehyde.

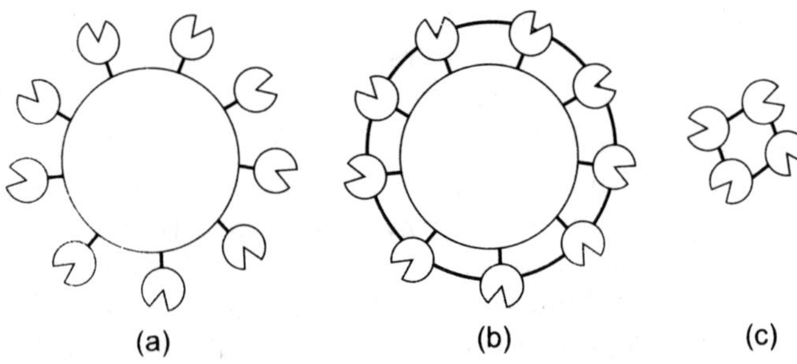

There are several different methods for producing immobilized enzymes with multifunctional reagents, as illustrated in Figure 3.1. Enzymes can be reacted with multifunctional reagent alone so that they are cross-linked intermolecularly by the reagent to form a water-insoluble derivative. Another method is to adsorb enzymes on a water-insoluble, surface-active support followed by intermolecular cross-linking with multifunctional reagents to strengthen the attachment. Multifunctional reagents can be also used to introduce functional groups into water-insoluble polymers, which then react covalently with water-soluble enzymes.

Fig. 3.1 Several different methods for producing immobilized enzymes with multifunctional reagents: (a) enzymes are adsorbed on the surface-active support followed by inter-molecular cross-linking, (b) functional groups are intro-duced on the support to react covalently with enzymes, and (c) enzymes are cross-linked intermolecularly.

3.1.2 Physical Method

Adsorption: This method is the simplest way to immobilize enzymes. Enzymes can be adsorbed physically on a surface-active adsorbent by contacting an aqueous solution of enzyme with an adsorbent. Commonly employed adsorbents are (Zaborsky, 1973): alumina, amon-exchange resins, calcium carbonate, carbon, cation-exchange resins, celluloses, clays, collagen, colloid-ion, conditioned metal, glass plates, diatomaceous earth, and hydroxyapatite. The advantages of adsorption techniques are as follows:

1. The procedure of immobilization is simple.
2. It is possible to separate and purify the enzymes while being immobilized.
3. The enzymes are not usually deactivated by adsorption.
4. The adsorption is a reversible process.

However, adsorption techniques also have several disadvantages:

1. The bonding strength is weak.
2. The state of immobilization is very sensitive to solution pH, ionic strength, and temperature.
3. The amount of enzymes loaded on a unit amount of support is usually low.

Entrapment: Enzymes can be entrapped within cross-linked polymers by forming a highly cross-linked network of polymer in the presence of an enzyme. This method has a major advantage in the fact that there is no chemical modification of the enzyme, therefore, the intrinsic properties of an enzyme are not altered. However, the enzyme may be deactivated during the gel formation. Enzyme leakage is also a problem. The most commonly employed cross-linked polymer is the polyacrylamide gel system. This has been used to immobilize alcohol dehydrogenase, glucose oxidase, amino acid oxidase, hexokmase, glucose isomerase, urease, and many other enzymes.

Microencapsulation: Enzymes can be immobilized within semipermeable membrane microcapsules. This can be done by the interfacial polymerization technique. Organic solvent containing one component of copolymer with surfactant is agitated in a vessel and aqueous enzyme solution is introduced. The polymer membrane is formed at the liquid-liquid interface while the aqueous phase is dispersed as small droplets. One example of this process is the polyamide nylon system, in which 1, 6-diaminohexane is the water-soluble diamme and 1,10-decanoyl chloride is the organic-soluble diacid halide. The organic solvent for this system is a chloroform-cyclohexane mixture (1:4 v/v) containing usually 1 percent (v/v) Span85 surfactant (Zaborsky, 1973). The immobilized enzyme produced by this technique provides an extremely large surface area.

3.2 EFFECT OF MASS-TRANSFER RESISTANCE

The immobilization of enzymes may introduce a new problem which is absent in free soluble enzymes. It is the mass-transfer resistance due to the large particle size of immobilized enzyme or due to the inclusion of enzymes in polymeric matrix. If we follow the hypothetical path of a substrate from the liquid to the reaction site in an immobilized enzyme, it can be divided into several steps (Figure 3.2): (1) transfer from the bulk liquid to a relatively unmixed liquid layer surrounding the immobilized enzyme; (2) diffusion through the relatively unmixed liquid layer; and (3) diffusion from the surface of the particle to the active site of the enzyme in an inert support. Steps

1 and 2 are the external mass-transfer resistance. Step 3 is the intraparticle mass-transfer resistance.

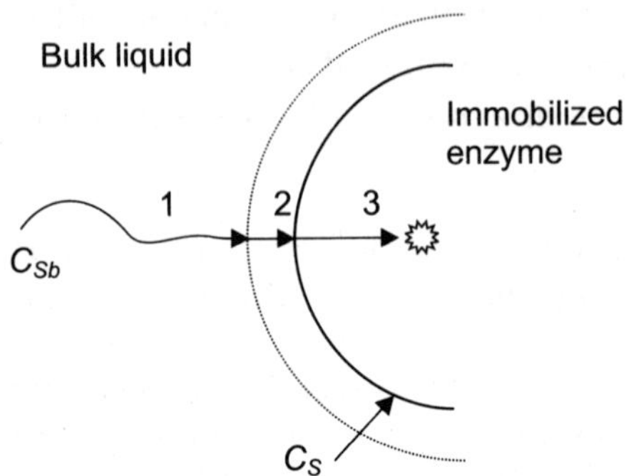

Fig. 3.2   Schematic diagram of the path of the substrate to the reaction site in an immobilized enzyme.

3.2.1   External Mass-Transfer Resistance

If an enzyme is immobilized on the surface of an insoluble particle, the path is only composed of the first and second steps, external mass-transfer resistance. The rate of mass transfer is proportional to the driving force, the concentration difference, as

$$N_s = k_s A (C_{Sb} - C_s) \tag{3.1}$$

where C_{Sb} and C_S are substrate concentration in the bulk of the solution and at the immobilized enzyme surface, respectively (Figure 3.2). The term k_S is the mass-transfer coefficient (length/time) and A is the surface area of one immobilized enzyme particle.

During the enzymatic reaction of an immobilized enzyme, the rate of substrate transfer is equal to that of substrate consumption. Therefore, if the enzyme reaction can be described by the Michaelis-Menten equation,

$$r_p = k_s a (C_{Sb} - C_S) = \frac{r_{max} C_S}{K_M + C_S} \tag{3.2}$$

where a is the total surface area per unit volume of reaction solution. This equation shows the relationship between the substrate concentration in the bulk of the solution and that at the surface of an immobilized enzyme. Eq. (3.2) can be expressed in dimensionless form as:

$$\frac{1 - X_S}{N_{Da}} = \frac{\beta x_S}{1 + \beta x_S} \tag{3.3}$$

where
$$x_S = \frac{C_S}{C_{Sb}}$$

$$N_{Da} = \frac{r_{max}}{k_s a C_{Sb}} \tag{3.4}$$

$$\beta = \frac{C_{Sb}}{K_M}$$

N_{Da} is known as *Damköhler number*, which is the ratio of the maximum reaction rate over the maximum mass-transfer rate. Dependingupon the magnitude of N_{Da}, Eq. (3.2) can be simplified, as follows:

1. If $N_{Da} \ll 1$, the mass-transfer rate is much greater than the reaction rate and the overall reaction is controlled by the enzyme reaction,

$$r_p = \frac{r_{max} C_{Sb}}{K_M + C_{Sb}} \tag{3.5}$$

2. If $N_{Da} \gg 1$, the reaction rate is much greater than the mass-transfer rate and the overall rate of reaction is controlled by the rate of mass transfer that is a first-order reaction,

$$r_p = k_s a C_{Sb} \tag{3.6}$$

To measure the extent which the reaction rate is lowered because of resistance to mass transfer, we can define the *effectiveness factor* of an immobilized enzyme, η, as

$$\eta = \frac{actual\ reaction\ rate}{rate\ if\ not\ slowed\ by\ diffusion} \tag{3.7}$$

The actual reaction rate, according to the external mass-transfer limitation model, is as given in Eq. (3.2). The rate that would be obtained with no mass-transfer resistance at the interface is the same as Eq. (3.5) except that C_S is replaced by C_{Sb}. Therefore, the effectiveness factor is

$$\eta = \frac{\dfrac{r_{max} C_S}{K_M + C_S}}{\dfrac{r_{max} C_{Sb}}{K_M + C_{Sb}}} = \frac{\dfrac{\beta x_S}{1 + \beta x_S}}{\dfrac{\beta}{1 + \beta}} \tag{3.8}$$

where the effectiveness factor is a function of x_S and β. If x_S is equal to 1, the concentration at the surface C_S is equal to the bulk concentration C_{Sb}.

Substituting 1 for x_S in the preceding equation yields $\eta = 1$, which indicates that there is no mass-transfer limitation. On the other hand, if x_S approaches zero, η also approaches zero, which is the case when the rate of mass transfer is very slow compared to the reaction rate.

3.2.2 Internal Mass-Transfer Resistance

If enzymes are immobilized by copolymerization or microencapsulation, the intraparticle mass-transfer resistance can affect the rate of enzyme reaction. In order to derive an equation that shows how the mass-transfer resistance affects the effectiveness of an immobilized enzyme, let's make a series of assumptions as follows:

1. The reaction occurs at every position within the immobilized enzyme, and the kinetics of the reaction are of the same form as observed for free enzyme.
2. Mass transfer through the immobilized enzyme occurs via molecular diffusion.
3. There is no mass-transfer limitation at the outside surface of the immobilized enzyme.
4. The immobilized enzyme is spherical.

The model developed by these assumptions is known as the *distributed model*.

First we derive a differential equation which describes the relationship between the substrate concentration and the radial distance in an immobilized enzyme. The material balance for the spherical shell with thickness dr as shown in Figure 3.3 is

$$\text{Input} - \text{Output} + \text{Generation} = \text{Accumulation} \tag{3.9}$$

$$4\pi \left\{ \begin{array}{c} (r + dr)^2 D_S \left[\dfrac{dC_S}{dr} + \dfrac{d}{dr}\left(\dfrac{dC_S}{dr}\right) dr \right] \\[2ex] - r^2 D_s \dfrac{dC_S}{dr} + \left(r^2\, dr\right) r_S \end{array} \right\} = 4\pi r^2 dr\, \dfrac{dC_S}{dt} \tag{3.10}$$

where D_S is diffusivity of the substrate in an immobilization matrix.

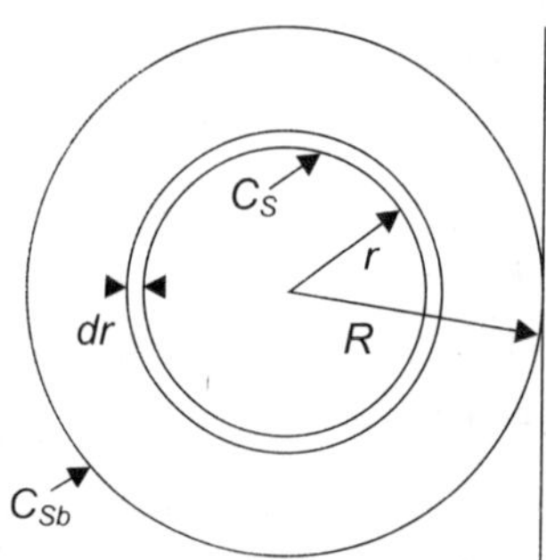

Fig. 3.3 Shell balance for a substrate in an immobilized enzyme.

For a steady-state condition, the change of substrate concentration, dC_S/dt, is equal to zero. After opening up the brackets and simplifying by eliminating all terms containing dr^2 or dr^3, we obtain the second order differential equation:

$$D_S \left(\frac{d^2 C_S}{dr^2} + \frac{2}{r} \frac{dC_S}{dr} \right) + r_S = 0 \tag{3.11}$$

Eq. (3.11) can be solved by substituting a suitable expression for r_S. Let's solve the equation first for the simple cases of zero-order and first-order reactions, and for the Michaelis-Menten equation.

Zero-order Kinetics: Let's assume that the rate of substrate consumption is constant (zero order) with respect to substrate concentration as

$$\begin{aligned} r_S &= 0 - k_0 & \text{if } C_S > 0 \\ r_S &= 0 & \text{otherwise} \end{aligned} \tag{3.12}$$

This is a good approximation when $K_M \ll C_S$ for Michaelis-Menten kinetics, in which case $k_0 = r_{\max}$.

By substituting Eq. (3.12) into Eq. (3.11), we obtain

$$\frac{d^2 C_S}{dr^2} + \frac{2}{r} \frac{dC_S}{dr} - \frac{k^0}{D_S} = 0 \qquad \text{for } C_S > 0 \tag{3.13}$$

The boundary conditions for the solution of the preceding equation are

$$\begin{aligned} \frac{dC_S}{dr} &\longrightarrow 0 & \text{as } r \longrightarrow R_C \\ C_S &= C_{Sb} & \text{as } r = R \end{aligned} \tag{3.14}$$

Eq. (3.13) becomes

$$\frac{d^2 \alpha}{dr^2} = \frac{k_0}{D_S} r \tag{3.15}$$

Integrating Eq. (3.15) twice with respect to r, we obtain

$$\alpha = \frac{1}{6} \frac{k_0}{D_S} r^3 + C_1 r + C_2 \tag{3.16}$$

Therefore,

$$C_S = \frac{1}{6} \frac{k_0}{D_S} r^2 + C_1 + \frac{C_2}{r} \tag{3.17}$$

Applying the boundary conditions (Eq. (3.14) on Eq. (3.17) yields

$$C_1 = C_{Sb} - \frac{1}{6} \frac{k_0}{D_S} R^2 - \frac{1}{3} \frac{k_0 R_C^3}{D_S R} \tag{3.18}$$

$$C_2 = \frac{1}{3}\frac{k_0}{D_S} R^3 \tag{3.19}$$

Therefore, the solution of Eq. (3.13) is

$$\frac{C_S}{C_{Sb}} = \frac{k_0}{6 C_{Sb} D_s}\left[\left(r^2 - R^2\right) - 2R_c^3\left(\frac{1}{R} - \frac{1}{r}\right)\right] + 1 \tag{3.20}$$

Eq. (3.20) is only valid when $C_S > 0$. The critical radius, below which C_S is zero, can be obtained by solving

$$\left(R_C^2 - R^2\right) - 2R_c^3\left(\frac{1}{R} - \frac{1}{R_C}\right) + \frac{6C_{Sb}D_S}{k_0} = 0 \tag{3.21}$$

The actual reaction rate according to the distribution model with zero order is $(4/3)\pi\,(R^3 - R_C^3)k_0$. The rate without the diffusion limitation is $(4/3)\pi\,R^3 k_0$. Therefore, the effectiveness factor, the ratio of the actual reaction rate to the rate if not slowed down by diffusion, is

$$\eta = \frac{(4/3)\pi\left(R^3 - R_C^3\right)k_0}{(4/3)\pi\,R^3\,k_0} = 1 - \left(\frac{R_C}{R}\right)^3 \tag{3.22}$$

First-order Kinetics: If the rate of substrate consumption is a first-order reaction with respect to the substrate concentration,

$$r_S = -kC_S \tag{3.23}$$

By substituting Eq. (3.23) into Eq. (3.11) and converting it to dimension-less form, we obtain

$$\frac{d^2 x_S}{d\dot{r}^2} + \frac{2}{\dot{r}}\frac{dx_S}{d\dot{r}} - 9\phi^2 x_S = 0 \tag{3.24}$$

where $\qquad x_S = \dfrac{C_S}{C_{Sb}}$

$$\dot{r} = \frac{r}{R} \tag{3.25}$$

$$\phi = \frac{R}{3}\sqrt{\frac{k}{D_S}}$$

and ϕ is known as *Thiele's modulus*, which is a measure of the reaction rate relative to the diffusion rate. Eq. (3.24) together with the boundary conditions

$$x_S \text{ is bounded} \qquad \text{as } \dot{r} \longrightarrow 0 \tag{3.26}$$

$$x_S = 1 \qquad\qquad \text{at } \dot{r} = 1$$

determines the function $C'_S(r')$.

In order to convert Eq. (3. 24) to a form which can be easily solved, we set $\dot\alpha = \dot r\, x_S$, so that the differential equation becomes

$$\frac{d^2\dot\alpha}{d\dot r^2} - 9\phi^2\,\dot\alpha = 0 \tag{3.27}$$

Now the general solution of this differential equation is

$$\dot\alpha = C_1\cosh 3\phi\dot r + C_2\sinh 3\phi\dot r \tag{3.28}$$

or
$$x_S = \frac{1}{\dot r}(C_1\cosh 3\phi\dot r + C_2,\sinh 3\phi\dot r) \tag{3.29}$$

Since x_S must be bounded as $\dot r$ approaches zero according to the first boundary condition, we must choose $C_1 = 0$. The second boundary condition requires that $C_2 = 1/\sinh 3\phi$, leaving

$$x_S = \frac{\sinh 3\phi\dot r}{\dot r\sinh 3\phi} \tag{3.30}$$

which shows how the substrate concentration changes as a function of the radial distance in an immobilized enzyme.

Figure 3.4 shows the substrate concentration profile with respect to the radial distance as predicted by Eq. (3.30). For a low value of Thiele's modulus ($\phi \le 1$), the rate of the enzyme reaction is slow compared to the diffusion rate.

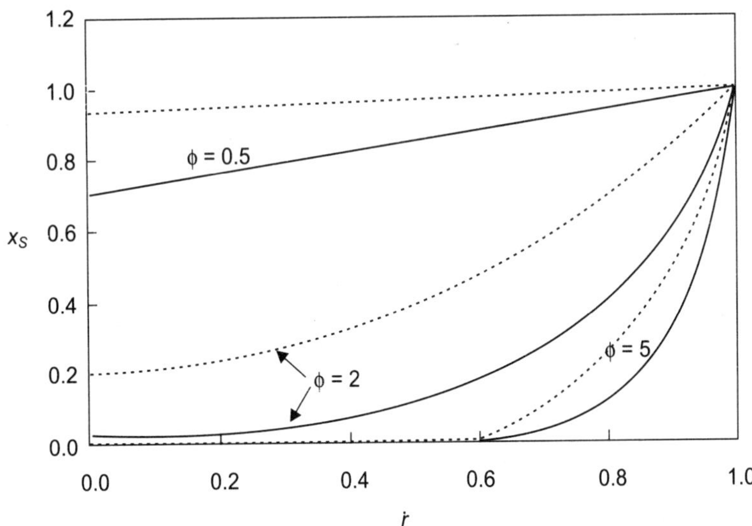

Fig. 3.4 The substrate concentration profile with respect to the radial distance predicted by: first-order kinetics (solid lines) and Michaelis-Menten kinetics with $\beta = 5$ (broken lines) for different values of Thiele's modulus (ϕ).

Therefore, the substrate diffuses into the core of the particle, which results in a fairly flat concentration distribution throughout the radial location of a particle. On the other hand, for higher values of the modulus ($\phi \geq 5$), the reaction rate is faster than the diffusion rate, so most of the substrate is consumed near the particle surface. When $\phi = 5$, the substrate concentration at $\dot{r} \leq 0.6$ is nearly zero.

The actual reaction rate with the diffusion limitation would be equal to the rate of mass transfer at the surface of an immobilized enzyme, while the rate if not slowed down by pore diffusion is kC_{Sb}. Therefore,

$$\eta = \frac{\dfrac{A_P}{V_P} D_S \dfrac{C_{Sb}}{R} \dfrac{dx_S}{d\dot{r}}\Big|_{\dot{r}=1}}{kC_{Sb}} \tag{3.31}$$

where A_p and V_p are the surface area and volume of an immobilized enzyme particle, respectively. Therefore, differentiating Eq. (3.30) with respect to $\dot{r}$ and substituting the resultant equation into Eq. (3.31) will yield

$$\eta = \frac{3\phi \coth 3\phi - 1}{3\phi^2} \tag{3.32}$$

Figure 3.5 shows how Thiele's modulus affects the effectiveness factor for spherical immobilized particles. When $\phi \leq 0.1$, the effectiveness factor is nearly equal to one, which is the case when the rate of reaction is not slowed down by the diffusion. On the other hand, when $\phi \geq 0.1$, the effectiveness factor is inversely proportional to the Thiele's modulus.

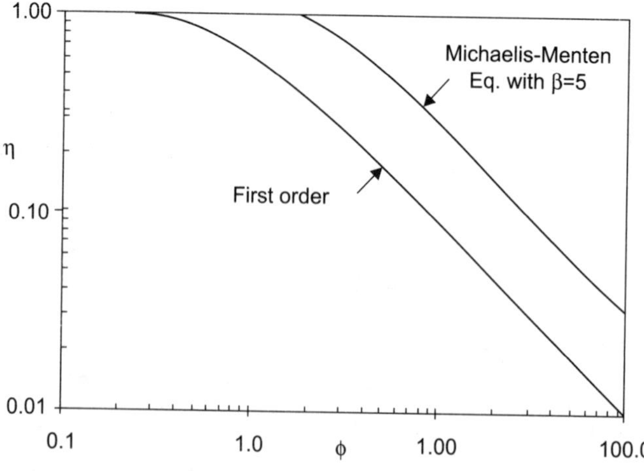

Fig. 3.5 Effect of Thiele's modulus on the effectiveness factor for spherical immobilized particles for first-order kinetics and Michaelis-Menten kinetics with $\beta = 5$. Thiele's modulus for first-order kinetics is $(R/3)\sqrt{k/D_S}$ and for Michaelis-Menten kinetics $(R/3)\sqrt{r_{\max}/D_S K_M}$

Michaelis-Menten Kinetics: If the rate of substrate consumption can be expressed by the Michaelis-Menten equation,

$$r_S = \frac{-r_{max}\,C_S}{K_M + C_S} \tag{3.33}$$

By substituting Eq. (3.33) into Eq. (3.11) and changing it to dimensionless form, we obtain

$$\frac{d^2 x_S}{d\dot{r}^2} + \frac{2}{\dot{r}}\frac{dx_S}{d\dot{r}} - 9\phi^2\,\frac{x_S}{1 + \beta x_S} = 0 \tag{3.34}$$

where β is C_{Sb}/K_M and Thiele's modulus (ϕ) is defined slightly differently from the first-order kinetics as

$$\phi = \frac{R}{3}\sqrt{\frac{r_{max}}{D_S K_M}} \tag{3.35}$$

Eq. (3.34) cannot be solved analytically because it is a nonlinear differential equation. It can be solved by various numerical techniques. Again Advanced Continuous Simulation Language (ACSL, 1975) can be used to solve the problem. Since Eq. (3. 34) is a second-order differential equation, it has to be changed to two simultaneous first-order differential equations to be solved by ACSL as

$$\frac{dy}{d\dot{r}} = \frac{2}{\dot{r}} Y + 9\phi^2\,\frac{x_S}{1 + \beta x_S} = 0 \tag{3.36}$$

$$\frac{dx_S}{d\dot{r}} = Y \tag{3.37}$$

The boundary conditions for the preceding equations are

$$Y = 0 \qquad \text{at } \dot{r} = \frac{R_C}{R}$$

$$x_S = 1 \qquad \text{at } \dot{r} = \frac{R_C}{R} \tag{3.38}$$

One of the simplest algorithms to solve a boundary value problem is the "shooting method." In this method, we assume initial values needed to make a boundary value problem into an initial value problem. We repeat this process until the solution of the initial value problem satisfies the boundary conditions. Therefore, proper initial conditions for the solution of the preceding problem are

$$Y = 0 \qquad \text{at } \dot{r} = \frac{R_C}{R}$$

$$x_S = x_{S0} \qquad \text{at } \dot{r} = \frac{R_C}{R} \tag{3.39}$$

where we have to guess the values of R_c and x_{S_0}. When guessing the values of R_c and x_{S_0}, we can think of two different cases:

1. When $R_c = 0$, assume the value of x_{S_0} at $\dot{r} = 0$. This is the case when the rate of enzyme reaction is slow compared to that of mass transfer which is represented by the low value of phi. As a result, the substrate reaches the center of sphere.

2. When $R_c > 0$, x_{S_0} is equal to zero at $\dot{r} = R_c/R$. Therefore, assume the value of R_c. This is the case when the rate of enzyme reaction is fast compared to that of mass transfer which is represented by the high value of ϕ. As a result, the substrate is consumed before it reaches the center of sphere.

Table 3.1 shows the ACSL program for the solution of Eq. (3. 34) when $\phi = 2$ and $\beta = 5$. This is the case when $R_c = 0$. Therefore, initially the substrate concentration is assumed to be zero, and it is checked that $x_S = 1$ at $\dot{r} = 1$ after solving the initial value problem. If it is not, a new value of xs is estimated using the equation:

$$x_{S0,new} = x_{S0,old} - 0.5\,(x_{S,\,at\,r=1} - 1) \tag{3.40}$$

Table 3.1   ACSL Program for the Solution of Eq. (3.34)

```
PROGRAM DIST ACSL
          INTEGER      N
          CONSTANT     PHI=2, BETA=5., ERR=0.001, CSO=0
          CINTERVAL    CINT=0.001
          VARIABLE     R=0.001
   INITIAL                  $ 'NOTE N IS HANDLED AS AN INTEGER'
          N = 0
      L1..N = N + 1     $ 'BUMP RUN COUNT¹
   END                  $ 'END OF INITIAL¹
   DERIVATIVE
          X = INTEG(Y, XO)
          Y = INTEG(-2*Y/R+9*PHI*PHI*X/(1+BETA*X),0)
          TERMT(R.GE.I)
   END                      $ 'END OF DERIVATIVE'
   TERMINAL                 $ 'IF CONVERGED OR TOO MANY TRIES'
      IF(ABS(X-1.).LT.ERR .OR. N.GT.10) GO TO L2
         XO=XO-0.5*(X-1)
         GO TO L1
   L2.. CONTINUE
         ETA=Y/((3"PH?-PHI)/(1+BETA))
         WRITE(6,L4)   PHI,ETA,Y,XO,X
   L4..  FORMATCPHI =',F10.2,' ETA =',F10.3,' Y =',F10.3,..
                    ' XO =',F10.3, ' X =',F10.3)
   END                      $ 'END OF TERMINAL'
   END                      $ 'END OF PROGRAM'
```

The solution is considered to be successful if $\left| x_{S,\,at\,\dot{r}=1} \right| \leq 1.001\,|$

Figure 3.6 shows the simple manual iteration for the solution of the boundary problem, Eq. (3.34), by MatchCad.

Figure 3.4 shows the substrate concentration profile when $\beta = 5$. The effectiveness factor when the reaction rate is expressed by the Michaelis-Menten equation can be calculated as

$$\eta = \frac{\dfrac{A_P}{V_P} D_S \dfrac{C_{Sb}}{R} \dfrac{dx_S}{d\dot{r}}\Big|_{\dot{r}=1}}{\dfrac{r_{max} C_{Sb}}{K_M + C_{Sb}}} \qquad (3.41)$$

Figure 3.5 shows the effect of the Thiele's modulus on the effectiveness factor.

First Solution:

$\phi := 5$

$\beta := 5 \qquad D(r, Y) := \left[\dfrac{Y_1}{\dfrac{-2\,Y_1}{r}} + 9 \cdot \phi^2 \dfrac{Y_0}{1 + \beta \cdot Y_0} cy \right]$

$r_1 := 1.0 \qquad N := 4$

Derivativefunction.

Note $\quad Y_1 = Y$

$\qquad\qquad Y_0 = X$

$r_0 := 0.4 \qquad X_0 := 0.002215 \qquad$ Guess values for initial r and X near R = Rc

Repeat this quesses until X=1 at r=1

$\text{Guess} := \begin{pmatrix} X_0 \\ 0 \end{pmatrix} \qquad \text{Sol} := \text{Rkadapt}(\textit{Guess}, r_0, R_1, N, D)$

$$\text{Sol} = \begin{pmatrix} 0.4 & 2.215 \times 10^3 & 0 \\ 0.55 & 8.828 \times 1-^{-3} & 0.112 \\ 0.7 & 0.059 & 0.727 \\ 0.85 & 0.308 & 2.91 \\ 1 & 0.999 & 6.405 \end{pmatrix}$$

Solved using Runge-Kutta adaptive method

The solution:

1st column: r

2nd column: X

3rd column: dX / dt = Y

Second Solution:

$\phi := 2 \qquad \beta := 5$

$D(r,Y) := \left(\dfrac{Y_1}{\dfrac{-2 \cdot Y_1}{r}} + 9 \cdot \phi^2 \cdot \dfrac{Y_0}{1 + \beta \cdot Y_0} \right)$

$r_1 := 1.0 \qquad N := 4$

$r_0 := \qquad X_0 = 0.2113$

$\text{Guess} := \begin{pmatrix} X_0 \\ 0 \end{pmatrix} \qquad \text{Sol} = \begin{pmatrix} 0 & 0.211 & 0 \\ 0.25 & 0.251 & 0.324 \\ 0.5 & 0.38 & 0.726 \\ 0.75 & 0.622 & 1.227 \\ 1 & 1 & 1.802 \end{pmatrix}$

$\text{Sol2} := \text{Rkadapt}(\text{Guess}, r_0\ r_1, N, D)$

Fig. 3.6 Solution of Eq. (3.34) by MathCad.

3.2.3 Effective Diffusivities in Biological Gels

The analysis of intraparticle mass-transfer resistance requires the knowledge of the *effective diffusivity* D_S of a substrate in an immobilized matrix, such as agarose, agar, or gelatin. Gels are porous semisolid materials, which are composed of macromolecules and water. The gel structure increases the path length for diffusion, and as a result decreases the diffusion rate.

Various techniques are available for determining the effective diffusivity of solute in gel (Itamunoala, 1988). One of the most reliable techniques is the *thin-disk method* which uses a diffusion cell with two compartments divided by a thin gel. Each compartment contains a well-stirred solution with different solute concentrations. Effective diffusivity can be calculated from the mass flux verses time measurement (Hannoun and Stephanopoulos, 1986). A few typical values of effective diffusivities are listed in Table 3.2.

Table 3.2   Typical Effective Diffusivities of Solutes in Biological Gels in Aqueous Solution

Solute	*Gel*	*Concentration wt%*	*Temp. °C*	*Diffusivity m^2/s*
Glucose[a]	Ca Alginate	2	25	6.10×10^{-10}
Ethanol[a]	Ca Alginate	2	25	1.00×10^{-9}
Sucrose[b]	Gelatin	0	5	2.85×10^{-10}
Sucrose[b]	Gelatin	3.8	5	2.09×10^{-10}
Sucrose[b]	Gelatin	5.7	5	1.86×10^{-10}
Sucrose[b]	Gelatin	7.6	5	1.35×10^{-10}
Lactose[b]	Gelatin	25	5	0.37×10^{-10}
L-Tryptophan[c]	Ca Algmate	2	30	6.67×10^{-10}

[a] Hannoun and Stephanopoulos (1986)
[b] Friedman and Kraemer (1930)
[c] Tanaka, etal. (1984)

Example 3.1

When the rate of diffusion is very slow relative to the rate of reaction, all substrate will be consumed in the thin layer near the exterior surface of the spherical particle. Derive the equation for the effectiveness of an immobilized enzyme for this diffusion limited case by employing the same assumptions as for the distributed model. The rate of substrate consumption can be expressed by the Michaelis-Menten equation.

Solution:

Since all substrate is consumed in the thin layer, we can assume that the layer has a flat geometry. Consider a control volume defined by the element at *r* with thickness *dr*, as shown in Figure 3.7.

The material balance for the control volume at steady state gives

$$AD_S\left[\frac{dC_S}{dr} + \frac{d}{dr}\left(\frac{dC_S}{dr}\right)dr\right] - AD_S\frac{dC_S}{dr} + A(dr)r_S = 0 \qquad (3.42)$$

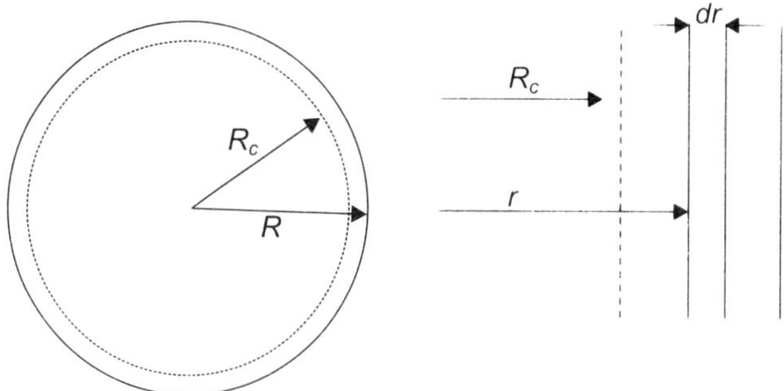

Fig. 3.7 Shell balance for the diffusion limited immobilized enzyme

After replacing r_S with the Michaelis-Menten equation in Eq. (3.42), it can be simplified to

$$D_S\frac{D^2C_S}{dr^2} = \frac{r_{max}\,C_S}{K_M + C_S} \qquad (3.43)$$

which can be solved numerically to obtain an equation for the substrate concentration profile. However, in this problem we do not need to solve this equation. What we need to know to derive an equation for η is dC_S/dr at $r = R$ because

$$\eta = \frac{\dfrac{A_P}{V_P}D_S\dfrac{dC_S}{dr}\bigg|_{r=R}}{\dfrac{r_{max}C_{Sb}}{K_M + C_{Sb}}} \qquad (3.44)$$

Eq. (3.43) is the same as

$$\frac{D_S}{2}\frac{d}{dC_S}\left(\frac{dC_S}{dr}\right)^2 = \frac{r_{max}C_{Sb}}{K_M + C_S} \qquad (3.45)$$

Therefore, by integrating Eq. (3.45) we obtain

$$\frac{dC_S}{dr}\bigg|_{r=R} = \left[\frac{2r_{max}}{D_S}\int_0^{C_{Sb}}\left(\frac{C_S}{K_M + C_S}\right)dC_S\right]^{1/2} \qquad (3.46)$$

$$= \frac{2r_{max}\,K_M}{D_S}\left[\beta - \ln(1+\beta)\right]^{1/2}$$

Substituting Eq. (3.46) into Eq. (3.44) and simplifying results in

$$\eta = \frac{\sqrt{2}}{\phi}\frac{(1+\beta)}{\beta}\left[\beta - \ln(1+\beta)\right]^{1/2} \qquad (3.47)$$

3.3. NOMENCLATURE

A	mterfacial area, m^2
a	interfacial area per unit volume, m^{-1}
$C_1\,C_2$	constants, dimensionless
C_S	substrate concentration, $kmol/m^3$
x_S	C_S/C_{Sb}, dimensionless
C_{Sb}	substrate concentration in the bulk solution, $kmol/m^3$
D_S	effective diffusivity of a substrate in an immobilized matrix, m^2/s
k	first-order rate coefficient, s^{-1}
K_M	Michaelis constant, $kmol/m$
k_S	mass-transfer coefficient, m/s
k_0	zero-order rate coefficient, $kmol/(m^3 sN)$
Da	Damköhler number, $r_{max}/k_S\,a\,C_{Sb}$, dimensionless
N_S	rate of substrate transfer, mol/s
R	radius of a spherical particle, m
R_c	critical radius below which C_S is zero, m
r	radius, m
$\dot{r}$	r/R, dimensionless
r_{max}	maximum rate of reaction per unit volume, $kmol/m^3 s$
r_P	rate of reaction for product per unit volume, $kmol/m^3 s$
r_S	rate of reaction for substrate per unit volume, $mol/m^3 s$
t	time, s
V	volume of reactant solution, m^3
α	rCs, $kmol/m^2$
$\dot{r}$	$r'C's$, dimensionless
β	C_{Sb}/K_M dimensionless
η	effectiveness factor defined in Eq. (3.47), dimensionless
ϕ	Thiele's modulus, $(R/3)\sqrt{k/D_S}$ for first-order kinetics, $(R/3)\sqrt{r_{max}/(D_S K_M)}$ for Michaelis-Menten kinetics, dimensionless

3.4 PROBLEMS

3.1 An enzyme is immobilized on solid surface. Assume that the external mass-transfer resistance for substrate is not negligible and that the Michaelis-Menten equation describes the intrinsic kinetics of enzyme reaction.

   a. Derive an expression which shows the relationship between the substrate concentration at the surface and that in the bulk solution.

   b. The following equation was found to fit the data for the immobilized enzyme reaction,

$$r_{obs} = \frac{r_{max}^{app}\, C_{Sb}}{K_M^{app} + C_{Sb}}$$

Do you expect that the values of the apparent values of r_{max} and K_M are similar to those of r_{max} and K_M for the free soluble enzyme? If they are different, what will be the meaning of K_{max}^{app} and K_M^{app}?

   c. You can pack the immobilized enzyme in a cylindrical column pass through substrate solution, which is known as a packed-column immobilized reactor. What will be the effect of the substrate flow rate on K_M^{app} for the packed-column reactor? Can you determine the intrinsic kinetic parameters (r_{max} and K_M for free enzyme) by using the packed-column immobilized reactor?

3.2 The Michaelis-Menten kinetic parameters for a soluble enzyme for a certain substrate are found to be

$$K_M = 0.05 \text{ mol/L}$$
$$r_{max} = 10 \text{ mol/L nin}$$

After immobilizing this enzyme on the surface of insoluble matrix by physical adsorption, it was found that the K_M^{app} value was increased to 0.08 mol/L whereas the r_{max}^{app} value stayed the same as r_{max}. What is the effectiveness factor of the immobilized enzyme when the substrate concentration is 1 mol/L?

3.3 The values of K_M and r_{max} for an enzyme (21°C and pH = 7.1) are 0.004 kmol/m³ and 10 kmol/m³s, respectively. We immobilized this enzyme by attaching it covalently to acrylamide-based polymers that can be assumed to have spherical shape (diameter = 1 mm). The effectiveness of the immobilized enzyme was found to be 70 percent of the free enzyme when the concentration of the substrate was 0.5 kmol/m³. The reaction was carried out in a stirred reactor with an agitation speed of 50 rpm. (a) Estimate the concentration of the substrate at the surface of the immobilized enzyme, (b) Estimate $k_S\, a$.

3.4 When the rate of diffusion is very slow relative to the rate of reaction, all substrate will be consumed in the thin layer near the exterior surface of the spherical particle. Derive the equation for the effectiveness of an immobilized enzyme for this diffusion limited case by employing the same assumptions as for the distributed model. The rate of substrate consumption can be expressed as a first-order reaction.

3.5 An enzyme which hydrolyzes the cellobiose to glucose, β-glucosidase is immobilized in a sodium alginate gel sphere (2.5 mm in diameter). Assume that the zero-order reaction occurs at every point within the sphere with $k_0 = 0.0795$ mol/sm^3, and cellobiose moves through the sphere by molecular diffusion with $D_S = 0.6 \times 10^{-5}$ cm^2/s (cellobiose in gel). Calculate the effectiveness factor of the immobilized enzyme when the cellobiose concentration in bulk solution is 10 mol/m^3.

3.6 Solve the boundary value problem

$$\frac{d^2 x_S}{d\dot{r}^2} + \frac{2}{\dot{r}}\frac{dx_S}{d\dot{r}} - 9\phi^2 \frac{x_S}{1 + \beta x_S} = 0$$

$$\frac{dx_S}{d\dot{r}} = 0 \qquad \text{at } \dot{r} = 0$$

$$x_S = 0 \qquad \text{at } \dot{r} = 1$$

where $\beta = 5$ and $\phi = 2$

3.7 An enzyme is immobilized by copolymerization technique. The diameter of the spherical particle is 2 mm and the number density of the particles in a substrate solution is 10,000/L. Initial concentration of substrate is 0.1 mole/L. A substrate catalyzed by the enzyme can be adequately represented by the first-order reaction with $k_0 = 0.002$ mol/Ls. It has been found that both external and internal mass-transfer resistance are significant for this immobilized enzyme. The mass-transfer coefficient at the stagnant film around the particle is about 0.02 cm/s and the diffusivity of the substrate in the particle is 5×10^{-6} cm^2/s.

   a. If the internal mass transfer resistance is negligible, what is the concentration of the substrate at the surface of the particle? What is the effectiveness factor for this immobilized enzyme?

   b. If the internal mass-transfer resistance can be described as the distributed model and the external mass transfer resistance is not negligible, what is the concentration of the substrate at the surface of the particle? What is the overall effectiveness factor considering both internal and external mass-transfer resistance?

3.8 An enzyme is immobilized uniformly in a gelatin slab (thickness L and area A). One side is in contact with substrate solution (C_{Sb}) and the other side is in contact with a glass plate. The mass transfer coefficient on the surface of the gelatin is k_S.

   a. Derive the equation for the substrate concentration with respect to x when the substrate is catalyzed by zero-order reaction (k_0). Assume that the substrate is transferred by molecular diffusion in the x direction only and the gelatin slab is thick enough to catalyze all the substrate while it diffuses into the slab. The substrate concentration at the surface of the slab in contact with the solution is C_{S_0}.

   b. What is the critical thickness (x_c) at which all substrate is consumed?

   c. What is the substrate concentration at the surface of the slab (Cs_0)?

3.5 REFERENCES

Advanced Continuous Simulation Language: User Guide Reference Manual. Concord, MA: Mitchell and Gauthier, Assoc., Inc., 1975.

Fnedman, L. and E. O. Kraemer, "The Structure of Gelatin Gels from Studies of Diffusion," *J. Am. Chem. Soc.* **52** (1930): 1295 – 1314.

Gutcho, S. J., *Immobilized Enzymes: Preparation and Engineering Techniques,* p. ~ 25. Park Ridge, NJ: Noyes Data Corp., 1974.

Hannoun, B. J. M. and G. Stephanopoulos, "Diffusion Coefficients of Glucose and Ethanol in Cell-free and Cell-occupied Calcium Algmate Membranes," *Biotech. Bioeng.* **28** (1986): 829 – 835.

Itamunoala, G. F., "Limitations of Methods of Determining Effective Diffusion Coefficients in Cell Immobilization Matrices," *Biotech. Bioeng.* **31** (1988): 714 – 717.

Tanaka, H., M. Matsumura, and I. A. Veliky, "Diffusion Characteristics of Substrates in Ca-Algmate Gel Beads, *Biotech. Bioeng.* **26** (1984): 52 – 58.

Zab orsky, O., *Immobilized Enzymes*, Cleveland, OH: CRC Press, 1973.

4

Industrial Applications of Enzymes

The applications of enzymes can be classified into three major categories: industrial enzymes, analytical enzymes, and medical enzymes. In this chapter, we review several industrial processes, utilizing industrial enzymes such as starch conversion and enzymatic hydrolysis of celluloses. Before we discuss the enzymatic hydrolysis of starch and cellulose, we review the organic chemistry of carbohydrates.

4.1 CARBOHYDRATES

Carbohydrates constitute a major class of naturally occurring organic compounds, including sugars, starches, and celluloses. They are essential to the maintenance of plant and animal life. Carbohydrates are classified into three major groups:monosaccharides, oligo-saccharides,[1] and polysaccharides. Monosacchandes are the simplest carbohydrate units. Oligosaccharides contain two or more of these simple mono sacchande units, and polysaccharides contain hundreds or thousands of them.

4.1.1 Monosaccharides

The basic carbohydrate molecules are simple sugars, or monosaccharides, which are polyhydroxy aldehyde, polyhydroxy ketone, and their derivatives. All simple monosaccharides have the general empirical formula, $(CH_2O)_n$, where n is the whole number ranging 3 to 8.

All monosaccharides can be grouped into two general classes as:
1. *aldoses:* contain a functional aldehyde grouping (–CHO), or
2. *ketoses:* contain a functional ketone grouping (> CO)
Oligo means few.

As indicated in Table 4.1, sugars can be further subclassified according to the number of carbons: trioses, tetroses, pentoses, and

---

[1] Oligo means few

[2] An asymmetric carbon atom has four different atoms or groups of atoms attached to it and may be a source of dissymmetry in the molecule. The asymmetric carbon atom may have two possible arrangements of the groups around it. The two structures may be nonsuperimposable mirror images, and can be expected to differ in the rotation of plane-polarized light to an equal extent but in the opposite direction.

Table 4.1 Classification of Monosaccharides

	$C_3H_6O_3$	$C_4H_8O_4$	$C_5H_{10}O_5$	$C_6H_{12}O_6$
Aldose family	CHO | CHOH | CH_2OH Aldotiose	CHO | CHOH | CHOH | CH_2OH Aldotiose	CHO | CHOH | CHOH | CHOH | CH_2OH Aldotiose	CHO | CHOH | CHOH | CHOH | CHOH | CH_2OH Aldotiose
Ketose family	CH_2OH | C=O | CH_2OH Ketotriose	CH_2OH | C=O | CHOH | CH_2OH ketotetrose	CH_2OH | C=O | CHOH | CHOH | CH_2OH Ketopentose	CH_2OH | C=O | CHOH | CHOH | CHOH | CH_2OH Ketohexose

hexoses. Monosaccharides have asymmetric carbon atoms[2]. The number of possible optical isomers for a compound can be determined by the formula 2^n, where n stands for the number of asymmetric carbons. As an example, aldohexose (see Table 4.1) has four asymmetric carbon atoms, second carbon through fifth carbon from the top. Therefore, it has 16 possible isometric forms, with eight L forms and eight D forms. The D form has OH on the right side of the highest-numbered asymmetric carbon (fifth carbon for aldohexos) and rotates polarized light in the +direction, while the L form has OH on the left side and rotates polarized light in the - direction.

Glucose (or dextrose) is one of aldohexoses which has two isometric forms (Figure 4.1): D and L. The D form predominates in the nature. Glucose is the most common and most important hexose and is found in most sweet fruits and in blood sugar.

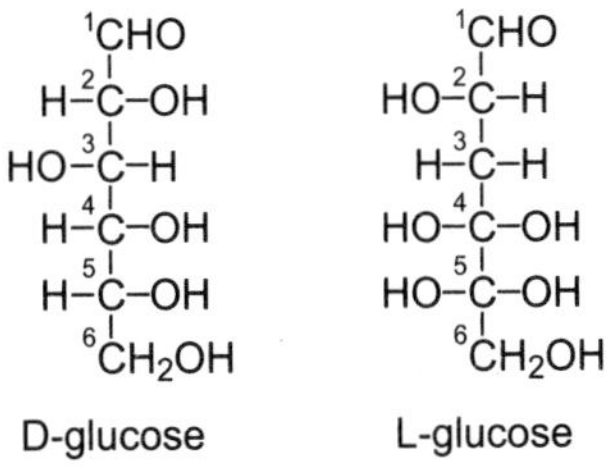

$$
\begin{array}{cc}
\text{D-glucose} & \text{L-glucose} \\
\end{array}
$$

^1CHO ^1CHO

$H-^2C-OH$ $HO-^2C-H$

$HO-^3C-H$ $H-^3C-H$

$H-^4C-OH$ $HO-^4C-OH$

$H-^5C-OH$ $HO-^5C-OH$

$^6CH_2OH$ $^6CH_2OH$

D-glucose   L-glucose

Fig. 4.1 Two isomeric forms of glucose.

In solution, very few sugar molecules exist with free aldehyde or ketone functional groups. Aldehydes and hydroxyls in a sugar molecule can react in a solution so that the H from the OH at the fifth carbon joins the aldehyde and the O from the same OH bonds to the first carbon, as shown in Figure 4.2.

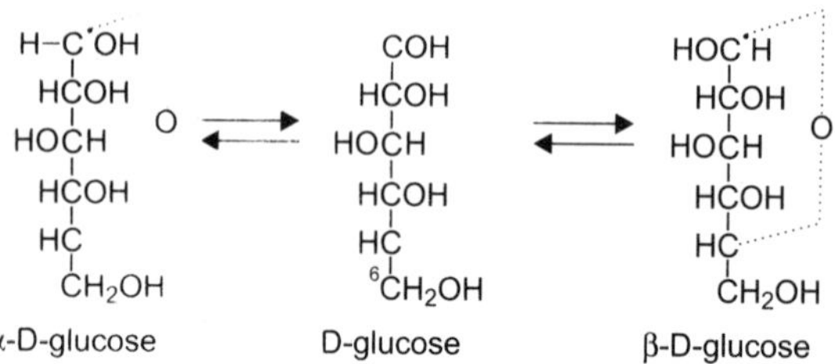

Fig. 4.2 Two stereoisomeric forms of D-glucose in solution: Fischer projection formulas.

The resulting structure has a ring form which is known as *cyclic hemiacetal.* There is an equilibrium between the ring and open forms. The open form allows the aldehyde or ketone group to react. The formation of a cyclic hemiacetal generates an additional asymmetric center at the original carbonyl atom. The new asymmetric center is known as the *anomeric carbon* (C* in Figure 4.2). In the linear *Fischer projection formulas,* the structure with the anomeric hydroxyl group oriented to the right is termed the α-form, and that to the left is β-form. Pure β-D-glucose equilibrates in water to give a mixture of 64 percent β- and 36 percent α-D-glucose.

A more realistic representation for the hemiacetal ring structure is the *Haworth projection formulas.* The formulas for α-D-glucose are shown in Figure 4.3. The shorthand form of the Haworth projection eliminates the Hs and indicates OHs by dashes. Five- and six-membered cyclic sugars are called *furanose and pyranose,* respectively.[3]

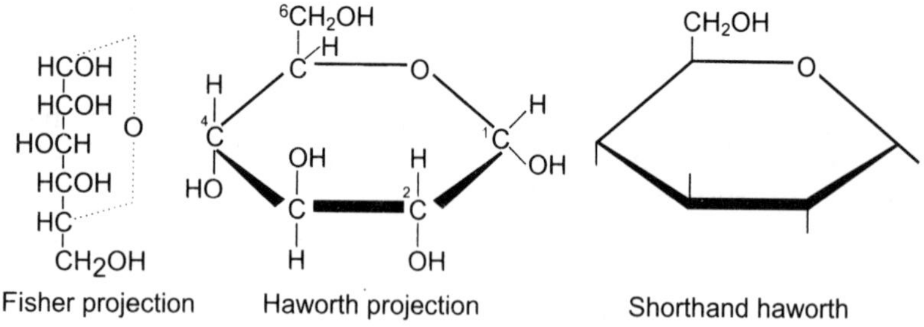

Fig. 4.3 Haworth projection formula of α-D-glucose (or α-D-glucopyranose)

Even though Haworth formulas give a sound representation of the ring structures of sugars, the real structure conformation can be most accurately represented by the *chair forms* of cyclohexane as shown in

---

[3] The terms furanose and pyranose arise from attachment of carbohydrate endings to the names of the cyclic compounds, furan and pyran.

Figure 4.4. However, despite the inaccuracy of the Haworth formulas, they are used more frequently than the chair conformation, because they are easier to draw and interpret.

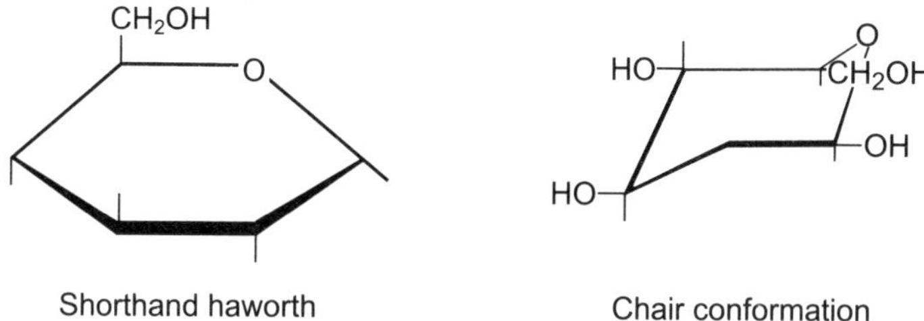

Fig. 4.4 Haworth and chair conformation of β-D-glucose (or β-D-glucopyranose).

Fructose is a keto sugar and is found in fruits and honey. Cyclization of fructose by formation of a hemiketal between the carbonyl group at the second carbon and the hydroxyl group at the fifth carbon gives a five-membered furanose ring, which can have two anomers, α- and β-D-fructose (or α- and β-D-fructofuranose), as shown in Figure 4.5. Fructose sweeter than other natural sugar. If we take the relative sweetness of cane sugar as one, glucose is measured to be 0.7 whereas fructose is 1.7. (Cheng and Lee, 1992)

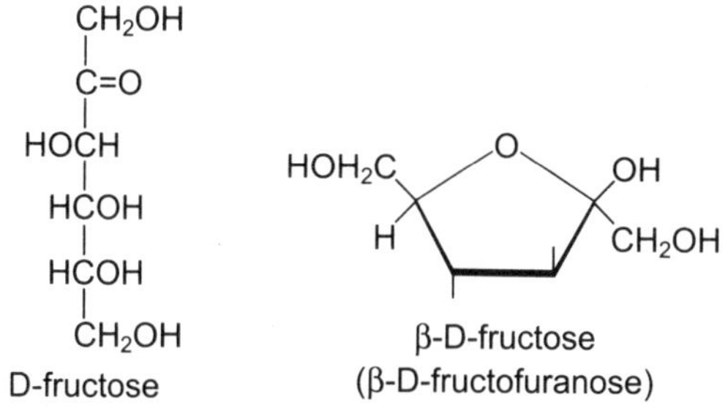

Fig. 4.5 Fructose

4.1.2 Disaccharides

Two sugars can link to each other by losing water from OHs to form disaccharides. Figure 4.6 shows the Haworth projection formulas of four important disaccharides: sucrose, lactose, maltose, and cellobiose, which all have the same molecular formulas, $C_{12}H_{22}O_{11}$. Sucrose and lactose are the most abundant and most important disaccharides of natural origin. Maltose and cellobiose are repeating units of polymeric starch and cellulose, respectively. Disaccharides may hydrolyze to form two monosaccharide molecules.

$$C_{12}H_{22}O_{11} + H_2O \longrightarrow C_6H_{12}O_6 + C_6H_{12}O_6$$

Sucrose, known as table sugar, is comprised of α-D-glucose and β-D-fructose. The aldehyde group (1′ carbon) of glucose is linked with the ketone group (2′ carbon) of fructose (1′ – 2′), so that no carbonyl group (–CO) from either monosaccharides portion is available as reducing agent. For this reason sucrose is termed a *nonreducing sugar.* Sucrose is the only nonreducing sugar among the four disaccharides.

Lactose, sugar present in milk, is a dimer of β-D-galactose bonded (1′ – 4′) with D-glucose. The aldehyde group of the left ring of lactose is used for linkage. However, the right ring of the lactose can be opened to react because its aldehyde group is not used for linkage. As a result, lactose is a reducing sugar.

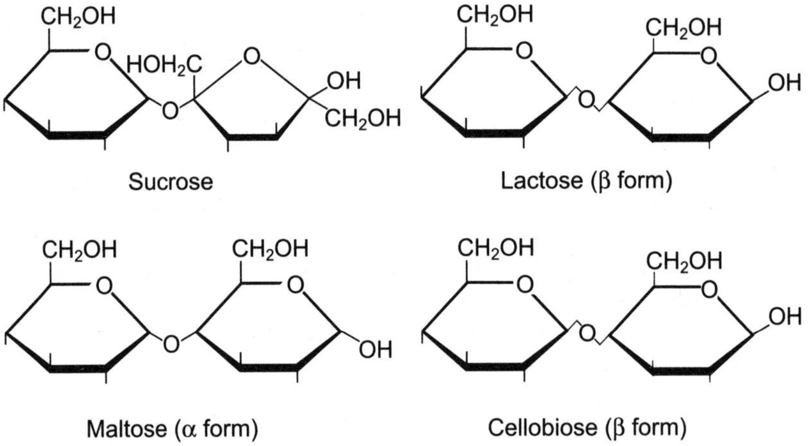

Sucrose

Lactose (β form)

Maltose (α form)

Cellobiose (β form)

Fig. 4.6 Important disaccharides

Maltose is repeating units of starch and can be obtained by the hydrolysis of starch using the diastase enzyme. Further hydrolysis of maltose yields two molecules of glucose. *Cellobiose,* a stereoisomer of maltose, is obtained by the partial hydrolysis of cellulose. Maltose and cellobiose are both reducing sugars, since the right rings may open to react, as reducing agents.

4.1.3 POLYSACCHARIDES

Polysacchandes consist of many simple sugar units linked together. One of the most important polysaccharides is starch, which is produced by plants for food storage. Animals produce a related material called glycogen.

Starch comprises a large percentage of cereals, potatoes, corn, and rice. Complete hydrolysis of starch yields glucose, but partial hydrolysis gives maltose as well. This shows that starch is a polymer of glucose units, joined by α-glycosidic linkage. Starch can be separated into two mam fractions by treatment with hot water. The insoluble component (10 percent to 25 percent) is *amylose,* the soluble component (75 percent to 90 percent) is *amylopectin.* The glucosyl residues (glucose minus water) of amylose are linked by α (1' – 4') glycosidic bonds in a single chain that contains up to 4,000 glucose units (Figure 4.7). The long linear molecule of amylose exists as a helix that contains six glucosyl residues per turn.

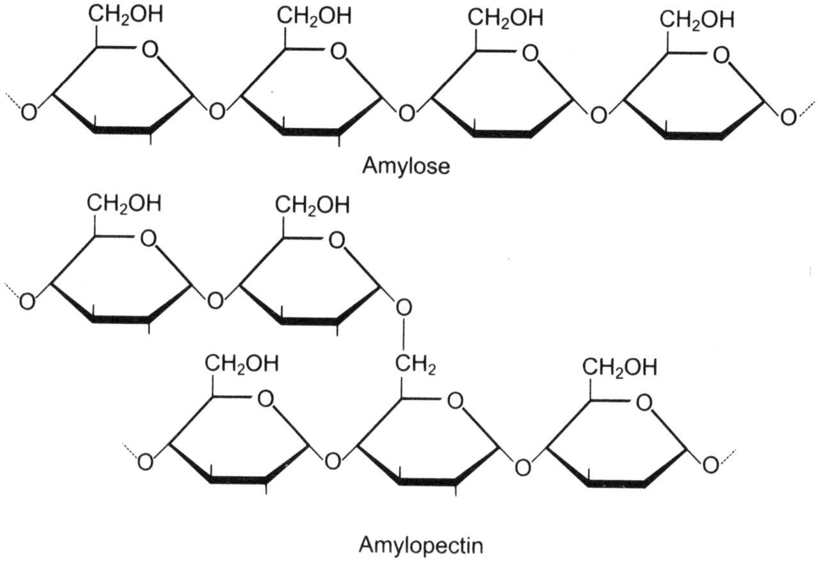

Fig. 4.7 Structure of amylose and amylopectin.

Amylopectin is a highly branched amylose. Various length of the linear chains, α (1' – 4') glucans containing 20 to 25 residues, are linked to a core chain by α (1' – 6') glycosidic bonds (Figure 4.7).

Amylose and amylopectin are degraded by α- and β-amylase, which are found in the pancreatic juice and saliva of animals, α-amylase is an *endoglycosidase* which attacks the amylose and amylopectin randomly along the α (1' – 4') bonds. β-amylase is an *exoglycosidase* which removes maltose units successively from the nonreducing end of the chain. Neither enzyme can hydrolyze the α (1' – 6') branch points, which can be degraded by other enzymes, called *debranching enzymes.* Another enzyme, *amylogucosidase* (also called glucoamylase), releases glucose units from the nonreducing end of the chain. Enzymatic hydrolysis of starch by sequential treatment with α-amylase and glucoamylase will produce glucose as the mam final product.

Dextrins, products of the partial hydrolysis of starch, are polysacchandes of lower molecular weight than starch. They are used in infant food because they are easier to digest than starches. Dextrins are sticky when wet and are used as mucilage on postage stamps and envelopes.

Cellulose is one of the three major structural components of all plant cell walls with two other components, hemicellulose and lignin. Cellulose is the most abundant organic compound of natural origin on the face of the earth. Complete hydrolysis of cellulose gives glucose. The cellulose molecule is comprised of long chains of cellobiose molecules joined together by β-1,4-glucosidic bonds as shown in Figure 4.8. The molecular weight of cellulose ranges from 300,000 to 500,000 (1,800 to 3,000 glucose units). The digestive systems of man and most other animals (except ruminants) do not contain the necessary enzymes (cellulase) for hydrolyzmg β-glucosidic linkages. However, cellulases are found in ruminants, various insects, fungi, algae, and bacteria.

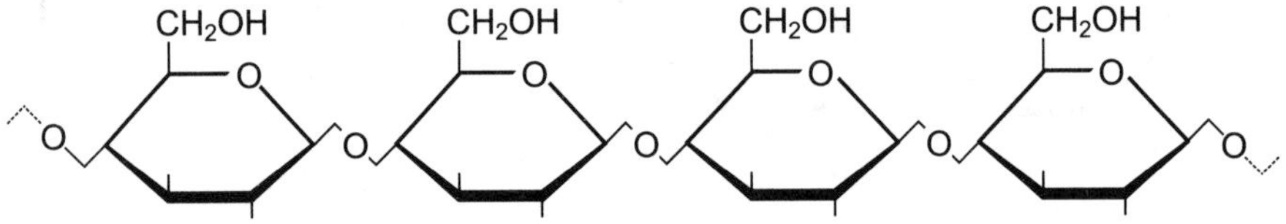

Fig. 4.8 Structure of cellulose

4.2 STARCH CONVERSION

In recent years, the conversion of starch to fructose has become a very important commercial process. *High-fructose corn syrup* (HFCS) is approximately twice as sweet as sucrose. It is used in soft drinks, canned fruits, lactic acid beverages, juice, bread, ice cream, frozen candies, and so on. HFCS can be obtained from a variety of cereals and vegetables, such as corn, wheat, rice, potatoes, and cassava. Corn is the most important source of HFCS because of low costs and excellent utilities of its by-products, corn meal, oil, gluten, germ, and fiber.

Corn Wet Milling: The first step of the HFCS process is the *com wet milling* (Joglekar et al., 1983). Corn is cleaned, shelled, and transferred to a large steep tank containing warm water (54°C) with 0.1 to 0.2 percent sulfur dioxide (pH 3–4). The steeping lasts about 40 hours. The sulfur dioxide inhibits fermentation and helps softening of the kernel.The steeped corn kernels are torn apart in a degerminating mill to free the germ (containing corn oil) and to loosen the hull. The germ is separated in a continuous liquid cyclone, washed, and dried for oil recovery. Starch and hull are ground and screened to eliminate the hull. The resulting mill starch contains 5 to 8 percent protein which is separated in a centrifuge. The separated-out starch is further purified in a hydroclone to reduce the protein content to a minimum level of 0.3 percent.

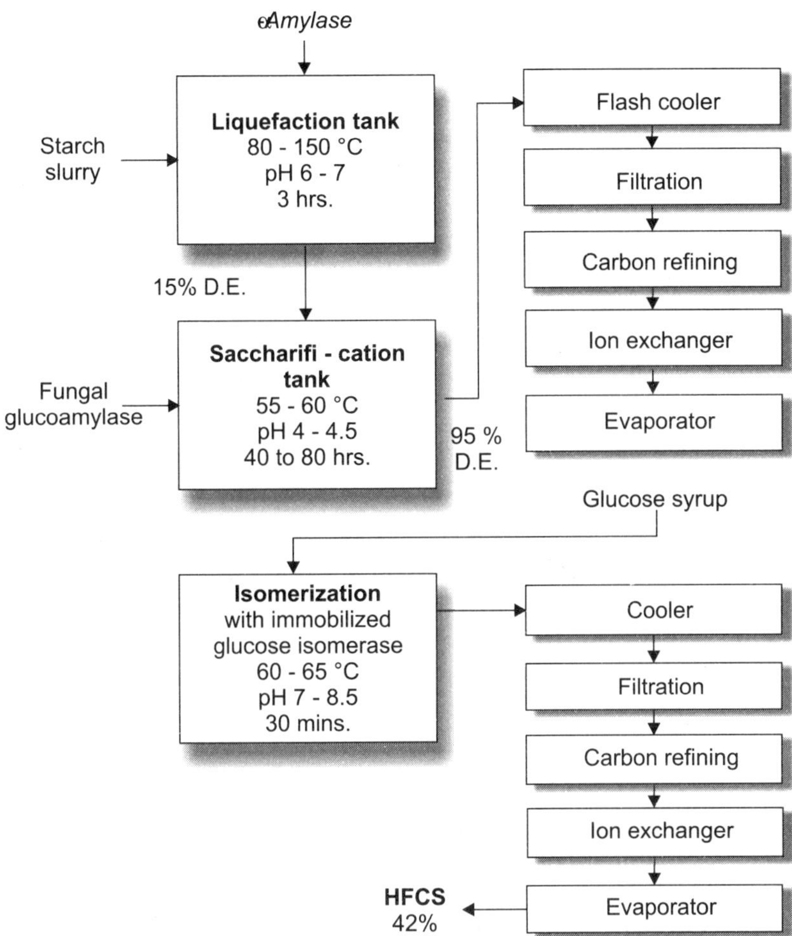

Fig. 4.9   Corn refining process (D.E.: dextrose equivalent)

Corn Refining Process: The starch slurry from the wet milling process is broken down to glucose and isomerized to fructose as shown in Figure 4.9. The starch slurry is gelatinized by cooking at high temperature (104 – 107°C) for 5 to 8 minutes and liquefied by α-amylase into low-molecular-weight dextrins and maltose. The syrup produced has D.E. (dextrose equivalent) of 15%. The enzyme used for this step, α-amylase, is thermostable and splits the starch at the interior of the molecule (Joglekar et al., 1983).

After liquefaction, the syrup is further hydrolyzed to glucose by the action of fungal glucoamylase, which acts on starch by splitting glucose units from the nonreducing end. It takes about 40 to 80 hours at a temperature between 55–60°C and pH of 4–4.5. After saccharification, the liquor (95 percent free glucose) is filtered, and passed through activated carbon, diatomaceous earth and ionexchange columns to remove impurities, color, and salts. It is then concentrated in an evaporator to 60 percent solids.

The glucose syrup obtained is isomerized to fructose by passing through an immobilized isomerase colum. The glucose isomerase is immobilized by an inert carrier, such as glass beads or DEAE-cellulose. A typical residence time is 30 minutes. The isomerization reaction is reversible with an equilibrium constant of about 1.0 at 60°C. Therefore, the expected final concentration of fructose will be less than half of the inlet glucose concentration. The finished product leaving the reactor contains 42 percent fructose, 50 percent glucose, and other saccharides. After isomerization, the syrup is purified by passing through a filter and ion-exchanger and is concentrated in an evaporator.

4.3 Cellulose Conversion

Cellulosic wastes have great potential as a feedstock for producing fuels and chemicals. Cellulose is a renewable resource that is inexpensive, widely available and present in ample quantities. Large amounts of waste cellulose products are generated by commercial and agricultural processes. In addition, municipal facilities must treat or dispose of tremendous quantities of cellulosic solid waste.

4.3.1 LIGNOCELLULOSIC MATERIALS

Lignocellulosic materials have a common basic structure, but vary greatly in chemical composition and physical structure.[4] Typically, these materials contain 30 percent to 60 percent cellulose, 10 percent to 30 percent hemicellulose (polyoses), and 10 percent to 20 percent ligmn. Cellulose provides strength and flexibility, while lignin supports and protects the cellulose from biological and chemical attack. Hemicellulose bonds lignin to cellulose.

Native cellulose is basically composed of *microfibrils*, which are bundles of lamellae containing an indefinite number of fibrillar units. Their schematic representation is shown in Figure 4.10. Cellulose molecules, hydrophilic linear polymers, are linked together to form *elementary fibrils* (or photofibrils), about 40Å wide, 30Å thick, and 100Å long. The linear polymers in an elementary fibril are oriented in a parallel alignment and are bounded by hydrogen bonds to form a crystalline region, which is surrounded by a disordered layer of cellulose molecules, an amorphous region or paracrystalline region (Rånby, 1969). Microfibrils in cell wall components are again surrounded by hemicelluose layer and lignin.

---

[4] Lignocellulose: The composition of woody biomass

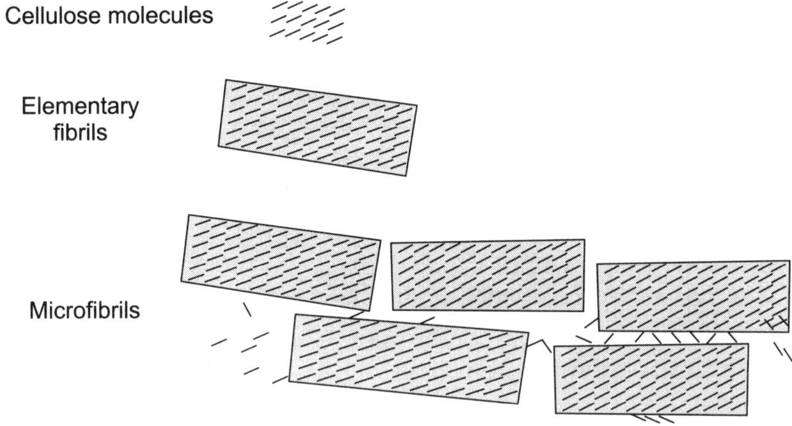

Fig. 4.10   Schematic cross section of a lamellae of cellulose microbibrils

Although cellulose and lignin are both polymers and major components of woody biomass, their chemical characteristics are totally different. *Lignin* is a complex aromatic biopolymer of high molecular weight and is formed by the polymerization of oxidatively formed radicals of p-hydroxycinnamyl alcohols (Hira et al, 1978). It should be noted that the term *lignin* cannot be regarded as one individually defined compound, but rather, as a collective term for a whole series of similar, large polymeric molecules which are closely related structurally to one another. The complexity of the chemical structure of lignin makes it very difficult to utilize except as a fuel.

Since isolated lignin is a by-product of the pulp industry, its economical utilization has been actively sought. Because of its relatively high calorific value (12,700 BTU/lb), most of waste lignin is being used as fuel in the chemical recovery processes of the pulp plants. Only a small part of lignin is utilized in adhesives, structural polymers, coating, dispersants, soil conditioner, pesticide carrier, and so on. Several processes for the conversion of polymeric lignin to simple chemical feedstock have been developed. However, only vanillin, dimethyl sulfide, and methyl mercaptan are produced in commercially significant quantities (Drew et al., 1978).

Hemicellulose (or polyose) is primarily composed of xylan, a branched polymer composed of five-carbon sugar, xylose. Typical polymerization degree of hemicellulose is 50 – 200, which is shorter than the cellulose molecules. The acid hydrolysis of hemicellulose, $(C_6H_{10}O_5)_n$, produces mainly xylose $(C_6H_{10}O_5)$, which can be converted to furfural, a chemical feedstock, or can be fermented to ethanol.

4.3.2 Cellulose Pretreatment and Hydrolysis

Major obstacles in the hydrolysis of cellulose are the interference of lignin (which cements cellulosic fibers together) and the highly ordered crystalline structure of cellulose. These obstacles necessitate a costly pretreatment step in which elementary cellulosic fibrils are exposed and separated.

Many pretreatments have been employed to enhance the degradation of lignocellulosic materials to glucose. The treatments fall into two general areas (Ryu and Lee, 1983):

1. *physical pretreatment* milling, irradiation, heating, and heating with other pretreatment, and

2. *chemical pretreatment* alkali treatments, acid treatments, deligmfication, and dissolving and reprecipitatmg.

Ball milling is the most commonly used pretreatment. It reduces crystallmity and particle sizes, while it increases surface area, bulk density, and the water soluble fraction. The major drawbacks of the milling are cost and the fact that noncellulosic substances are not removed.

Common chemical pretreatments of alkali and acid contacting improve hydrolysis by breaking down the lignin, hemicellulose, and cellulose. However, chemical hydrolysis is not specific and a variety of products are formed. A balance must be met between the enhanced hydrolysis and production of undesirable by-products. Deligmfication treatment, such as Kraft and sulfite pulping used in the pulp-and-paper industry, is too expensive to be considered as an economical pretreatment.

The rate and extent of enzymatic hydrolysis was found to be increased significantly by combining the pretreatment and reaction steps into one process (Kelsey and Shafizadeh, 1980; Ryu and Lee, 1983; Deeble and Lee, 1985; Jones and Lee, 1988). The separate processes can be combined by using an attrition bioreactor, which is a stirred reactor with stainless-steel balls (Ryu and Lee, 1983; Deeble and Lee, 1985; Jones and Lee, 1988). By using this reactor, the amount of time required for the hydrolysis of newsprint or sawdust can be reduced to hours as compared to days which are necessary in a regular stirred reactor. Enhanced conversion of cellulose in the attrition bioreactor is due to a combination of fact ors, including a reduction in crystallmity, an increase in pore volume and surface area, and an increase in the accessibility of glucosidic bond sites to the cellulase complex. It was also found that enzyme deactivation in the attrition bioreactor is not significant, since interfacial forces, not shear forces, cause the most deactivation. Elimination of the air-liquid interface by covering the reactor substantially increased the enzyme stability.

4.3.3 Cellulases

Many fungi are capable of producing extracellular enzymes that can degrade cellulose. They are *Trichoderma (T) reesei, T. viride, T. koningii, T. lignorum, Penicillium funiculosum, Fusarium solani, Sclerotium rolfsii,* and so on. Bacterial species such as *Cellulomonas* along with *Clostridium thermocellum* can also produce cellulases (Marsden and Gray, 1986).

The cellulase enzymes produced by submerged fermentation of the cellulolytic microfungus *Trichoderma* are usually classified into three general categories: endo-β,1,4 glucanases (endoglucanases), cellobiohydrolases, and β-1,4 glucosidases (cellobiase). Typical proportions of the endoglucanases and cellobiohydrolases in the extracellular protein from *Trichoderma reesei* grown in cellulose are 15 percent to 20 percent and 35 percent to 85 percent, respectively. The cellobiase makes up less than 1 percent of the protein (Marsden and Gray, 1986).

The endoglucanases hydrolyze cellulose molecules randomly along the molecule and are more effective in combination with cellobiohydrolases. Cellobiohydrolases cleave cellobiose and glucose from the nonreducmg end of the cellulose molecule. β-Glucosidases convert cellobiose to glucose, with some indication of the ability to degrade oligosaccharides. Endoglucanases and cellobiohydrolases must be absorbed by the cellulose before a reaction can occur, producing soluble sugars and oligomers. β-Glucosidases, however, remain in solution and hydrolyze the cellobiose in solution.

Various enzymes have been reported to be susceptible to deactivation upon shearing due possibly to the disturbance of their tertiary structure. Several investigators have studied the interfacial deactivation of *T. reesei* enzymes (Kim et al., 1982; Reese and Mandels, 1980). The addition of a surfactant has been found to substantially reduce enzyme deactivation. The surfactant impedes the migration of enzyme to the air-liquid interface.

Consequently, less enzyme reaches the interface, where it will deactivate by unfolding when subjected to surface tension forces. The cellulase deactivation due to this interfacial effect combined with the shear effect was found to be far more severe and extensive than that due to the shear effect alone (Kirn et al., 1982).

4.3.4 Kinetics of Enzymatic Hydrolysis of Cellulose

The development of a reliable kinetic model for the enzymatic hydrolysis of cellulose is helpful for the understanding of its mechanism and subsequent reactor design. Several kinetic models for the enzymatic hydrolysis of pure cellulose have been proposed by many investigators. The basic assumptions for the derivation of models are summarized in Table 4.2. Cellulose materials are composed of a highly ordered crystalline phase and a permeable amorphous phase. Some considered this multiplicity of cellulose structure and others ignored it and assumed that cellulose is a homogeneous material.

Concerning the enzyme acted on the substrate, some researchers considered all three components of the cellulase enzyme system (endoglucanases, cellobiohydrolases, and cellobiase). Since the kinetic behavior of the three enzyme system are different and not fully understood, simplified models were also suggested by assuming the cellulase system can be represented quantitatively by a single enzyme.

To illustrate the procedure for the development of kinetic models of the cellulose hydrolysis, let's examine the model proposed by Ryu et al. (1982). A kinetic model for the enzymatic hydrolysis of cellulose was derived based on the following assumptions:

1. The cellulosic material S_o is composed of amorphous matter S_a, crystalline matter S_c, and nonhydrolyzable merts S_x, and their rates of enzymatic degradation are different.

2. The cellulase system can be represented quantitatively by a single enzyme E.

3. The cellulase enzyme is first adsorbed E*on the surface of cellulose, followed by enzyme-substrate formation E*S and hydrolysis to release both product and enzyme.

4. The products (glucose and cellobiose) inhibit the cellulase enzyme competitively.

The reaction scheme based on the preceding assumption can be written as follows:

$$E \underset{k_d}{\overset{k_{ad}}{\rightleftharpoons}} E^* \tag{4.1}$$

$$E^* + S_a \underset{k_{2a}}{\overset{k_{1a}}{\rightleftharpoons}} E^*S_a \tag{4.2}$$

$$E^* + S_c \underset{k_{2c}}{\overset{k_{1c}}{\rightleftharpoons}} E^*S_c \tag{4.3}$$

$$E^* + S_X \underset{k_{2x}}{\overset{k_{1x}}{\rightleftharpoons}} E^*S_X \tag{4.4}$$

$$E + P \underset{k_{2p}}{\overset{k_{1p}}{\rightleftharpoons}} EP \tag{4.5}$$

$$E^*S_a \xrightarrow{k_{3g}} E^* + P \tag{4.6}$$

$$E^*S_c \xrightarrow{k_{3c}} E^* + P \tag{4.7}$$

Table 4.2 Comparison of Various Kinetic Models of Enzymatic Hydrolysis of Cellulose.

State of Substrate	Enzyme System[1]	Kinetic Approach[2]	Product Inhibition	Reference
Homogeneous Material	E_{12}	PSS	Product Competitive	Howell & Stuck (1975)
Homogeneous Material	E_{123}	MM	Competitive	Huang (1975)
Degree of Polymerization	E_1, E_2, E_3	MM	Non-competitive	Okazaki et al. (1978)
Homogeneous Material		PSS	Competitive	Howell & Mangat (1978)
Crystalline Amorphous	E_{123}	MM		Peitersen Ross (1979)
Crystalline Amorphous	E_{123}	PSS	Competitive	Ryu et al. (1982)
Homogeneous Material	E_{12},E_3		Non. competitive	Fan (1983

[1] E_1 = Endoglucanase, E_2 = cellobiohydrolase, E_3 = cellobiase E_{12} = Combmation of E_1 and E_2, E_{123} = Combination of E_1, E_2, and E_3

[2] PSS = Pseudo steady state, MM = Michaelis-Menten

The adsorption of the enzyme, Eq. (4.1), can be described by the Langmuir-type adsorption isotherm:

$$\frac{C_{E*}}{C_{E^*_{max}}} = \frac{K_e C_E}{1 + K_e C_E} \tag{4.8}$$

where $C_{E^*_{max}}$ is the concentration of the maximum adsorbed enzyme and K_e the adsorption constant. At low enzyme concentration ($K_e C_E \ll 1$), Eq. (4.8) reduces to

$$C_{E*} = K_e C_{E^*_{max}} \; C_E = K_d C_E \tag{4.9}$$

The rate of product formation is

If we assume that the total enzyme content is constant

$$\frac{dC_P}{dt} = K_{3a} C_{E*Sa} + k_{3c} C_{E*Sc} \tag{4.10}$$

If we assume that the total enzyme content is constant

$$C_{E0} = C_E + C_{E*} + C_{E*S_a} + C_{E*S_c} + C_{E*S_x} + C_{EP} \tag{4.11}$$

To derive a rate equation, let's follow the Briggs-Haldane approach as explained in Chapter 2, which assumes that the change of the intermediate concentration with time is negligible. This is known as pseudo-steady-state assumption. The change of the intermediate concentrations are:

$$\frac{dC_{E^*Sa}}{dt} = k_{1a}C_{E^*}C_{Sa} - (k_{2a} + k_{3a})\,C_{E^*Sa} \cong 0 \quad (4.12)$$

$$\frac{dC_{E^*Sc}}{dt} = k_{1c}C_{E^*}C_{Sc} - (k_{2c} + k_{3c})C_{E^*Sc} \cong 0 \quad (4.13)$$

$$\frac{dC_{E^*Sx}}{dt} = k_{1x}C_{E^*}C_{Sx} - k_{2x}C_{E^*Sx} \cong 0 \quad (4.14)$$

$$\frac{dC_{EP}}{dt} = k_{1p}C_{E}C_{P} - k_{2p}C_{EP} \cong 0 \quad (4.15)$$

From Eqs. (4.12) and (4.13),

$$dC_{E^*} = \frac{K_a}{C_{S_a}}C_{E^*S_a} + \frac{K_c}{C_{S_c}}C_{E^*S_c} \quad (4.16)$$

where
$$K_a = \frac{k_{2a} + k_{3a}}{k_{1a}} \quad \text{and} \quad K_c = \frac{k_{2c} + k_{2c}}{k_{1c}}$$

Substituting Eq. (4.16) into Eq. (4.10) yields

$$\frac{dC_P}{dt} = \left(k_{3a} + k_{3c}\frac{K_aC_{Sc}}{K_cC_{Sa}} \right)C_{E^*Sa} \quad (4.17)$$

To obtain an expression for $C_{E^*S_a}$, substitute Eqs. (4.9), 4.14, 4.15 and 4.16 into Eq. (4.11) and rearrange for $C_{E^*S_a}$,

$$C_{E^*Sa} = \frac{C_{E_0}}{\dfrac{K_a}{C_{S_a}}\left(1 + \dfrac{1}{K_d} + \dfrac{C_{S_a}}{K_a} + \dfrac{C_{S_c}}{K_c} + \dfrac{C_{S_x}}{K_x} + \dfrac{C_P}{K_pK_d} \right)} \quad (4.18)$$

where
$$K_x = \frac{k_{2x}}{k_{1x}} \quad \text{and} \quad K_p = \frac{k_{2p}}{k_{1p}}$$

Substitution of Eq. (4.18) into Eq. (4.11) yields

$$\frac{dC_P}{dt} = \frac{\left(k_{3a}\dfrac{C_{Sa}}{K_a} + k_{3c}\dfrac{C_{S_c}}{K_c} \right)C_{E_0}}{1 + \dfrac{1}{K_d} + \dfrac{C_{S_a}}{K_a} + \dfrac{C_{S_c}}{K_c} + \dfrac{C_{S_x}}{K_x} + \dfrac{C_P}{K_pK_d}} \quad (4.19)$$

If we define

$$\phi = \frac{C_{S_c}}{C_{S_a} + C_{S_c}} \tag{4.20}$$

and
$$\gamma = \frac{C_{S_x}}{C_{S_a} + C_{S_c} + C_{S_x}} = \frac{C_{S_x}}{C_S} \tag{4.21}$$

Eq. (4.19) can be rewritten as

$$\frac{dC_P}{dt} = \frac{\left[k_{3a}(1-\phi) + k_{3c}\phi\dfrac{K_a}{K_c}\right]C_{E_0}(1-\gamma)C_S}{K_a\left(1 + \dfrac{1}{K_d} + \dfrac{C_P}{K_p K_d}\right) + \left\{(1-\gamma)\left[(1-\phi) + \phi\dfrac{K_a}{K_c}\right] + \gamma\dfrac{K_a}{K_x}\right\}C_S} \tag{4.22}$$

The preceding equation can be simplified for initial reaction rate by substituting $C_S = C_{S_0}$, $C_P = 0$, and $\phi = \Phi$(4.23)

$$\left.\frac{dC_P}{dt}\right|_{t=0} = \frac{\left[k_{3a}(1-\Phi) + k_{3c}\Phi\dfrac{K_a}{K_c}\right]C_{E_0}(1-\gamma)C_{S_0}}{K_a\left(1 + \dfrac{1}{K_d}\right) + \left\{(1-\gamma)\left[(1-\Phi) + \Phi\dfrac{K_a}{K_c}\right] + \gamma\dfrac{K_a}{K_x}\right\}C_{S_0}} \tag{4.23}$$

which can be further simplified if we use pure cellulose as a substrate, then $\gamma = 0$

$$\left.\frac{dC_P}{dt}\right|_{t=0} = \frac{\left[k_{3a}(1-\Phi) + k_{3c}\Phi\dfrac{K_a}{K_c}\right]C_{E_0}C_{S_0}}{K_a\left(1 + \dfrac{1}{K_d}\right) + \left[(1-\Phi) + \Phi\dfrac{K_a}{K_c}\right]C_{S_0}} \tag{4.24}$$

Eq. (4.24) can be simplified to a Michelis-Menten equation as

$$\left.\frac{dC_P}{dt}\right|_{t=0} = \frac{r_{max}^{app}C_{S_0}}{K_M^{app} + C_{S_0}} \tag{4.25}$$

where r_{max}^{app}. and K_M^{app} represent the apparent maximum reaction rate and apparent Michaelis constants which are equivalent to

$$r_{max}^{app} = \frac{\left[k_{3a}(1-\Phi) + k_{3c}\Phi\dfrac{K_a}{K_c}\right]C_{E_0}}{(1-\Phi) + \Phi\dfrac{K_a}{K_c}} \tag{4.26}$$

$$\text{and} \qquad K_M^{app} = \frac{K_a\left(1 + \dfrac{1}{K_d}\right)}{(1 - \Phi) + \Phi\dfrac{K_a}{K_c}} \qquad (4.27)$$

The apparent kinetic parameters are expressed as the summation of the terms corresponding to amorphous and crystalline fractions. The r_{max}^{app} and K_M^{app} can be estimated by the Lmeweaver-Burk plot or the Langmuir plot of experimental data as explained in Chapter 2.

4.4 EXPERIMENT

4.4.1 Reducing Sugar Analysis

The total reducing sugars produced from the enzymatic hydrolysis ofcellulose can be measured by the dimtrosalicylic acid (DNS) method (Miller, 1959; Andreotti, 1980), as follows:

Materials:

1. 3,5-dimtrosalicylic acid
2. NaOH
3. distilled water
4. Rochelle salts (sodium potassium tartrate)
5. phenol (melt at 50°C)
6. sodium metabisulfite
7. phenolphthalein solution
8. 0.1NHCL
9. 1 g/ell α-D(+)glucose (Dextrose) standard solution
10. spectrophotometer

Dns Reagent Preparation:

1. Dissolve the 10.6 g 3,5-dinitrosalicylic acid and 19.8 g NaOH into 1416 mL distilled water in a stirred beaker, then add 306 g Rochelle salts, 7.6 mL (8.132 g) phenol, and 8.3 g sodium metabisulfite.
2. Titrate 3 mL sample with phenolphthalein with 0.1 N HC1. It should take 5-6 mL of 0.1 N HC1. Add NaOH into the preceding DNS solution if required (2 g NaOH for the additional 1 mL 0.1 N HC1).

Assay Procedures:

1. Dilute the standard glucose solution to make 0.2, 0.4, 0.6, 0.8, and 1.0 g/ell solutions. Use these known glucose solutions as samples for step 3 and 4. Prepare a calibration curve which shows the absorbance versus glucose concentration.
2. Dilute a sample so that the expected concentration is in the range of the calibration curve.
3. Place 1 mL sample in a test tube and add 3 mL DNS reagent. Place in boiling water for 5 minutes. Cool to room temperature.
4. Read light absorbance at 550 nm with a water blank.
5. Read the value of glucose concentration corresponding to the absorbance from the calibration curve.

4.4.2   Enzyme Assay: Filter Paper Activity

According to the International Union of Biochemistry, one enzyme international unit (IU) was defined as the enzyme strength which can catalyze 1 μmole of substrate per minute. In describing the activity of the cellulase, one IU is equivalent to the strength to release 1 μmoles of glucose per minute, because the molecular weight of the substrate, cellulose polymer, is not well defined.

Frequently we use filter paper as a substrate for the measurement of cellulase activity because it is well defined and yields reproducible results. The activity (IU) we obtain with filter paper is referred to as filter paper unit (FPU). The experimental procedure to measure the FPU of cellulase (Mandels et al., 1976) is as follows:

Materials:

1. Whatman No. 1 filter paper cut into 1 times 6 cm strips (about 50 mg)
2. 0.1 M sodium acetate buffer solution (pH 4.7)
3. cellulase: any commercial products, such as Type II cellulase (Sigma Chemical Co., St. Louis, MO)
4. DNS reagent

Assay Procedures:

1. Dissolve 0.1 g of cellulase in 10 mL of 0.1 M NaAc buffer. You are going to measure the enzyme activity of this solution.
2. Coil the filter paper strip and put it in a small test tube. Add 0.5 mL of enzyme solution and 1.0 mL of the buffer solution. Mix on a vortex mixer briefly and incubate for 1 hour at 50°C. Stop the reaction by immersing the test tube into an ice bath.

3. Take 0.1 mL (increase or decrease this amount depending on the enzyme strength) of sample and add water to make 1 mL. (If the sample amount is 0.1 mL, the dilution rate is 10.) Add 3 mL of DNS reagent to stop the reaction. Place in boiling water for 5 minutes. Cool to room temperature and read light absorbance at 550 nm using water as a blank. Read the glucose concentration corresponding with the absorbance from the calibration curve.

4. Calculate the FPU/mL of enzyme solution as follows:

$$\text{FPU/mL} = \frac{(1.5 \text{ mL}) \, (10)(\text{glucose conc. in mg / mL}}{(1 \text{ hr})(0.180 \text{ mg / } \mu\text{mole})(0.5 \text{ mL})(60 \text{ min / hr})}$$

$$= 2.778 \times (\text{glucose conc. inmg / mL}$$

4.4.3   Enzymatic Hydrolysis of Cellulose (Materials)

1. cellulose: Newfsprint (cut into 1.0 cm sections) or pure cellulose such as filter paper (cut into small pieces) and Solka Floe SW40 (James River Corp., Berlin, NH), and hammer-milled sulfite pulp
2. 2.5 g cellulase: Any commercial products, such as Type II Cellulase (Sigma Chemical Co., St. Louis, MO)
3. 600-mL glass tempering beaker (jacketed) (Cole-Parmer Instrument Co., Chi-ca-go, IL)
4. magnetic stirrer
5. water bath to control the temperature of the jacketed vessel
6. 0.1 M sodium acetate buffer solution (pH 4.7)

Procedures:

1. Set the temperature of the water bath at 50°C and circulate water through the water jacket to control the temperature.
2. Pour 300 mL 0.1M Sodium acetate buffer solution (pH 4.7) into the vessel and start the stirrer and add 2.5 g of cellulase. (Typical FPU of commercially prepared enzyme is about 100 FPU/g.)
3. Start the hydrolysis reaction by adding 6 g of cellulose into the vessel.
4. Take 3 mL sample once every hour and continue the experiment several hours.
5. Determine the total reducing sugar content in the sample by using the DNS method and prepare a curve to show the change of the reducing sugar concentration with time.

4.5   NOMENCLATURE

C	concentration, $kmol/m^3$
k	rate constant
K_M^{app}	apparent Michaelis constant, $kmol/m^3$
K	dissociation constant, dimensionless
K_d	defined in Eq. (4.9)
K_e	adsorption constant
r_{max}^{app}	apparent maximum rate of reaction per unit volume, $kmol/m\ s$
t	time, s
ϕ	defined in Eq. (4.20), dimensionless
Φ	ϕ at the beginning of the hydrolysis reaction, dimensionless
γ	defined in Eq. (4.21), dimensionless

Subscript

E	enzyme
EP	enzyme-product complex
ES	enzyme-substrate complex
P	product
S	substr

4.6   PROBLEMS

4.1 By employing the Michelis-Menten approach, derive the rate equation, Eq. (4.19), for the enzymatic hydrolysis of cellulose with the kinetic mechanisms described as Eqs. (4.1) through (4.7).

4.2 Write the mechanism of enzymatic hydrolysis of cellulose based on the following assumptions:

   a. The substrate is pure cellulose, which is assumed to be a material with a uniform quality.

   b. the cellulase system can be represented quantitatively by a single enzyme.

   c. The cellulase enzyme is first adsorbed on the surface E^* of cellulose, followed by the enzyme-substrate formation E^*_S and hydrolysis to release the product and the enzyme.

   d. The products (glucose and cellobiose) inhibit the cellulase enzyme competitively.

4.3 Derive the rate equation dC_p/dt for the enzymatic hydrolysis using the mechanism described in Problem (4.2), and explain how you can determine kinetic parameters.

4.4 Write the mechanism of enzymatic hydrolysis of cellulose based on the following assumptions:

  a. The cellulosic material S_o is composed of amorphous S_a, crystalline S_c, and nonhydrolyzable inert part S_x, and their rates of enzymatic degradation are different.

  b. The cellulose is first hydrolyzed at its surface to cellobiose by the synergistic action of enzymes, endoglucanases and cellobiohydrolases, which are denoted as E_{12}. The cellobiose is then hydrolyzed to glucose by cellobiase E_3 in the aqueous phase.

  c. The products (glucose and cellobiose) inhibit the cellulase enzyme competitively.

4.7 REFERENCES

Andreotti, R. E., "Laboratory Experiment for High Yield Cellulose Fermentation," *II International Course-ciim-Symposium on Byconversion and Biochemical Engineering*, New Delhi, India (1980).

Cheng, Y. L. and T. Y. Lee, "Separation of Fructose and Glucose Mixture by Zeolite Y, "*Biotech. Bioeng.* **40** (1992):498-504.

Deeble, M. F. and J. M. Lee, "Enzymatic Hydrolysis of Cellulosic Substance in an Attrition Bioreactor," *Biotech. Bioeng. Symp.* No. 15 (1985): 277-293.

Drew, S. W., K. L. Kadam, S. P. Shoemaker, W. G. Glasser, and P. Hall, "Chemical Feedstocks and Fuels from Lignin," *AIChE Symp. Ser.* **74**, 181 (1978):21-25

Fan, L. T. and Y.-H. Lee, "Kinetic Studies of Enzymatic Hydrolysis of Insoluble Cellulose: Derivation of a Mechanistic Kinetic Model," *Biotech. Bioeng.* **25** (1983):2707-2733.

Fengel, D., "Ideas on the Ultrastructural Organization of the Cell Wall Components, "*J. Polymer Sci. Part C* **36** (1971):383-392.

Hira, A., S. M. Barnett, C. H. Shieh, and J. Montecalvo, Jr., "An Extracellular Lignase: A Key to Enhanced Cellulose Utilization," *AIChE Symp. Ser.* **74**, 181 (1978): 17-20.

Howell, J. A. and M. Mangat, "Enzyme Deactivation during Cellulose Hydrolysis," *Biotechnol. Bioeng.* **20** (1978):847-863.

Howell, J. A. and J. D. Stuck, "Kinetics of Solka Floe Cellulose Hydrolysis by *Trichoderma viride* Cellulase," *Biotechnol. Bioeng.* **17** (1975): 873-893.

Huang, A. A., "Kinetic Studies on Insoluble Cellulose-Cellulase System," *Biotechnol. Bioeng.* **17** (1975): 1421-1433.

Joglekar, R., R. J. Clerman, R. P. Ouellette, and P. N. Cheremisinoff, *Biotechnology in Industry*, pp. 73-97. Ann Arbor, **MI**: Ann Arbor Sci. Pub., 1983.

Jones, E. O. and J. M. Lee, "Kinetic Analysis of Byconversion of Cellulose in Attrition Bioreactor," *Biotech. Bioeng.* **31** (1988):35-40.

Kelsey, R. G. and F. Shafizadeh, "Enhancement of Cellulose Accessibility and Enzymatic Hydrolysis by Simultaneous Wet Milling," *Biotechnol. Bioeng.* **22** (1980): 1025-1036.

Kirn, M. H., S. B. Lee, D. D. Y. Ryu, and E. T. Reese, "Surface Deactivation of Cellulase and its Prevention," *Enzyme Microb. Technol.* **4**(1982):99-103.

Mandels, M., R. Andreotti, and C. Roche, "Measurement of Saccharifying Cellulase,"*Biotechnol. Bioeng. Symp.* **6** (1976):21-33.

Marsden, W. L. and P. P. Gray, "Enzymatic Hydrolysis of Cellulose in Lignocellulosic Materials," *CRC Critical Review Biotech.* **3** (1986):235–276.

Miller, G. L., "Use of Dinitrosalicylic Acid Reagent for

Determination of Reducing *Sugar" Analytical Chem.* **31** (1959):426-428.

Okazaki, M. and M. Moo-Young, "Kinetics of Enzymatic Hydrolysis of Cellulose: Analytical Description of a Mechanistic Model," *Biotechnol. Bioeng.* **20** (1978):637-663.

Peitersen, N. and E. W. Ross, "Mathematical Model for Enzymatic Hydrolysis and Fermentation of Cellulose by *Trichoderma" Biotechnol. Bioeng.* 21 (1979):997-1017.

Ranby, B., "Recent Progress on the Structure and Morphology of Cellulose," *Adv. Chem. Ser.* **95** (1969): 139-148.

Reese, E. T. and M. Mandels, "Stability of the Cellulase of *Trichoderma reesei* under Use Conditions," *Biotech. Bioeng.* **22** (1980):323-335.

Ryu, D. D. Y. , S. B. Lee, T. Tassman, and C. Macy, "Effect of Compression Milling on Cellulose Structure and on Enzymatic Hydrolysis Kinetics," *Biotech. Bioeng.* **26** (1982): 1047-1067.

Ryu, S. K. and J. M. Lee, "Byconversion of Waste Cellulose by Using an Attrition Bioreactor,"*Biotech. Bioeng.* **25** (1983):53-65.

5

Cell Cultivations

The purpose of this chapter is to give an introduction to the basics of microbial, animal, and plant cells and their cultivation techniques and applications. It covers only what is necessary for understanding the terminologies and procedures introduced in this book.[1]

5.1 MICROBIAL CELL CULTIVATIONS

Even though the role of microorganisms in biotransformation was not recognized until the nineteenth century, microorganisms have been used by humans since prehistoric times in the preparation of food, alcoholic beverages, milk products, textiles, and so on. Today, the use of microorganisms is even more widespread than before. They are not only used for the traditional microbial processes but also for new processes such as the production of pharmaceuticals, industrial chemicals, enzymes, agricultural chemicals, waste water treatments, mineral leaching, and recombinant DNA technologies.

5.1.1 Microbial Cells

All living organisms were classified by Haekel in 1866 as the animal kingdom, the plant kingdom, and the *protist* as shown in Table 5.1. The protists refer relatively simple biological organisms compared to plants and animals and include algae, protozoa, fungi, and bacteria. The development of the electro microscope allowed the scientists to recognize that unit structure of all living organisms are divided into two category, prokaryotes and eukaryotes.

The *prokaryotic cell* is the unit of structure in two microbial groups: bacteria and blue-green algae. The prokaryotic cell is small and simple, as shown in Figure 5.1, which is not compartmentalized by unit membrane systems. The cell has only two structurally distinguishable internal regions: *cytoplasm* and *nuclear region* (or nucleoplasm). The cytoplasm has grainy dark spots as a result of its

---

[1] It is recommended that readers obtain any college-level textbooks in biology and micro-biology for further reading and reference as the need occurs.

content of *ribosomes,* which are composed of protein and ribonucleic acid (RNA). The ribosome is the site of important biochemical reactions for protein synthesis. The nuclear region is of irregular shape, sharply segregated even though it is not bounded by membrane. The nuclear region contains deoxyribonucleic acid (DNA), which contains genetic information that determines the production of proteins and other cellular substances and structures.

Table 5.1   Classification of living organisms

Multicellular	*Animals*			Eukaryotes
	Plants			
Unicellular	Protists	Algae Protozoa Fungi Molds Yeasts		
		Bacteria		Prokaryotes

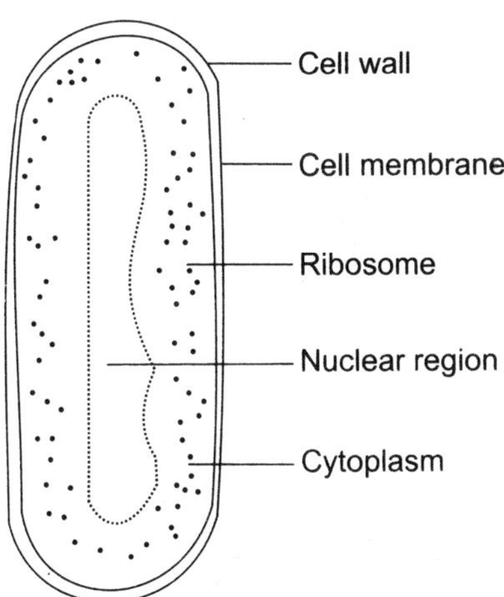

Fig. 5.1   Sketch of a typical prokaryotic cell.

The prokaryotic cell is surrounded with a cell wall and a cell membrane. The cell wall, considerably thicker than the cell membrane, protects the cell from external influences. The cell membrane (or cytoplasmic membrane) is a selective barrier between the interior of the cell and the external environment. The largest molecules known to cross this membrane are DNA fragments and low-molecular-weight proteins. The cell membrane can be folded and extended into the cytoplasm or internal membranes. The cell membrane serves as the surface onto which other cell substances attach and upon which many important cell functions take place.

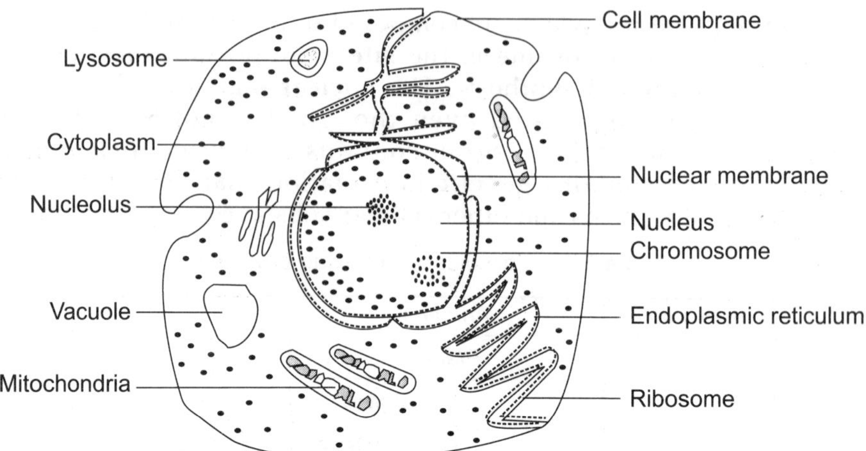

Fig. 5.2 Sketch of a typical eukaryotic cell.

The more complex *eukaryotic cell* is the unit structure in plants, animals, protozoa, fungi, and algae. The eukaryotic cell has internal unit membrane systems that segregate many of the functional components of the cell, as shown in Figure 5.2.

They are 1,000 to 10,000 times larger and more complex than prokaryotic cells. The nucleus is surrounded by a double membrane with pores 40 to 70 mµ wide, containing cytologically distinguishable chromosomes. The nucleus controls hereditary properties and all vital activities of the cell. The *chromosomes* are long and threadlike bodies and are found in the nuclei of cells, which contain the genes arranged in linear sequence in nucleoproteins (proteins plus nucleic acid). The cytoplasm contains large numbers of granules called *ribosome*, which are involved in continuous reactions to synthesize cell materials. The ribosome is especially concentrated along the rough surface of the *endoplasmic reticulum*, an irregular network of interconnected membrane-delimited channels. The *mitochondria* contain the electron transport enzymes that utilize oxygen in the process of energy generation. *Vacuole* and *lysosome* are organelles that serve to isolate various chemical reactions in a cell.

Microbial Nomenclature: Microbiologists use the *binomial system,* in which each organism has two names, for example, *Bacillus subtilis.* Proper names of organisms are always italicized. The first word is the name of the *genus* (plural, *genera)* and is capitalized. The genus name is a Latin or Greek word. Examples of genus names and their meanings are:

Bacillus: a small rod

Lactobacillus: a small milk rod

Micrococcus: a small grain

Clostridium. a small spindle

Pasteurella: after Louis Pasteur, Latinized

Salmonella: after Daniel E. Salmon, Latinized

Saccharomyces: sugar fungus

The second word in the name of a microorganism is the species name and is not capitalized. There may be several species with the same genus name, for example, *Bacillus subtilis, B. albus,* and *B. coagulans.* Note that when the same genus name is repeated several times, it is abbreviated

For industrial applications of microorganisms, bacteria and fungi are especially important. Therefore, they are discussed in more detail in the following sections.

5.1.2 Bacteria

Shapes: Bacteria are unicellular microscopic organisms. There are about 1,500 known species occurring in practically all natural environments. The typical diameter of the cell ranges from 0.5 to 1 μm. The lengths of bacterial cells vary greatly. Bacteria occur in a variety of shapes such as:

cocci: spherical or ovoid

bacilli: cylindrical or rod shaped

spirilla: helically coiled

Growth Pattern: Bacteria reproduce predominantly by a process known as *binary fission* as illustrated in Figure 5.3. This process involves several steps: cell elongation, invagination of the cell wall, distribution of nuclear material, formation of the transverse cell wall, distribution of cellular material into two cells, and separation into two new cells. This is an asexual reproductive process.

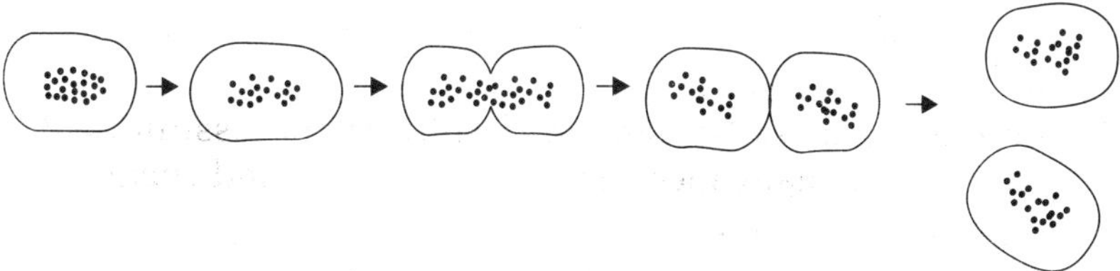

Fig. 5.3 Bacterial multiplications by binary fission.

Nutritional Requirements: It is important to be able to cultivate bacteria under laboratory conditions to study their characteristics. To be able to do this, one must know what food material and physical conditions are required. Bacterial cells that are actively growing are about 90 percent water. The elemental composition of bacteria is listed in Table 5.2. The nutritional medium to cultivate bacteria should contain those basic components listed.

All biological systems, from microorganisms to man, share a set of nutritional requirements, which are:

1. Sources of energy
   a. *phototrophs:* organisms which are capable of employing radiant energy.[2]
   b. *chemotrophs'.* organisms which obtain the energy for their activities and self-synthesis from chemical reactions that can occur in the dark.
2. Sources of carbon
   a. *autotrophs'.* organisms which can thrive on an entirely inorganic diet, using CO_2 or carbonates as a sole source of carbon.
   b. *heterotrophs:* organisms which cannot use CO_2 as a sole source of carbon but require, in addition to minerals, one or more organic substances, such as glucose or amino acids, as sources of carbon.
3. Sources of nitrogen: atmospheric nitrogen, inorganic nitrogen compounds, or other derived nitrogen.

Table 5.2 Elemental Composition of Bacteria

Components	Dry Weight %
Carbon	50
Oxygen	20
Nitrogen	14
Hydrogen	8
Phosphorus	3
Sulfur	1
Metals	4

4. Sources of sulfur and phosphorus: elementary sulfur, inorganic sulfur, or organic sulfur.
5. Sources of metallic elements: sodium, potassium, calcium, magnesium, manganese, iron, zinc, copper, and cobalt.
6. Sources of vitamins.

Physical Conditions: After determining the proper nutrients for the cultivation of bacteria, it is necessary to determine the physical environment in which the organisms will grow best. Three major physical factors to be taken into consideration are temperature, the gaseous environment, and pH.

Since microbial activity and growth are manifestations of enzymatic action, and since the rates of enzyme reactions increase

---

[2] Troph is derived from the Greek word *trophe* which means nourishing.

with increasing temperatures, the rate of microbial growth is temperature dependent. Depending on the temperature range over which they grow, bacteria are called *psychrophiles,*[3] *mesophiles,* or *thermophiles.* The temperature ranges in which each group are capable of growth and the optimum temperatures are summarized in Table 5.3.

The principal gases in the cultivation of bacteria are oxygen and carbon dioxide. There are four types of bacteria, according to their response to oxygen:

Table 5.3 Approximate Temperature Range for Growth of Various Bacteria

Types of Bacteria	Temperature Range for Growth	Optimum Temperature
Psychrophiles	−7 ~ 35°C	20 ~ 30°C
Mesophiles	7~ 45°C	30 ~ 40°C
Thermophiles	40 ~ 75°C	45 ~ 60 °C

1. *Aerobic* bacteria grow in the presence of free oxygen.
2. *Anaerobic* bacteria grow in the absence of free oxygen.
3. *Facultatively anaerobic* bacteria grow in either the absence or the presence of free oxygen.
4. *Microaerophilic* bacteria grow in the presence of minute quantities of free oxygen.

For most bacteria the optimum pH for growth lies between 6.5 and 7.5. Although a few bacteria can grow at the extremes of the pH range, for most species the minimum and maximum limits fall somewhere between pH 4 and pH9.

Spore Formation: Some bacteria form spores when growth ceases due to starvation or other causes. Spores are more resistant than normal cells to heat, drying, radiation, and chemicals. Spores can remain alive for many years; however, they can convert back to normal cells at proper conditions. Spore-forming bacteria are found most commonly in the soil.

Gram Reaction: *Gram staining* is one of the most widely used differential staining technique that separates bacteria into two groups: gram-positive and gram-negative.[4] The procedure is as follows:

1. Cover bacterial smear on a slide with crystal violet and let stand for 20 seconds and wash off the stain. All cells stain violet.

---

[3] Psychro is derived from the Greek word *psychros* which means cold.

[4] Differential staining technique: Staining procedures that elicit differences between bacterial cells.

2. Cover the smear with Gram's iodine solution and let stand for 1 minute. Crystal violet will form a complex with iodine within cells. All cells will remain violet.

3. Flood the smear with 95 percent ethyl alcohol for 10 to 20 seconds and rinse the slide with water. Gram-positive cells will remain violet, but gram-negative cells will be colorless.

4. Cover the smear with safranin, a red counterstain, for 20 seconds, wash for a few seconds, and dry. Gram-positive cells will remain violet, but gram-negative cells will become red.

There are characteristic differences between most gram-positive and gram-negative bacteria, which can be summarized in Table 5.3.

Table 5.3 Some Differences between Gram-positive and Gram-negative Bacteria.

Property	Gram-positive	Gram-negative
Result of Gram staining	Purple	Red
Susceptibility to penicillin	Marked	Much less
Digestibility by lysozyme	Many species	Requires pretreatment
Digestibility by trypsin	Resistant	Susceptible
Number of layers in wall	1	2
Thickness of wall	20–80 nm	10 nm
Lipid content of wall	0–3%	11–22%
Susceptibility to shear	Marked	Much less
Susceptibility to detergent	Marked	Much less

5.1.3 Fungi

Fungi are plants devoid of *chlorophyll* and are therefore unable to synthesize their own food.[5] They range in size and shape from single-celled yeasts to multicellular mushrooms. Among them, yeasts and molds are industrially important.

Yeasts: Yeasts are widely distributed in nature. They are found in fruits, grains and other food containing sugar. They are also in the soil, in the air, on the skin and in the intestines of animals. Since yeasts do not have chlorophyll, they depend on higher plants and animals for their energy. Yeasts are generally unicellular organisms and their shape is spherical to ovoid. Their size is 1 to 5 μm in width and from 5 to 30 μm in length. The cell wall is quite thin in young cells but thickens with age.

---

[5] Chlorophyll is a complex molecule which strongly adsorbs visible light.

The most common growth pattern for yeasts is budding, which is an asexual process as illustrated in Figure 5.4. A small bud (or daughter cell) is formed on the surface of a mature cell. The bud grows and is filled with nuclear and cytoplasmic material from the parent cell. When the bud is as large as the parent, nuclear apparatus in both cells is reoriented and the cells are separated. The daughter cell may cling to the parent cell, often even after the cells are divided.

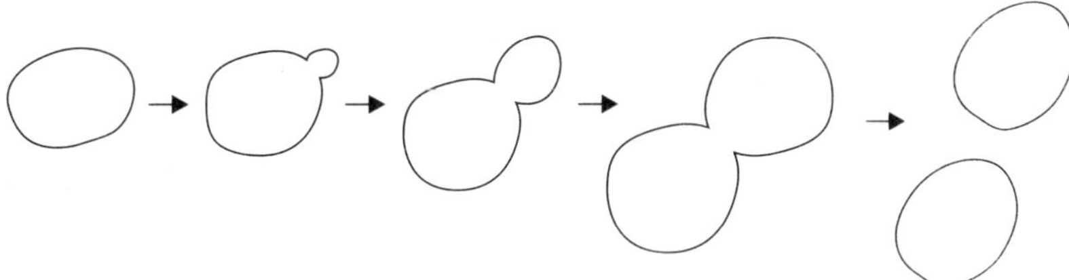

Fig. 5.4 A typical growth pattern of yeast by budding.

The most important yeasts are strains of *Saccharomyces cerevisiae*, which are used in the manufacture of wine and beer and in the leavening of bread.

Molds: Molds are filamentous fungi (Figure 5.5). A single reproductive cell or spore (conidia) is germinated to form a long thread, *hyphae*, which branches repeatedly as it elongates to form a vegetative structure called a *mycelium*. This consists of a multinucleate mass of cytoplasm within a rigid, much-branched system of tubes. Since a mycelium is capable of growing indefinitely, it can attain macroscopic dimensions. The most important classes of molds industrially are *Aspergillus* and *Penicillium*. Molds are used in the production of antibiotics, industrial chemicals, enzymes, and food additives.

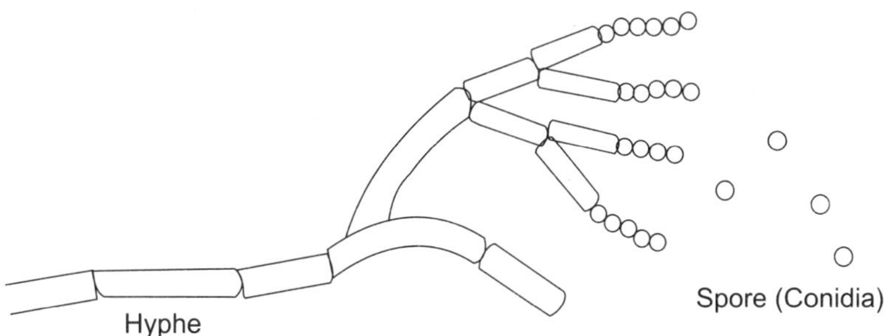

Fig. 5.5 Growth patern of ffilamentous fungus by branching

5.1.4 Culture Media

The growth of microbial population in artificial environments is called *cultivation*. A culture that contains only one kind of microorganism is *a pure culture*. A *mixed culture* is one that contains more than one kind of microorganism. The necessary steps for cultivating microorganisms are:

1. Preparing a culture medium in which a microorganism can grow best.
2. Sterilizing in order to eliminate all living organisms in the vessel.
3. Inoculating the microorganism in the prepared medium.

To cultivate microorganisms, culture medium has to be prepared in one of the commonly employed culture vessels: a test tube, a flask, a Petri dish, or a fermenter. There are two main types of culture media: *natural* (or empirical, or complex) and *synthetic* (or chemically defined) media. They vary widely in form and composition, depending on the species of organism to be cultivated and the purpose of the cultivation.

Natural media are those used on the basis of experience and not on the basis of exact knowledge of their composition and action. Natural or complex media usually contain peptones, beef extract, or yeast extract. When a solid medium is desired, a solidifying agent such as gelatin or agar may be incorporated into the medium. Examples of a relatively simple liquid and a solid medium that support the growth of many common heterotrophs are nutrient broth and nutrient agar. Their composition is as follows:

- *Nutrient broth:* 3 g of beef extract, 5 g of peptone, 5 g of yeast extract, and water to make 1 L.

Table 5.4 Typical Growth Medium for Yeasts

Glucose	100 g
Yeast extract	8.5 g
NH_4Cl	1.32g
$MgSO_4$	0.11 g
$CaCl_2$	0.06 g
Antifoam	0.2 mL
Water to make	1L

- *Nutrient agar.* the same ingredient as Nutrient broth, 15 g of agar and water to make 1 L.

Typical growth medium for yeasts is listed in Table 5.4.

Synthetic media consist of dilute, reproducible solutions of chemically pure, known inorganic and/or organic compounds. They are often required for research purposes. The medium may be as simple as inorganic ammonium salt plus minerals and a sugar, or as complex as purified casein with added vitamins, minerals, and a sugar. These media have the added advantage that they can be produced with a constant composition year after year.

Sterilization: After a suitable culture medium is selected for the cultivation of a specific microorganism, it is poured into a culture vessel. If you use test tubes or flasks as your culture vessel, the ends of test tubes or flasks should be covered with a suitable closure to allow for the exchange of gases with the atmosphere, yet to keep foreign organisms out of the media. Various types of closures are used in the modern laboratory including cotton plugs, plastic foam, screw caps, metal caps, and aluminum foil.

The medium is then sterilized to eliminate all living organisms in the vessel. The most common method of sterilization is by moist heat (steam under pressure) in an autoclave. Generally, the autoclave is operated at approximately 15 psi at 121°C. The time of sterilization depends on the nature of the material, the type of container, and the volume. For example, test tubes of liquid media can be sterilized in 15 to 20 minutes at 121°C.

Inoculation: Inoculation is the seeding of a culture vessel with the microbial material (inoculum). The inoculum is introduced with a metal wire or loop which is rapidly sterilized just before its use by heating it in a flame. Transfers of liquid culture are often made by using a sterilized pipette. The inoculation is usually done in a laminar flow hood to minimize the risk of contamination. It is important to know proper pipetting techniques for inoculating or sampling during cultivation.

5.1.5 Example of Penicillin Production

The production of penicillin represents one of the most remarkable examples of the successful scale-up of a laboratory-scale cultivation of a microorganism into a profitable industrial-scale process.

Between 1941 and 1945, U.S. production of penicillin rose from virtually nothing to more than 650 billion units per month, whereas its cost dropped from $20 per 100,000 units to roughly 60 cents (Sturchio, 1988).[6] The yield of penicillin was also increased about a thousand times, which was due to the following (Pelczar and Reid, 1972):

1. Improvements in composition of the medium - corn steep liquor as a growth promoter, substitution of lactose for glucose, and addition of phenylacetic acid.
2. Development of the submerged culture technique.

---

[6] One unit is equivalent to about 0.6 μg of the international standard for active penicillin

3. The production of mutant strains of *Penicillium chrysognum* by X-ray and ultraviolet radiation.

4. Refinements of downstream separation techniques.

Today, the yield is even higher than ever by selecting better mutant strains, and improving medium and fermentation techniques. The major steps in penicillin production are as follows (Queener and Swartz, 1979):

1. Preparation and sterilization of medium: Typical medium consists of corn steep liquor (4 percent to 5 percent dry weight); an additional nitrogen source such as soy meal, yeast extract, whey; a carbon source such as lactose; and various buffers.

2. Inoculation: Lyophilized spores are grown in an agar slant culture, which is inoculated into a shake flask culture, followed by primary, secondary seed culture and large-scale fermenter in increasing volume. This gradual increase of the seed culture volume is used in order to make the inoculum size large enough so that each step is reasonably short and large-scale equipment is used efficiently.

3. Cultivation: A stirred fermenter is employed in fed-batch mode by feeding glucose and nitrogen during cultivation. Typical size of the vessel is 40,000 to 200,000 liters. Oxygen is supplied by sparging air at a rate of 0.5 to 1.0 volumes of air per fluid volume per min. The power input by the turbine agitator and sparged air is about 1 to 4 watts per liter. The pH is maintained at 6.5. In a typical penicillin fermentation, most of the cell mass necessary is obtained during the first 40 hours. The penicillin starts to be produced at the exponential growth phase and continues to be produced until it reaches the stationary phase. The growth must continue at a certain minimum rate to maintain the high penicillin productivity. This is why glucose and nitrogen are fed continuously during the fermentation instead of being added at the beginning. Penicillin is excreted into the medium.

4. Downstream processing: After removing the mold mycelium, the penicillin is separated from the broth by means of a two-stage continuous countercurrent extraction with amyl or butyl acetate.

5.2 ANIMAL CELL CULTIVATIONS

Cells removed from animal tissue can be cultivated in nutritional medium outside the donor's body. Cultured cells grow by increasing in number and size. Tissue culture methodology has given researchers the opportunity to study cancer cells, to classify malignant tumors, to determine tissue compatibility in transplantation, and to study specific cells and their interactions.

The mammalian cell culture technique can be employed to produce clinically important biochemicals such as human growth hormones, interferon, plasminogen activator, viral vaccines, and monoclonal antibodies. Traditionally, these biochemicals had been produced using living animals or extracted from human cadavers. As examples, monoclonal antibodies can be produced by cultivating hybridoma cells in the peritoneal cavity of mice, and the human growth hormone to cure dwarfism can be extracted from human cadavers. However, the quantity obtained from these methods is quite limited for the wide clinical usages of the products.

Some mammalian gene products can be also produced by bacterial systems using recombinant DNA technology. Fast growth rate and inexpensive medium requirements of bacterial cells make them an economical alternative to the mammalian cell culture. However, bacteria lack the capability of the *post-translational modifications*, which involve proteolytic cleavage, subunit association, or a variety of additional reactions such as glycosylation, methylation, phosphorylation, or acylation. These modifications are important for proper biological activity of a product. For example, glycosylation can protect proteins against proteolytic breakdown, maintain structural stability, and alter antigenicity (Butler, p. 43, 1987). Therefore, the inactive polypeptides have to be isolated and refolded into proper structures of active proteins *in vitro* which can be difficult and expensive procedures.

However, it is difficult to cultivate large quantities of mammalian cells because of the following reasons (Feder and Tolbert, 1983):

1. They are larger and more complex than most microorganisms.
2. Their growth rate is very slow compared to the micro-organisms. Therefore, the productivity is low and the maintenance of sterility is difficult.
3. They are enclosed with a delicate plasma membrane without the tough cell wall normally found in microorganisms or plant cells. As a result, they are fragile.
4. Their nutritional requirements are not fully defined yet, requiring expensive blood serum for medium.
5. They are part of an organized tissue rather than an individual cellular organism.
6. Most animal cells only grow when attached to a surface.

5.2.1 Animal Cells

Animal cells are eukaryotic cells. They are bound together by intercellular material to form tissue. *Tissue* is customarily divided into four categories: epithelium, connective tissue, muscle, and nerve. *Epithelial tissue* forms the covering or lining of all free body surfaces, both external and internal. In *connective tissue*, the cells are always embedded in an extensive intercellular matrix, which may be liquid, semisolid, or solid. *Muscle cells* are usually elongate and bound together into sheets or bundles by connective tissue. Muscles are responsible for most movement in higher animals. *Nerve cells* are composed of a cell body, containing the nucleus, and one or more long thin extensions called fibers. Nerve cells are easily stimulated and can transmit impulses very rapidly.

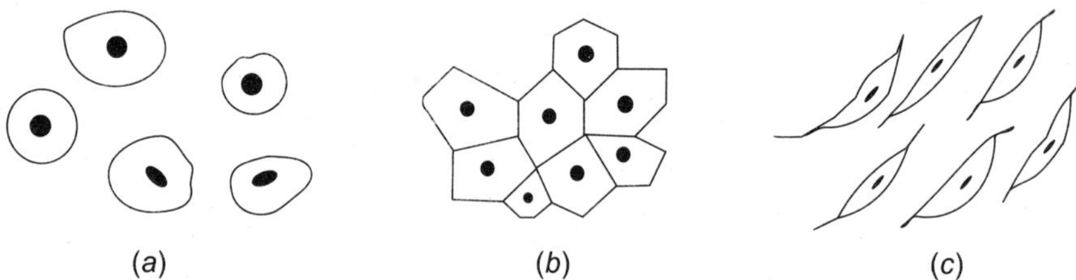

(a)   (b)   (c)

Fig. 5.6 Animal cells commonly used in culture: (a) lymphocytes, (b) epithelial cells, and (c) fibroblast cells.

Suspension Cells: Blood and lymph are rather atypical connective tissues with liquid matrices. Cells from blood or lymph fluids are suspension cells, or nonanchorage dependent when grown in culture. Nonanchorage-dependent cells do not require a surface to grow on.

As an example, lymphocytes (Figure 5.6a) derived from the lymphoid tissue are nonanchorage-dependent cells and spherical with a diameter of 10–20 μm. They can be cultivated in a liquid suspension medium in a similar way to bacteria.

Anchorage-Dependent Cells: Most normal mammalian cells are anchorage-dependent, that is, they require a surface for attachment and growth. The most widely used anchorage-dependent cell types are epithelial or fibroblast (broadly classified as connective) cells (Figure 5.6 b and c). Anchorage-dependent cells require wettable surface to grow on, such as glass or plastic. Petri dishes and roller bottles are the most commonly used devices. Bottles are laid on a slowly rotating roller in an incubator. A one-liter bottle typically contains 100 mL of medium to facilitate cells both to grow on the wall and to be exposed to medium and gas. However, roller bottles are suitable only for small-scale laboratory use because of small surface area per unit vessel volume (about 500 cm^2 per 1 L).

The area-to-volume ratio can be increased by growing cells on spongy polymers, on a ceramic matrix, in hollow fibers, in microcapsules, or on microscopically small beads called microcarriers.

5.2.2 Growth Media

The nutritional requirements of mammalian cells are more stringent than those of microorganisms because, unlike microorganisms, animals do not metabolize inorganic nitrogen. Therefore, many amino acids and vitamins should be provided. Typical medium contains amino acids, vitamins, hormones, growth factors, mineral salts, and glucose. Furthermore, the medium needs to be supplemented with two to twenty percent (by volume) of mammalian blood serum. The serum provides components that have not yet been identified but have been shown to be necessary for culture viability. Table 5.5 shows the contents of one of the commonly used media, known as Eagle's medium (Eagle, 1959).

The serum in the medium is not only expensive but also can be the source of virus or mycoplasma contamination. Since the chemical nature of serum is not well defined, its contents may vary batch after batch, which can affect the result of culture. The presence of many different proteins in serum can also complicate the downstream separation processes. For these reasons, many attempts have been made to formulate serum-free media. These formulations contain purified hormones and growth factors which can substitute for serum supplements (Butler, p.11, 1987).

5.2.3 Monoclonal Antibodies

Lymphocytes are specialized white blood cells involved in the immune response. B-lymphocytes, the producer of antibodies, are present in the spleen, lymph nodes, and blood. When a foreign substance enters the body of vertebrate animals, lines of B-lymphocytes proliferate and secrete protein molecules called immunoglobulin or antibodies. The antibodies have combining sites that recognize the shape of particular determinants on the surface of the foreign substance, or *antigens*. As a result, the antibodies can bind to the antigens, and neutralize and eliminate the foreign substances. Due to their specificity to identify particular molecules or cells, the antibodies have been important tools for researchers and clinicians in detecting the presence and level of drugs, bacterial and viral products, hormones, and other antibodies in blood samples.

There are five recognized classes of antibodies: immunoglobulin G, A, M, D, and E. Figure 5.7 illustrates the structure of the antibody, immunoglobulin G (IgG), which has a Y-shaped protein structure comprised of one pair of heavy chain and light chain polypeptides, linked by disulfide bonds. Each chain has two regions: (1) variable region which is different for each antibody and provides differently shaped combining sites that bind specifically to different antigens, and (2) constant region for all antibodies of a given subclass. The

Table 5.5   Composition of Eagle's Minimum Essential Medium (Eagle, 1959)

Compound	mg/L	Compound	mg/L
L-Amino Acids		**Vitamins**	
Arginine	105	Choline	1
Cystine	24	Folic acid	1
Glutamine	292	Inositol	2
Histidine	31	Nicotinamide	1
Isoleucine	52	Pantothenate	1
Leucine	52	Pyridoxal	1
Lysine	58	Riboflavin	0.1
Methionine	15	Thiamine	1
Phenylalanine	32		
Threonine	48	**Salts**	
Tryptophan	10	NaCl	6800
Tyrosine	36	KC1	400
Valine	46	CaC12	200
Carbohydrate		$MgCl_2\ 6H_2O$	200
Glucose	1000	$NaH_2PO_4\ 2H_2O$	150
Serum	5-10%	$NaHCO_3$	2000

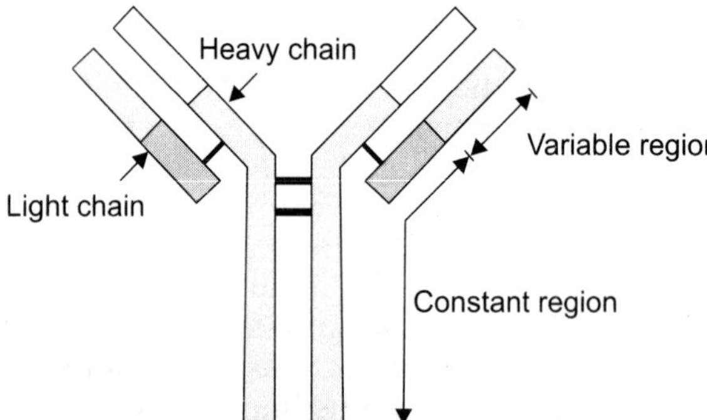

Fig. 5.7   Structure of an antibody immunoglobulin G (IgG).

nature of antigen-antibody binding is analogous to that of the enzyme-substrate complex discussed in Chapter 2.

There are many different lines of B-lymphocytes and each produce different antibodies that recognize specific antigenic determinants. Therefore, when an animal is injected with an immunizing agent, it responds by making diverse mixtures of antibodies, which are virtually impossible to separate. To produce large amounts of identical antibody (*monoclonal antibodyl*) which recognize only one chemical structure, we should be able to grow a specific cell line of

B-lymphocyte. However, it was found that antibody-secreting cells cannot be maintained in a culture medium.

Cell Fusion: Unlike antibody-secreting cells, *myeloma cells,* malignant tumor cells of the immune system, can be cultured continuously. Kohler and Milstein (1975) developed a method to fuse (hybridize) B-lymphocytes from the mouse spleen with mouse myeloma cells, so that the fused cell, hybrid-myeloma (or *hybridoma)* cell, can have the characteristic of the both cell lines: that is, the production of specific antibodies and the immortality. Since the hybridoma is derived from a single B-lymphocyte, it produces only one kind of antibody, thus a monoclonal antibody.

A typical procedure for the cell fusion is as follows (Figure 5.8):

1. Inject a chosen antigen into a mouse. The immune system in the mouse responds by proliferating B-lymphocyte cells that secrete antibodies.

2. Remove the spleen of the mouse and separate B-lymphocyte cells.

3. Cultivate a suitable malignant myeloma cells deficient in HPGRT (hypoxanthine guanine phosphoribosyl transferase), which is a genetic marker for the selection of the hybrid cells after fusion.

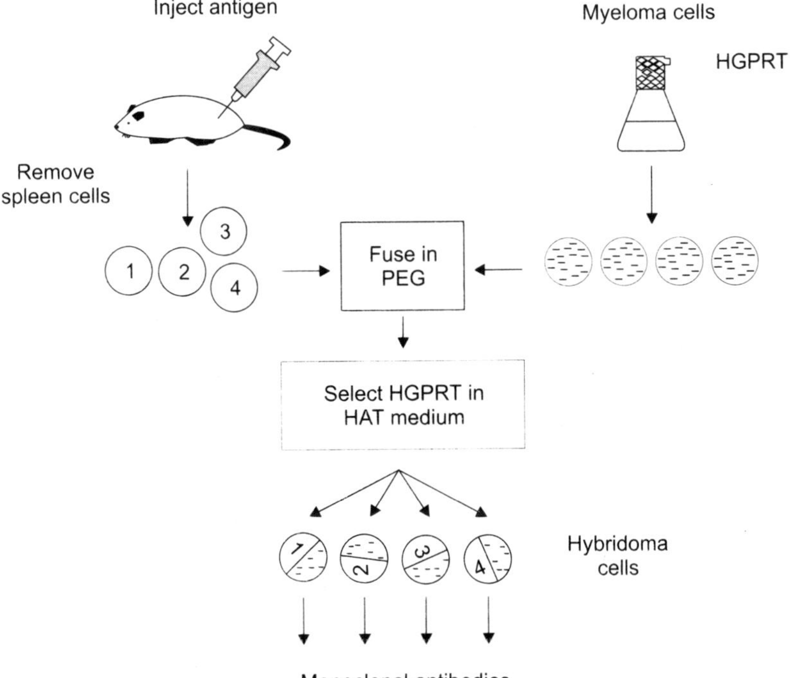

Fig. 5.8 The procedure for the fusion of B-lymphocytes

4. Fuse the B-lymphocyte cells with myeloma cells by mixing them in a medium containing 40 percent to 50 percent polyethylene glycol (PEG). The medium will contain the mixtures of B-lymphocytes, myelomas, and hybrid-myeloma cells. The B-lymphocytes contain

   HPGRT and also the hybridoma cells do. Therefore, the myeloma cells are noted as HPGRT-, whereas B-lymphocytes and hybridoma are as HPGRT$^+$.

5. Select HPGRT$^+$ cells by culturmg the mixture in a medium containing HAT (hypoxanthine, aminopterin, and thymidine), which is growth inhibitory to HPGRT$^+$ cells. Therefore, the myeloma cells will die in this medium while the hybridoma cells proliferate. The unfused lymphocytes will die due to their limited lifespan.

Antibody Assays: Monoclonal antibodies can be analyzed by the solid-phase assay techniques such as enzyme linked immunosorbent assays (ELISA) or radioimmuno assays (RIA). The typical assay procedures are as follows (Figure 5.9)

1. Apply a specific antigen (Ag) solution to a microtitre plate capable of adsorbing antigens by nonspecific hydrophobic interaction. Add a non-interfering protein such as bovine serum albumin (BSA) to occupy any remaining attachment site on the plate.

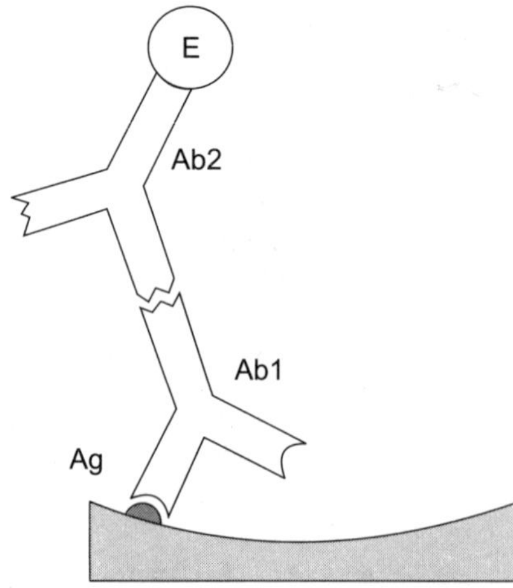

Fig. 5.9 Schematic diagram for the ELISA antibody assay. The enzyme will be replaced by a radioactive label in the case of RIA assay.

2. Add the monoclonal antibody (Abl) solution to be analyzed to the plate. The antibody will be bound to the antigen on the solid surface.

3. Add a second monoclonal antibody (Ab2) that has specificity against the total immunoglobulin of the species from which the hybridoma was originated. An enzyme (E) for the ELISA or a radioactive label for the RIA is covalently attached to the second antibody.

4. RIA: Measure the radioactivity by autoradiography or by scintillation counting of the solid support.

   ELISA: Add an appropriate substrate and measure its product by spectrophotometer.

5.3. PLANT CELL CULTIVATIONS

Plants are the valuable source of various chemical compounds such as pharmaceuticals, flavors, pigments, fragrances, and agrochemicals. These products, known as *secondary metabolites,* are usually produced in trace quantities in plants and have no obvious metabolic function.[7] They seem to serve as a chemical interface between the producing plants and their surrounding environment, such as adaptations to environmental stresses or chemical defenses against microorganisms, and higher predators.

Despite substantial advances in synthetic organic chemistry, many secondary metabolic compounds are either too difficult or too costly to synthesize. Some complex mixtures like rose oil cannot be constituted by manufacture. The majority of the plants that produce commercially useful substances are grown in tropical and subtropical regions of the world. As a result, the availability and costs of these materials depend on the political and economic circumstances of the countries of origin.

In order to produce secondary metabolic products from a plant, exogenous plant tissue instead of a whole plant, may be cultivated as a suspension culture in an aseptic condition. The technical rationale for using plant tissue is based on the unique biochemical *totipotency* of plant cells.[8]

The cultivation of plant cells instead of the whole plant for the production of metabolites can provide several advantages:

1. Plant cells can be cultivated where they are needed regardless of weather and geographical conditions. As a result, there is no need to ship or store bulky raw materials.

---

[7] Primary metabolites are produced in larger quantities than the secondary metabolites and have specific metabolic functions. Primary metabolites obtained from higher plants are used as foods, food additives, and industrial raw materials, such as carbohydrates, vegetable oils, protein, and fatty acids. They are generally high-volume, low-value bulk materials.

[8] The totipotency is the ability to generate or regenerate a whole organism from a part.

2. The product quality and yields can be also well controlled by eliminating the problems encountered in the processing of botanicals, such as the quality of the raw material, uniformity within and among lots, and damage in shipment and storage.

3. Some metabolic products can be produced from suspension culture in higher quantities than that observed in whole plants.

It is a biochemical engineer's challenge to be able to cultivate plant cells on a large scale and to maximize the production and accumulation of secondary metabolites. This can be accomplished by selecting proper genotypes and high-yielding cell clones, formulating suitable media to cultivate the cells, and designing and operating effective cell culture systems. We can utilize our experience and knowledge that we have obtained from the microbial cultivation. However, plant cells and microbial cells are so different that we need to modify and adjust the culture conditions and reactor configuration to meet the specific requirements for the plant cell culture.

Some of the major differences between plant and microbial cells are as follows:

1. Plant cells are 10 to 100 times larger than bacterial and fungal cells.

2. The metabolism of plant cells is slower than microbial cells in the one order of magnitude, which requires the maintenance of sterility for a longer period of time.

3. Plant cells tend to grow in clumps which cause sedimentation, poor mixing, plugging the inlet and outlet lines, wall growth, and so on.

4. Plant cells are more sensitive to shear than microbial cells.

5. Metabolic production in plant cells is subject to more complex regulatory mechanisms than metabolic production in microbial cells.

6. Plant cells are more genetically unstable than microbial cells.

Therefore, there is no question that only high value-added, low market-volume products are economically feasible to produce from plant cell culture techniques due to the previously listed hindrances. Examples of potential products from plant cell culture are listed in Table 5.6.

5.3.1. Plant Cells

A generalized plant cell is shown in Figure 5.10. The size of a single plant cell is usually within the range of 20-40 μm in diameter and 100-200 μm long (Wilson, 1980). The major structure of the cell is similar to that of typical eukaryotic cells as discussed in the previous section on microbial cultivation. However, plant cells have distinctive features such as a rigid wall, a large vacuole, and the presence of chloroplasts.

The plant cell is surrounded by a cell wall, which determines many features of the plant. The outer layer of the cell wall are called middle lamella because it contains heavy layer of pectin (a polygalacturonan) that serves as the glue to hold one plant cell firmly to an adjacent cell. The inner layer of the wall is cell membrane. The cell membrane is completely different from the cell wall in form, composition, and function. Whereas the wall is a rigid, relatively thick structure, the cytoplasmic membrane is thin (approximately 75 Å) and flexible. The membrane is composed of protein and lipid, whereas the wall is carbohydrate in nature. The wall provides support, whereas the membrane regulates the movement of substances into and out of the cell.

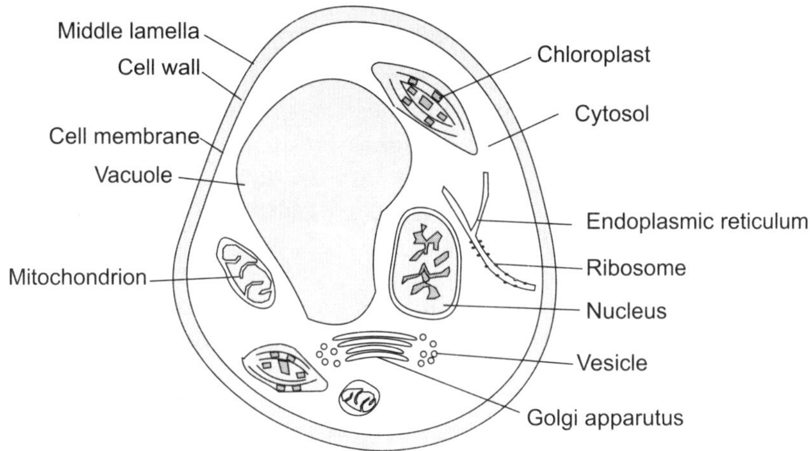

Fig. 5.10 Generalized plant cells

The *vacuole* serves as a receptacle for waste metabolic products or secondary plant substances. In young cells, the vacuoles are small and numerous. As the cell becomes larger and older, the vacuoles enlarge and coalesce. In a mature plant cell, vacuoles may occupy 90 percent of the volume of the cell. The vacuole is surrounded by a plasma membrane. The major component of large vacuoles is water, which contains dissolved solutes, such as inorganic ions, amino acids, organic acids, water-soluble pigments (anthocyanins), and insoluble materials in the form of crystals and needles. In addition, the vacuole also contains proteins such as hydrolases, catalase, and phosphatases. *Cytosol* refers the liquid around all of the floating structures between the nucleus and cell membrane.

Chloroplast is the site of photosynthesis in the plant cell, which contains the green pigment chlorophill that is responsible for trapping the light for the production of carbohydrates. *Nucleus* is the cell's control center and contains DNA for the protein translation and transcription. The proteins synthesized are sorted and packaged into vesicles in *Golgi apparatus. Endoplasmic reticulum* is the network of tiny tubes that interconnect the different parts of the cells. *Ribosome* is concentrated on the surface of endoplasmic reticulum and is actively involved int the synthesis of proteins. *Mitochondrion* contains genetic material and many enzymes important for cell metabolism.

5.3.2. Types of Plant Tissue Culture

Although *plant tissue culture* should refer only to those of unorganized aggregates of cells, it is commonly used as a collective term to describe all types *of in vitro* plant cultures (George and Sherrington, 1984).

Unorganized growth occurs frequently when pieces of whole plants are cultured *in vitro*. The tissue lacks any recognizable structure of the original intact plant.

1. *Callus cultures* are amorphous cell aggregates arising from the unorganized growth of explants on an aseptic solid nutrient medium. The *explants* are the small organs or tissue sections. The cell aggregates do not correspond with any particular tissue of the whole plant. The term *callus culture* was chosen because cell proliferation was thought to be induced by injury of the explant during excision. However, it has been found to be induced by plant growth regulators in the solid nutrient medium.

2. *Suspension (or cell) cultures* consist of cells and cell aggregates, growing dispersed in liquid medium. They are usually initiated by placing pieces of a friable callus culture in moving liquid medium. Suspension cultures are therefore the result of the progression from plant to explant, to callus, and finally to suspension. Suspension culture is more suitable for mass propagation of plant cells than callus culture because the former can be maintained and manipulated similar to submerged microbial fermentations.

3. *Protoplast cultures* involve the growth of protoplasts in solid or liquid medium. Protoplasts can be prepared by mechanically or enzymatically removing the cell wall. The isolated protoplasts can be used: (1) to modify the genetic information of plant cells, (2) to create plant hybrids through protoplast fusion, (3) to study plant viral infections, and so on. Another promising application of protoplast cultures is the micropropagation of plants. After the divisions of protoplasts, cell walls can be regenerated to give rise to callus and subsequently to whole

plants, which provides one route whereby plants may be multiplied.

Table 5.6 Plant Products of Commercial Interest

Food Products	Color Flavors	anthocyanms, betacyanins, saffron apricot, banana, apple, cherry, grape, peach, pineapple, rasberry, strawberry, asparagus, capsicum, celery, tomato, vanilla, cocoa
	Oils	garlic, jasmine, lemon, mint, onion, patchouly, rose, vetiver
	Sweeteners	miraculin, monellin, stevioside, thaumatin
	Spices	cardamom, cinnamon, rosemary, sage, turmeric
Pharma-ceuticals	Alkaloids	ajmalacine, atropine, berberine, camptothecin, ceuticals codeine, hyoscyamine, quinine, morphine, scopolamine, serpentine, vinblastine, vincristine
	Steroids	digitoxm, digoxin, diosgenin
	Others	L-Dopa, ginsengoside, shikonin, rosmarinic acid, saponin, ubiqumone-10, diosgenin
	Foreign Proteins	monoclonal antibody, interleukins, GM-CSF, various enzymes
Agricultural	Chemicals	pyrethrins, rotenone, azadirachtin, neriifolin, salannin, alleopathic chemicals

Organized growth occurs when organized plant parts are transferred to culture medium where they may continue to grow with their structure preserved. Differentiated plant organs can be grown in culture without loss of integrity, which is known as *organ culture.*

1. *Root cultures* can be established from root tips taken from many plants. Fast growing root cultures can be obtained from dicotyledonous plant species by infection with the microorganism *Agrobacterium rhizogens.*

   The "hairy root" clones that are produced can be cultivated to produce metabolites (Hjortso and Flores, 1987).

2. *Embryo cultures* may be established from embryos removed from sterilized seeds, ovules, or fruits. The embryos produced from cell culture technique, known as *somatic embryos,* can be isolated and germinated to provide one plant per explant. Embryo culture can be employed for the rapid production of seedlings from seeds which have a protracted dormancy period. The method has many potential advantages over traditional propagation systems, such as fast turnaround, genetic uniformity, mass production, and propagation of disease-free plants.

5.3.3 Culture Media

Although the nutritional requirements of various tissue cultures vary widely, typical plant tissue culture media contain the following basic components (George and Sherrington, 1984):

1. *Major nutrients:* Salts of nitrogen, potassium, calcium, phosphorus, magnesium, and sulfur, which are six major elements for the growth of higher plants.

2. *Minor nutrients:* Salts of iron, manganese, zinc, boron, copper, molybdenum, and cobalt in trace amounts.

3. *Organic supplements:* Small amount of vitamins (myo-inositol, thiamine, nicotinic acid, pyridoxine, and so on.), amino acids (usually omitted but sometimes used with advantage), and other undefined supplements (meat, malt, and yeast extract, and protein hydrolysates, and so on.).

4. *Plant growth regulators:* Auxins (usually provided with the synthetic analogue, 2,4-dichlorophenoxyacetic acid) and cytokinins (commonly used synthetic analogue, kinetin) which regulate growth and morphogenesis in plant tissue and organ culture.

5. *Carbohydrate source:* Usually sucrose to replace the carbon which the plant normally fixes from the atmosphere by photosynthesis, since most plant cell culture lacks photosynthetic capability.

6. *Solidifying agent for semisolid medium:* Agar.

Table 5.7 Compositions of the Murashige and Skoog (1962) Medium.

Mineral Salts	mg/L	Organic Constituents	mg/L
NH_4NO_3	1650	Sucrose	30,000
KNO_3	1900	Agar	10,000
$CaCl_2 \cdot 2H_2O$	440	Edamin(optional)	1,000
$MgSO_4 \cdot 7H_2O$	370	Glycine	2
KH_2PO_4	170	Indoleaceticacid	1–30
Na_2-EDTA	37.3	Kinetin	0.04–10
$FeSO_4 \cdot 7H_2O$	27.8	*myo*-inositol	1
H_3BO_3	6.2	Nicotinicacid	0.5
$MnSO_4 \cdot 4H_2O$	22.3	Pyridoxine-HCL	0.5
$ZnSO_4 \cdot 4H_2O$	8.6	Thiamine-HCl	0.1
KI	0.83		
$Na_2MoO_4 \cdot 2H_2$	0.25		
$CuSO_4 \cdot 5H_2O$	0.025		
$CoCl_2 \cdot 6H_2O$	0.025		

A significant contribution to the formulation of a defined growth medium was made by Murashige and Skoog (1962). The medium that they formulated (known as the MS medium) has since proved to be one of the most widely used in plant cell culture work. The recipe for MS medium is given in Table 5.7.

5.3.4   Secondary Metabolite Production

Secondary metabolites can be classified into three major categories (Shuler, 1981): alkaloids, essential oils, and glycosides.

Alkaloids are crystalline, nitrogen-containing compounds which can be extracted by the use of acidic solutions.[9] Alkaloids are physiologically active on all animals and used in the pharmaceutical industry. Familiar alkaloids include codeine, nicotine, caffeine, and morphine. *Essential oils* consist of mixtures of terpenoids and used as flavorents, fragrances, and solvents. *Glycosides* include phenolics, tannins and flavonoids, saponins, and cyanogenic glycosides, some of which can be utilized as dye, food flavors, and pharmaceuticals.

One of the important factors influencing the productions of secondary metabolites from plant cells is morphological *differentiation*, which is the process for individual cells to undergo changes that result in the localized differences in biochemical and metabolic activity and in structural organization. Many secondary metabolites are known to be produced during cell differentiation processes. Therefore, they are often found in highly specialized tissues such as roots, leaves, and flowers. Since the morphological differentiation and maturation are largely absent in plant cell culture, the productions of secondary metabolites tend to be suspended in plant cell culture. As a result, only a limited number of plant cell culture systems can produce reasonable amounts of the secondary metabolites, even though the natural plant, from which the cells were obtained, is able to produce them. However, undifferentiated cells in suspension cultures often form aggregates made of a few to several hundred cells due to the stickiness of the cell surface, from the excretion of polysaccharides and high cell concentrations (Mavituna and Park, 1987). Because of concentration gradients and cellular interaction, cells at the center of an aggregate will be exposed to an environment that is different from cells on the periphery. Consequently, differentiation may occur to some extent in aggregates, favoring the formation of secondary metabolites.

However, there have been reports that some suspension cultures are capable of synthesizing specific products to levels higher than the plant from which they were derived. As an example, Schulte et al.

---

[9] The name *alkaloid* means like an alkali.

(1984) have reported that secondary product (anthraquinones) formation in optimized cell cultures surpassed that of differentiated plants in 17 out of 19 different species belonging to the genera *Asperula, Galium, Rubia,* and *Sherardia.* The highest anthraquinone yield was reported with *Galium verum* (1.7 g/L) and the highest concentration achieved with *Rubia fruticosa* (20 percent of dry weight).

There is also strong evidence indicating that there is an inverse relationship between growth rate and secondary metabolite production (Lindsey and Yeoman, 1983). When growth is intense, the primary processes of the cell are cell division and production of cell mass. In the stationary phase, when growth is minimal, conditions favor the production and accumulation of secondary metabolites.

The medium composition also has a significant influence on the quantity of secondary metabolites produced (Zenk et al., 1977). The basic requirement for the media formulation is to fulfill cell growth.

After cells reach a certain population, the modification of the medium may affect product accumulation. Fujita et al. (1981) improved the yield of shikonin derivatives from the suspension cultures of *Lithospermum erythrorhizon* by using a production medium. The production medium usually contains more sucrose but less inorganic components and vitamins than the growth medium.

Product accumulation by plant cell culture may be stimulated by biotic or abiotic *elicitors.* Biotic elicitors are compounds or substances of microbial origin and abiotic elicitors are stress agents such as UV irradiation, osmotic shock, or heavy-metal ions. Biotic elicitors are usually produced by homogenizing fungal mycelium and sterilizing the homogenate. The effects of the biotic elicitors on the accumulation of secondary metabolites depend on the specificity and the concentration of an elicitor, duration of elicitor contact, and the growth stage of the plant cell culture (Constabel, 1988).

5.4 CELL GROWTH MEASUREMENT

In any biological system, growth can be defined as the orderly increase of all chemical components. Increase of mass might not really reflect growth because the cells could be simply increasing their content of storage products such as glycogen or poly-β-hydroxybutyrite. *Balanced growth* is defined as growth during which a doubling of the biomass is accompanied by a doubling of all other measurable properties of the population such as protein, DNA, RNA, and intracellular water. In other words, cultures undergoing balanced growth maintain a constant chemical composition. In an adequate medium to which they have become adapted, bacteria are in a state of balanced growth.

To follow the course of growth, it is necessary to make quantitative measurements. Cell growth can be determined by measuring cell number, cell mass, or cell activity.

5.4.1 Measurement of Cell Number

Microscopic Counts: The number of cells in a population can be measured under a microscope by counting cells placed in special counting chambers. There are two types of chambers used for counting cell number in liquid samples:

1. *hemocytometer.* a blood cell counting chamber for use with organisms of 3 μm in diameter or larger.

2. *Petroff-Hausser counting chamber,* for use primarily with bacteria.

Both chambers have a special square grid marked on the surface of the glass slide. A ridge on each side of the grid holds a cover slip off of the grid by a known distance so that the volume of a square is precisely known. A sample of cell suspension to be counted is allowed to flow under the cover slip and to fill the counting chamber. Then the number of cells per unit area of grid can be counted under the microscope. Very dense suspensions can be counted if they are diluted appropriately.

The desirable features of direct-counting methods are:

1. Minimal equipment is required.
2. Results are obtained rapidly.
3. The morphological characteristics of the organisms can be observed.

The disadvantages are:

1. Dead cells cannot usually be distinguished from live cells.
2. The method is not suitable for cell suspensions of low density.
3. Small cells are difficult to see under the microscope and can be missed when counting.
4. The actual counting procedure is tiresome and may cause considerable eyestrain.
5. It is not suitable for highly flocculating cells such as mycelium.

Viable Plate Count: A viable cell is defined as one that is able to divide and form a colony. There are two ways of performing a plate count—the spread plate method and pour plate method. With the *spread plate method,* a volume of no larger than 0.1 mL is spread over the agar surface. With *the pour plate method,* the sample is mixed with melted agar and poured into a sterile plate. The plate is then incubated until the colonies appear, and the number of colonies is counted. It is important that the number of colonies developing on the plates should be neither too large nor too small. To obtain the appropriate number of cells per unit volume, the sample has to be diluted. For the larger dilution, it is common to use the serial dilution technique. For example, to make $1/10^6$ dilution, three successive $1/100$ dilutions or six successive $1/10$ dilutions can be made.

Coulter Counter: To avoid the tedium of direct microscopic counting, a Coulter counter can be employed. By using this technique, not only the cell number, but the cell size can be measured. The disadvantage of this technique is that it cannot distinguish between cells and any impure particles. The technique is also difficult to use with organisms in chains and is useless with mycelial organisms.

5.4.2 Measurement of Cell Mass

Cell Dry Weight: Cell dry weight can be measured directly by taking an aliquot of cell suspension and centrifugivng it. After the supernatant is discarded, the cells are thoroughly washed with distilled water to eliminate all soluble matter. The suspension is recentrifuged and the settled cells are dried in an oven and weighed. This is the most direct approach for the quantitative measurement of a cell mass and is probably the most reliable and reproducible. However, such determinations are time consuming and relatively insensitive to small changes of the cell mass. This technique can only be used with dense cell suspensions, and the cells must be washed completely free of all extraneous matter.

Turbidity: The cell mass can be measured optically by determining the amount of light scattered by a suspension of cells. The technique is based on the fact that small particles scatter light proportionally, within certain limits, to their concentration. When a beam of light is passed through a suspension of organisms, the reduction in the amount of light transmitted as a consequence of scattering is thus a measure of the cell density. Such measurements are usually made in a spectrophotometer, which reads in *absorbency (A)* units. The *absorbency* is defined as the logarithm of the ratio of the intensity of light striking the suspension (I_0) to that transmitted by the suspension (I):

$$A = \log \frac{I_0}{I} \tag{5.1}$$

A calibration curve can be obtained by measuring the absorbance of the samples with known cell concentration. The measurements are usually made at a wavelength of 600 ~ 700 nm.

5.4.3. Indirect Methods

The indirect methods for measuring cell mass are based on the overall stoichiometry for growth and product formation, which may be written in the general form:

$$\text{C-source} + \text{N-source} + \text{phosphate} + O_2 \longrightarrow$$
$$\text{cell mass} + CO_2 + H_2O + \text{product} + \text{heat}$$

The change of the cell mass can be monitored indirectly by measuring nutrient consumption, product formation, cell components, heat evolution, or other physical properties of broth.

Nutrient Consumption: It is necessary to choose a nutrient which is not likely to be used to synthesize a metabolic product. Phosphate, sulfate, or magnesium can be good candidates. When cell mass is the major product, the concentration of a carbon source can be measured to estimate the cell mass.

Product Formation: It is important to check that the product formed is growth associated. Some products are formed after cell mass reaches the stationary phase of the growth cycle and, therefore, are not growth associated. The evolution of carbon dioxide can be measured and related stoichiometrically to the cell growth. One product that is quite general to many fermentations is the hydrogen ion. The amount of alkali added to the fermentation broth to maintain the pH may be proportional to the growth.

Cell Components: For cultures undergoing balanced growth, the macro-molecular cell components such as protein, RNA, and DNA can be measured instead of cell mass. However, care is needed because the proportion of these materials in a cell can change with time if the culture does not undergo balanced growth.

Heat Evolution: Cell growth evolves heat. The heat of combustion of organisms is fairly constant with a typical value of 5 kcal/g. The amount of heat evolved depends on the efficiency of the carbon energy utilization. Therefore, the measurement of the heat of fermentation can be related indirectly to cell growth. However, the total heat accumulated in a fermentation system is the combined effect of various sources of heat generation and disappearance such as the heat from agitation and evaporation, the heat dissipated to the surroundings through the fermenter walls, and the sensible heat in the air stream. Therefore, to measure the growth by heat evolution, a complete energy balance of a fermentation system has to be made, which may not be an easy task.

Viscosity: Mycelial growth or polysaccharide formation increase the viscosity of the fermentation broth. Therefore, the measurement of broth viscosity is useful in many commercial fermentations. Frequently, the fermentation broth shows non-Newtonian behavior, which makes the measurement of actual viscosity very complicated. However, apparent viscosity measured at a fixed shear rate can be used to estimate cell or product concentration.

5.5 CELL IMMOBILIZATION

As in the case of enzymes, whole cells can be immobilized for several advantages over traditional cultivation techniques. By immobilizing the cells, process design can be simplified since cells attached to large particles or on surfaces are easily separated from product stream. This ensures continuous fermenter operation without the danger of cell washout. Immobilization can also provide conditions conducive to cell differentiation and cell-to-cell communication, thereby encouraging production of high yields of secondary metabolites. Immobilization can protect cells and thereby decrease problems related to shear forces.

Table 5.8 Cell Immobilization Methods

Methods	Materials
Attachment to a Surface	wood chips, collagen, microcarriers, ion exchange resins, metal oxides
Entrapment within Porous Matrices	Cordierite, pore glass, acrylamide, alginate, collagen, K-carrageenan
Containment behind a Barrier	polylysine, ethylcellulose, emulsion, synthetic membrane
Self-aggregation	mycelial pellets, microbial floes, polyelectrolytes, cross-linking agents

Cell immobilization methods can be divided into four major categories: attachment to a surface, entrapment within a porous matrix, containment behind a barrier, and self-aggregation as summarized in Table 5.8 (Karel et al., 1985).

Attachment to a Surface: Cells can be attached on the surface of wood chips, collagen, microcarriers, or ion-exchange resin. One example of this type of immobilization is the use of *microcarriers* for the anchorage dependent animal cells. The main advantage of the microcarriers is that they provide a large surface area for cell attachment. The materials for a microcarrier include ion-exchange resins, dextran-based beads coated with gelatin, polyacrylamide beads, polystyrene beads, hollow glass beads, cylindrical cellulose beads, and fluorocarbon droplets stabilized with polylysine. At present, dextran-based microcarriers are the most widely used for the immobilization of mammalian cells (Hu and Dodge, 1985).

Entrapment Within a Porous Matrix: Cells are allowed to diffuse into preformed porous matrices such as bricks, cordierite, and pore glass, in which they will grow and be trapped. The main advantages of this method are that the preformed support materials are more resistant to disintegration in packed beds or stirred vessels than other support materials, and the entrapment is not usually harmful to the cells. However, it is difficult to reach a high cell concentration due to the limited pore volume usually available for entrapment within typical support materials.

Another way is to entrap cells within porous matrices formed *in situ*. Various gelatinous materials such as acrylamide, alginate, collagen and κ-carrageenan, can be mixed with cell suspensions and be gelled into various shape and sizes.

One simple procedure to make calcium alginate gel sphere is as follows: Concentrated suspension cells are mixed with alginate to give a final alginate concentration between 1 percent to 3 percent (w/v) and the cell-alginate mixture is pumped through a hypodermic needle drop-wise into a calcium chloride solution. The beads form instantly with diameters between 1 mm to 5 mm depending on the alginate and cell contents of the liquid and the needle size. Extreme caution needs to be taken to maintain sterility during this immobilization process.

The major disadvantage of using alginate immobilization is the leakage of cells from cell division occurring within the individual beads. Cell leakage can be minimized either by increasing the alginate or calcium chloride concentrations in beads or by making the beads small. However, the increase of the alginate and calcium chloride concentration in the beads can decrease the substrate diffusion rate through the gel and may affect the viability of entrapped cells (Cheetham et al., 1979).

Containment Behind a Barrier: Cells can be immobilized within micro-capsules that have either a permanent or nonpermanent semipermeable membrane. The advantage of encapsulation techniques is the large surface area for contact of substrate and cells. The semipermeable membrane also selectively passes only low-molecular weight components.

The cells can be entrapped by inclusion within membrane filter devices, such as hollow-fiber, flat plate, and spiral wound units.

Hollow-fiber membranes form a tubular structure which is usually arrayed as a parallel fiber bundle within a cylindrical container. The cells are trapped on the shell side of the hollow fibers while aerated nutrient medium is rapidly recirculated through the fibers. This type of membrane support can provide extra protection against contamination. However, the major disadvantages of membrane systems are high cost, membrane fouling resulting in added mass-transfer resistance, and aeration difficulties.

Self-aggregation: Self-aggregated or flocculated cells also can be regarded as immobilized cells because their large size provides similar advantages as immobilization by other methods. While molds will form pellets naturally, some bacteria or yeast cells require flocculation. Artificial flocculating agents or cross-linkers can be added to enhance the process.

5.6   EXPERIMENTS

5.6.1   Yeast Gro wth Curve

In this experiment, a strain of yeast will be cultivated in an Erlenmeyer flask, and the change of the cell concentration will be monitored by using three different techniques: microscopic count, dry weight measurement, and turbidity measurement.

Materials:

1. Any strain of yeast grown as a suspension culture. You can obtain a strain of yeast from a microbiology laboratory. You can buy a specific strain from American Type Culture Collection (ATCC), Rockville, MD. Recommended strain is *Saccharomyces cerevisiae* ATCC 4126.
2. glucose, yeast extract, NH_4Cl, $MgSO_4$, $CaCl_2$, and antifoam to make medium
3. two 125 mL Erlenmeyer flasks
4. sterile pipette
5. Bunsen burner
6. autoclave
7. incubator
8. hemocytometer
9. centrifuge
10. chemical balance
11. spectrophotometer

Procedures:

1. Prepare culture media according to 5.4 and pour 50 mL each into two 125-mL Erlenmeyer flasks.
2. Plug the Erlenmeyer flasks with one of the followings: plastic, aluminum, stainless steel cap, or cotton.
3. Autoclave the flasks 15 to 18 psi for 20 minutes.

4. Inoculate 1 mL yeast culture into the flask containing the sterilized medium. Be careful not to contaminate the pipette, plug, or the flask during the inoculation. To minimize the chance of contamination, flame the plug and the neck of the flask after removing the plug for the inoculation.

5. Place the flasks in an incubator at 37°C.

6. Take a 2-mL sample at predetermined intervals and measure the cell concentration by employing microscopic count, dry weight measurement, and turbidity measurement (see Section 5.4). The sampling interval should be arranged so that a good growth curve showing all three phases of growth can be shown. Before sampling, mix the contents of the flask by shaking.

5.6.2 Plant Cell Growth Curve Materials;

1. A well-established plant suspension cell line. The following procedures are based on the cultivation of tobacco cells. If another cell line is chosen, the medium can be modified.

2. Murashige and Skoog salt mixture (Gibco), 2,4-dichlophenoxy-acetic acid (2,4-D), KH_2PO_4, inositol, thiamine HC1, and sucrose to make medium

3. two 125-mL Erlenmeyer flasks

4. sterilized wide-mouth serological pipette (10 mL)

5. incubator shaker

6. chemical balance

7. centrifuge

8. Eppendorf tube

Procedures:

1. Prepare modified Linsmaier and Skoog medium (Linsmaier and Skoog, 1965) which contains 4.3 g/L of Murashige and Skoog salt mixture, 0.2 mg/L 2,4-D, 0.18 g/L KH_2PO_4, 0.1 g/L inositol, 1 mg/L thiamine HCL, and 30 g/L sucrose. Adjust pH to 5.8. with 1 N KOH and dispense the medium into two 125 mL Erlenmeyer flasks so that each flask contains 30 mL each.

2. Plug the Erlenmeyer flasks and autoclave them at 15 to 18 psi for 20 minutes.

3. Inoculate the flask containing 30 mL of medium with 1.5 mL of seven-day-old suspension culture and put it in a shaker incubator (150 rpm) at 27°C. Keep the cells dark by blocking the glass window of the incubator.

4. Take 1.5-mL sample each day from the well-mixed culture and place it in a tarred Eppendorf tube to weigh. Centrifuge the tube at 9,000 rpm for 4 to 5 minutes and remove the

supernatant. Reweigh the tube and calculate the wet cell weight percentage. Put the tube in an oven (70°C) for two days and reweigh to calculate the dry weight percentage.

5. Plot the change of wet and dry cell concentrations with respect to time.

5.6.3 Plant Cell Immobilization Materials:

1. 30 mL of tobacco suspension cells grown by the method described in the previous experiment
2. alginate
3. calcium chloride
4. peristaltic pump
5. autoclave
6. chemical balance

Procedures:

1. Mix 0.875 g alginate, 5 mL plant culture medium, and 25 mL of water and sterilize.
2. Wait for the 30 mL of suspension plant culture to settle and remove the supernatant. It normally takes about 10 minutes.
3. Add the alginate mixture and mix with concentrated cells.
4. Pump the cell-alginate mixture through a sterilized silicone tube (1.6 mm ID) and feed drop-wise into a flask containing sterilized 200 mL of 0.12 M $CaCl_2$. The droplets instantaneously will react with $CaCL_2$ to form spherical beads (3.75–4.5 mm in diameter).
5. Keep the beads in the solution for 1 hour to ensure that precipitation reaction reaches completion.
6. Inoculate the flask containing 30 mL of medium with about 30 immobilized plant cell beads and put it in a shaker incubator (150 rpm) at 27°C. Keep the cells dark by blocking the glass window of the incubator.
7. You can measure the wet and dry cell concentration of the free suspended cells and immobilized cells during the batch cultivation of immobilized cells. To determine the cell concentration of immobilized cells, you should dissolve the beads in 1 M potassium phosphate for 24 hours.

5.7 REFERENCES

Butler, M., *Animal Cell Technology: Principles and Products.* Milton Keynes, England: Open University Press, 1987.

Cheetham, P. S. J., K. W. Blunt, and C. Bucke, "Physical Studies on Cell Immobilization Using Calcium Alginate Gels," *Biotech. Bioeng.* **21** (1979):2155-2168.

Constabel, F., "Principles Underlying the Use of Plant Cell Fermentation for Secondary Metabolite Production," *Biochem. Cell. Biol.* **66** (1988):658-664.

Eagle, H., "Amino Acid Metabolism in Mammalian Cell Cultures," *Science* **130** (1959):432-437.

Feder, J. and W. R. Tolbert, "The Large-Scale Cultivation of Mammalian Cells," *Scientific American* **248** (1983):36-43.

Fujita, Y., Y. Kara, C. Suga, and T. Morimoto, "Production of Shikonin Derivatives by Cell Suspension Cultures of *Lithospermum erythrorhizon:* II. A New Medium for the Production of Shikonin Derivatives," *Plant Cell Reports* 1(1981):61-63.

George, E. F., and P. D. Sherrington, *Plant Propagation by Tissue Culture.* Hants, England: Exegetics Ltd., 1984.

Hjortso, M. A. and H. E. Flores, "Root Culture as an Experimental System for the Production of Plant Chemicals," Paper presented at *AIChE 1987 Annual Meeting,* New York, NY, November 15-20, 1987.

Hu, W. and T. C. Dodge, "Cultivation of Mammalian Cells in Bioreactors," *Biotech. Prog.* 1 (1985):4-10.

Karel, S. F., S. B. Libicki, and C. R. Robertson, "The Immobilization of Whole Cells: Engineering Principles," *Chem. Eng. Sci.* **40** (1985): 1321-1354.

Kohler, G. and C. Milstein, "Continuous culture of fused cells secreting antibody of predefined specificity," *Nature* **256** (1975):495-497.

Lindsey, K., M. M. Yeoman, G. M. Black, and F. Mavituna, "A novel method for the immobilization and culture of plant cells," *FEBS Letters* 155 (1983): 143-149.

Linsmaier, E. M. and F. Skoog, "Organic Growth Factor Requirements of Tobacco Tissue Cultures," *Physiol. Plant* **18** (1965): 100-127.

Mavituna, F. and J. M. Park, "Size Distribution of Plant Cell Aggregates in Batch Culture," *Chem. Eng. J.* 35 (1987):B9-B14.

Murashige, T. and F. Skoog, "A Revised Medium for Rapid Growth and Bioassays with Tobacco Tissue Cultures," *Physiologia Plantarum* **15** (1962):473-497.

Pelczar, M. J. and R. D. Reid, *Microbiology,* pp. 820-823. New York, NY: McGraw-Hill Book Co., 1972.

Queener, S. and R. W. Swartz, "Penicillins: Biosynthetic and Semisynthetic," in *Economic Microbiology,* A. H. Rose, ed., 3 (1979):35-122.

Sahai, O. and M. Knuth, "Commercializing Plant Tissue Culture Processes: Economics, Problems and Prospects," *Biotech. Progress* 1 (1985):1-9.

Schulte, U., H. El-Shagi, and M. H. Zenk, "Optimization of 19 *Rubiaceae* species in cell culture for the production of anthraquinones," *Plant Cell Reports* 3 (1984):51-54.

Shuler, M. L., "Production of Secondary Metabolites from Plant Tissue Cultures - Problems and Prospects," *Ann. N.Y. Acad. Sci.* **369** (1981):65-79.

Sturchio, J. L., "Chemistry in Action: Penicillin Production in World War II," *Today's Chemist,* (February 1988):20-22.

Wilson, G., "Continuous Culture of Plant Cells Using the Chemostat Principle," in *Advances in Biochemical Engineering,* Vol. 16, Ed. A. Fiechter, New York: Springer-Verlag, 1980, pp. 1-25.

Zenk, M. H., H. El-Shagi, H. Arens, J. Sto\"ckigt, E. W. Weiler, and B. Deus, "Formation of the Indole Alkaloids Serpentine and Ajmalicine in Cell Suspension Cultures of *Catharanthus roseus,*" in *Plant Tissue Culture and Its Biotechnological Application,* Eds. W. Barz, E. Reinhard, M. H. Zenk, New York: Springer-Verlag, 1977, pp. 27-43.

Suggested Reading

Benson, H. J., *Microbiological Applications: A Laboratory Manual in General Microbiology* (4th ed.). Dubuque, IA: Wm. C. Brown Publishers, 1985.

Brock, T. D. and K. M. Brock, *Basic Microbiology with Applications.* Englewood Cliffs, NJ: Prentice-Hall, Inc., 1973.

Butler, M., *Animal Cell Technology: Principles and Products.* Milton Keynes, England: Open University Press, 1987.

Frobisher, M., R. D. Hmdsdill, K. T. Crabtree, and C. R. Goodheart, *Fundamentals of Microbiology* (9th ed.). Philadelphia, PA: W. B. Saunders Co., 1974.

Levy, J., J. J. R. Campbell, and T. H. Blackburn, *Introductory Microbiology.* New York, NY: John Wiley \& Sons, Inc., 1973.

Milstein, C., "Monoclonal Antibodies," *Scientific American* **243** (1980):66-74.

Pelczar, M. J. and R. D. Reid, *Microbiology.* New York, NY: McGraw-Hill Book Co., 1972.

Stanier, R. Y., E. A. Adelberg, and J. Ingraham, *The Microbial World* (4th ed.). Englewood Cliffs, NJ: Prentice-Hall, Inc., 1976.

6

Cell Kinetics and Fermenter Design

6.1 INTRODUCTION

Understanding the growth kinetics of microbial, animal, or plant cells is important for the design and operation of fermentation systems employing them. Cell kinetics deals with the rate of cell growth and how it is affected by various chemical and physical conditions.

Unlike enzyme kinetics discussed in Chapter 2, cell kinetics is the result of numerous complicated networks of biochemical and chemical reactions and transport phenomena, which involves multiple phases and multicomponent systems. During the course of growth, the heterogeneous mixture of young and old cells is continuously changing and adapting itself in the media environment which is also continuously changing in physical and chemical conditions. As a result, accurate mathematical modeling of growth kinetics is impossible to achieve. Even with such a realistic model, this approach is usually useless because the model may contain many parameters which are impossible to determine.

Therefore, we must make assumptions to be able to arrive at simple models which are useful for fermenter design and performance predictions. Various models can be developed based on the assumptions concerning cell components and population as shown in Table 6.1 (Tsuchiya et al., 1966). The simplest model is the *unstructured, distributed model* which is based on the following two assumptions:

1. Cells can be represented by a single component, such as cell mass, cell number, or the concentration of protein, DNA, or RNA. This is true for balanced growth, since a doubling of cell mass for balanced growth is accompanied by a doubling of all other measurable properties of the cell population.

2. The population of cellular mass is distributed uniformly throughout the culture. The cell suspension can be regarded as a homogeneous solution. The heterogeneous nature of cells can be ignored. The cell concentration can be expressed as dry weight per unit volume.

Table 6.1 Various Models for Cell Kinetics

Population	Cell Components	
	Unstructured	Structured
Distributed	Cells are represented by a single component, which is uniformly distributed throughout the culture	Multiple cell components, uniformly distributed throughout the culture interact with each others
Segregated	Cells are represented by a single component, but they form a heterogeneous mixture	Cells are composed of multiple components and form a heterogeneous mixture

Besides the assumptions for the cells, the medium is formulated so that only one component may be limiting the reaction rate. All other components are present at sufficiently high concentrations, so that minor changes do not significantly affect the reaction rate. Fermenters are also controlled so that environmental parameters such as pH, temperature, and dissolved oxygen concentration are maintained at a constant level.

In this chapter, cell kinetic equations are derived from the unstructured, distributed model, and those equations are applied for the analysis and design of ideal fermenters. More realistic models which consider the multiplicity of cell components, structured model, are introduced at the end of the chapter.

6.2 DEFINITIONS

First, let us define the terminologies we use for microbial growth. If we mention the cell concentration without any specification, it can have many different meanings. It can be the number of cells, the wet cell weight, or the dry cell weight per unit volume. In this text, the following nomenclature is adopted:

C_X Cell concentration, dry cell weight per unit volume

C_N Cell number density, number of cells per unit volume

ρ Cell density, wet cell weight per unit volume of cell mass

Accordingly, the growth rate can be defined in several different ways, such as:

dC_X/dt Change of dry cell concentration with time

r_X Growth rate of cells on a dry weight basis

dC_N/dt Change of cell number density with timernGrowth rate of cells on a number basis

δ Division rate of cells on a number basis, $d \log_2 C_N/dt$

It appears that dCx/dt and r_X are always the same, but this is not true. The former is the change of the cell concentration in a fermenter, which may include the effect of the input and output flow rates, cell recycling, and other operating conditions of a fermenter. The latter is the actual growth rate of the cells. The two quantities are the same only for batch operation.

The growth rate based on the number of cells and that based on cell weight are not necessarily the same because the average size of the cells may vary considerably from one phase to another. When the mass of an individual cell increases without division, the growth rate based on cell weight increases, while that based on the number of cells stays the same. However, during the exponential growth period, which is the phase that we are most interested in from an engineer's point of view, the growth rate based on the cell number and that based on cell weight can be assumed to be proportional to each other.

Sometimes, the growth rate can be confused with the *division rate*, which is defined as the rate of cell division per unit time. If all of the cells in a vessel at time $t = 0$ ($C_N = C_{N_0}$) have divided once after a certain period of time, the cell population will have increased to $C_{N_0} \times 2$. If cells are divided n times after the time t, the total number of cells will be

$$C_N = C_{N_0} \times 2^N \tag{6.1}$$

and the average division rate is

$$\bar{\delta} = \frac{n}{t} \tag{6.2}$$

Since $N = \log_2 C_N - \log_2 C_{N_0}$ according to Eq. (6.1), the average division rate is

$$\bar{\delta} = \frac{1}{t} (\log_2 C_N - \log_2 C_{N_0}) \tag{6.3}$$

and the division rate at time t is

$$\delta = \frac{d\log_2 C_N}{dt} \tag{6.4}$$

Therefore, the growth rate defined as the change of cell number with time is the slope of the C_N versus t curve, while the division rate is the slope of the $\log_2 C_N$ versus t curve. As explained later, the division rate is constant during the exponential growth period, while the growth rate is not. Therefore, these two terms should not be confused with each other.

6.3   GROWTH CYCLE FOR BATCH CULTIVATION

If you inoculate unicellular microorganisms into a fresh sterilized medium and measure the cell number density with respect to time and plot it, you may find that there are six phases of growth and death, as shown in Figure 6.1. They are:

1. **Lag phase:** A period of time when the change of cell number is zero.

2. **Accelerated growth phase:** The cell number starts to increase and the division rate increases to reach a maximum.

3. **Exponential growth phase:** The cell number increases exponentially as the cells start to divide. The growth rate is increasing during this phase, but the division rate which is proportional to $d\ln C_{N_1}/dt$, is constant at its maximum value, as illustrated in Figure 6.1.

4. **Decelerated growth phase:** After the growth rate reaches a maximum, it is followed by the deceleration of both growth rate and the division rate.

5. **Stationary phase:** The cell population will reach a maximum value and will not increase any further.

6. **Death phase:** After nutrients available for the cells are depleted, cells will start to die and the number of viable cells will decrease.

6.3.1 Lag Phase

The lag phase (or initial stationary, or latent) is an initial period of cultivation during which the change of cell number is zero or negligible. Even though the cell number does not increase, the cells may grow in size during this period.

The length of this lag period depends on many factors such as the type and age of the microorganisms, the size of the inoculum, and culture conditions.

The lag usually occurs because the cells must adjust to the new medium before growth can begin. If microorganisms are inoculated from a medium with a low nutrient concentration to a medium with a high concentration, the length of the lag period is usually long. This is because the cells must produce the enzymes necessary for the metabolization of the available nutrients. If they are moved from a high to a low nutrient concentration, there is usually no lag phase.

Another important factor affecting the length of the lag phase is the size of the inoculum. If a small amount of cells are inoculated into a large volume, they will have a long lag phase. For large-scale operation of the cell culture, it is our objective to make this lag phase as short as possible. Therefore, to inoculate a large fermenter, we need to have a series of progressively larger seed tanks to minimize the effect of the lag phase.

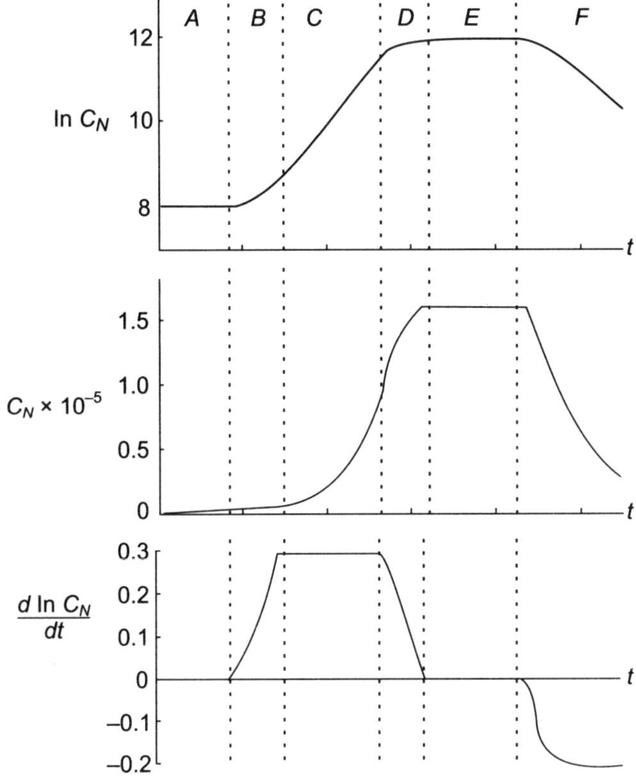

Fig. 6.1 Typical growth curve of unicellular organisms: (A) lag phase; (B) accelerated growth phase; (C) exponential growth phase; (D) decelerated growth phase; (E) stationary phase; (F) death phase.

At the end of the lag phase, when growth begins, the division rate increases gradually and reaches a maximum value in the exponential growth period, as shown by the rising inflection at B in Figure 6.1. This transitional period is commonly called the accelerated growth phase and is often included as a part of the lag phase.

6.3.2   Exponential Growth Phase

In unicellular organisms, the progressive doubling of cell number results in a continually increasing rate of growth in the population. A bacterial culture undergoing *balanced growth* mimics a first-order autocatalytic chemical reaction (Carberry, 1976; Levenspiel, 1972). Therefore, the rate of the cell population increase at any particular time is proportional to the number density (C_N) of bacteria present at that time:

$$r_N = \frac{C_N}{dt} = \mu\, C_N \tag{6.5}$$

where the constant μ is known as the *specific growth rate* [hr^{-1}]. The specific growth rate should not be confused with the growth rate, which has different units and meaning. The growth rate is the change of the cell number density with time, while the specific growth rate is

$$\mu = \frac{1}{C_N}\frac{dC_N}{dt} = \frac{d\ln C_N}{dt} \tag{6.6}$$

which is the change of the natural log of the cell number density with time. Comparing Eq. (6.4) and Eq. (6.6) shows that

$$\mu = \frac{d\ln C_N}{dt} = \ln 2\left(\frac{d\log C_N}{dt}\right) = \delta \ln 2 \tag{6.7}$$

Therefore, the specific growth rate μ is equal to ln2 times of the division rate, δ.

If μ is constant with time during the exponential growth period, Eq. (6.5) can be integrated from time t_1 to t as

$$\int_{C_{N_0}}^{C_N}\frac{dC_N}{C_N} = \int_{t_0}^{t}\mu\,dt \tag{6.8}$$

to give

$$C_N = C_{N_0}\exp\left[\mu\,(t - t_0)\right] \tag{6.9}$$

where C_{N_0} is the cell number concentration at t_1 when the exponential growth starts. Eq. (6.9) shows the increase of the number of cells exponentially with respect to time.

The time required to double the population, called the doubling time (t_d), can be estimated from Eq. (6.9) by setting $C_N = 2C_{N_0}$ and $t_1 = 0$ and solving for t:

$$t_d = \frac{\ln 2}{\mu} = \frac{1}{\delta} \tag{6.10}$$

The doubling time is inversely proportional to the specific growth rate and is equal to the reciprocal of the division rate.

---

[1] This equation has the same form as the rate equation for an enzyme catalyzed reaction, the Michaelis-Menten equation:

$$r_P = \frac{r_{max}\,\dot{C}_S}{K_M + C_S}$$

and also as the Langmuir adsorption isotherm:

$$\theta = \frac{C_A}{K + C_A}$$

where θ is the fraction of the solid surface covered by gas molecules and K is the reciprocal of the adsorption equilibrium constant.

6.3.3   Factors Affecting the Specific Growth Rate

Substrate Concentration: One of the most widely employed expressions for the effect of substrate concentration on μ is the *Monod equation,* which is an empirical expression based on the form of equation normally associated with enzyme kinetics or gas adsorption:[1]

$$\mu = \frac{\mu_{max}C_S}{K_S + C_S} \qquad (6.11)$$

where C_S is the concentration of the limiting substrate in the medium and K_S is a system coefficient. This relationship is shown graphically in Figure 6.2. The value of K_S is equal to the concentration of nutrient when the specific growth rate is half of its maximum value (μ_{max})

While the Monod equation is an oversimplification of the complicated mechanism of cell growth, it often adequately describes fermentation kinetics when the concentrations of those components which inhibit the cell growth are low.

According to the Monod equation, further increase in the nutrient concentration after μ reaches μ_{max} does not affect the specific growth rate, as shown in Figure 6.2. However, it has been observed that the specific growth rate decreases as the substrate concentration is increased beyond a certain level.

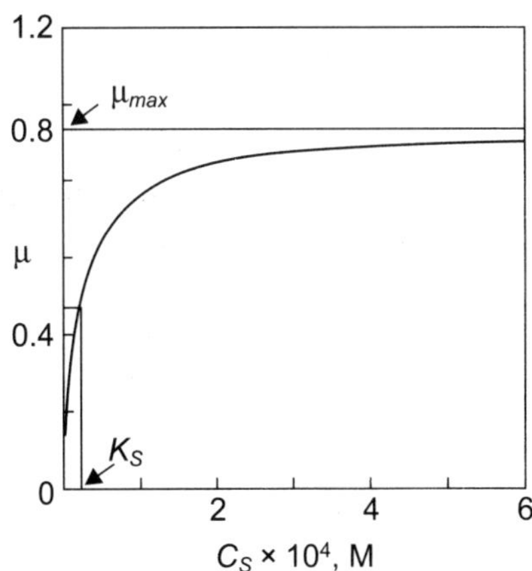

Fig. 6.2   Dependence of the specific growth rate on the concentration of the growth limiting nutrient: , μ_{max} = 0.935 hr^{-1}, K_s = 0.22 × 1(T^{-4} mol/L.

Several other models have been proposed to improve the Monod equation. They are:

$$\mu = \frac{\mu_{max}C_S}{K_{I1} + C_S + (C_{I2}C_S)^2} \qquad (6.12)$$

$$\mu = \mu_{max}[1 - \exp(-C_S/K_S]\tag{6.13}$$

$$\mu = \frac{\mu_{max}}{(1 + K_S C_S^{-\lambda})}\tag{6.14}$$

$$\mu = \frac{\mu_{max} C_S}{\beta C_N + C_S}\tag{6.15}$$

If several substances can limit the growth of a microorganism, the following model can be employed

$$\mu = \mu_{max} \frac{C_1}{K_1 + C_1} \frac{C_2}{K_2 + C_2} \cdots \tag{6.16}$$

If the limiting nutrient is the energy source for the culture, a certain amount of substrate can be used for purposes other than growth. Some models include a term, k_e, which accounts for the maintenance of cells as

$$\mu = \frac{\mu_{max} C_S}{K_S + C_S} - k_e \tag{6.17}$$

When C_S is so low that the first term of the right-hand side of Eq. (6.17) is less than or equal to k_e, the specific growth rate is equal to zero. These alternative models give a better fit for the growth of certain microorganisms, but their algebraic solutions are more difficult than for the Monod equation.

Product Concentration: As cells grow they produce metabolic by-products which can accumulate in the medium. The growth of microorganisms is usually inhibited by these products, whose effect can be added to the Monod equation as follows:

$$\mu = \mu_{max} \left(\frac{C_S}{K_S + C_S} \right)\left(\frac{K_P}{K_P + C_P} \right)\tag{6.18}$$

or

$$\mu = \mu_{max} \left(\frac{C_S}{K_S + C_S} \right)\left(1 - \frac{C_P}{C_{Pm}} \right)^{n}\tag{6.19}$$

The preceding equations both describe the product inhibition fairly well. The term C_{Pm} is the maximum product concentration above which cells cannot grow due to product inhibition.

Other Conditions: The specific growth rate of microorganisms is also affected by medium pH, temperature, and oxygen supply. The optimum pH and temperature differ from one microorganism to another.

6.3.4   Stationary and Death Phase

The growth of microbial populations is normally limited either by the exhaustion of available nutrients or by the accumulation of toxic products of metabolism. As a consequence, the rate of growth declines and growth eventually stops. At this point a culture is said to be in the *stationary phase*. The transition between the exponential phase and the stationary phase involves a period of unbalanced growth during which the various cellular components are synthesized at unequal rates. Consequently, cells in the stationary phase have a chemical composition different from that of cells in the exponential phase.

The stationary phase is usually followed by a death phase in which the organisms in the population die. Death occurs either because of the depletion of the cellular reserves of energy, or the accumulation of toxic products. Like growth, death is an exponential function. In some cases, the organisms not only die but also disintegrate, a process called *lysis*.

6.4   STIRRED-TANK FERMENTER

As defined in Chapter 2, a bioreactor is a device within which biochemical transformations are caused by the action of enzymes or living cells. The bioreactor is frequently called a fermenter whether the transformation is carried out by living cells or *in vivo* cellular components (enzymes). However, in this text, we call the bioreactor employing living cells a *fermenter* to distinguish it from the bioreactor which employs enzymes, the *enzyme reactor*. In laboratories, cells are usually cultivated in Erlenmeyer flasks on a shaker. The gentle shaking in a shake flask is very effective to suspend the cells, to enhance the oxygenation through the liquid surface, and also to aid the mass transfer of nutrients without damaging the structure of cells.

For a large-scale operation, the stirred-tank fermenter (STF) is the most widely used design in industrial fermentation. It can be employed for both aerobic or anaerobic fermentation of a wide range of cells including microbial, animal, and plant cells. Figure 6.3 shows a diagram of a fermenter used for penicillin production (Aiba et al., 1973). The mixing intensity can be varied widely by choosing a suitable impeller type and by varying agitating speeds. The mechanical agitation and aeration are effective for the suspension of cells, oxygenation, mixing of the medium, and heat transfer. The STF can be also used for high viscosity media. It was one of the first large-scale fermenters developed in the pharmaceutical industries. Its performance and characteristics are extensively studied. Since a stirred-tank fermenter is usually built with stainless steel and operated in mild operating conditions, the life expectancy of the fermenter is also long.

The disadvantage of the STF stems from its advantages. The agitator is effective in mixing the fermenter content, but it consumes a large amount of power and can damage a shear-sensitive cell system such as mammalian or plant cell culture. Fluid shear in mixing is produced by velocity gradients from tangential and radial velocity components of fluids leaving the impeller region. The velocity profile of a regular flat-bladed, disk turbine is parabolic in shape, with the highest observed velocity occurring at the centerline of each blade. Moving away from the centerline, the resultant velocity decreases by 85 percent at one blade-width distance above or below, creating a high shear region (Bowen, 1986). As the blade width-to-diameter ratio increases, the velocity profile becomes a more blunt, less parabolic shape, which yields a lower amount of shear due to a more gradual velocity gradient. Therefore, by increasing the blade width, a STF can be employed successfully in cultivating animal cells (Feder and Tolbert, 1983) or plant cells (Hooker et al., 1990).

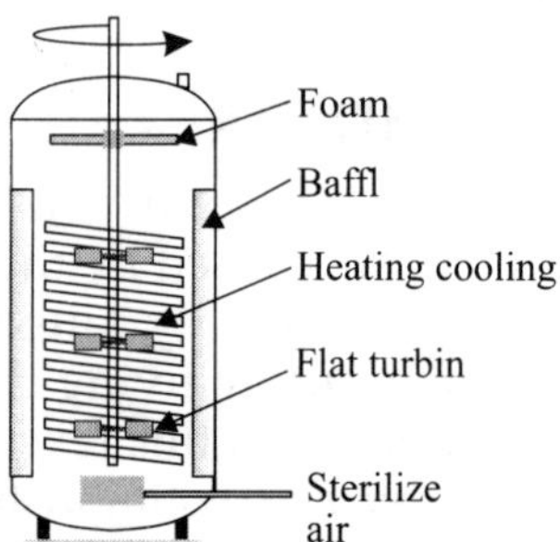

Fig. 6.3 Cutaway diagram of a fermenter used for penicillin production.

Many bench-scale fermenters are made of glass with a stainless steel top plate. As mentioned earlier, larger fermenters are made of stainless steel (Type 316). The height-to-diameter ratio of the vessels is either two-to-one or three-to-one. It is usually agitated with two or three turbine impellers. The impeller shaft enters the vessel either from the top or the bottom of the vessel through a bearing housing and mechanical seal assembly. The impeller diameter to tank diameter ratio (D_I/D_T) is generally 0.3 to 0.4. In the case of a two-impeller system, the distance between the first impeller and the bottom of the vessel and between the two impellers is 1.5 impeller diameters. The distance is reduced to one impeller diameter in the case of a three-impeller system. Four equally spaced baffles are usually installed to prevent a vortex formation which reduces the mixing efficiency. The width of the baffle is usually about one tenth of the tank diameter. For aerobic fermentation, a single orifice sparger or ring sparger is used to aerate the fermenter. The sparger is located between the bottom impeller and the bottom of the vessel. The pH in a fermenter can be maintained by employing either a buffer solution or a pH controller. The temperature is controlled by heating or cooling as the system requires.

6.4.1 Batch or Plug-Flow Fermenter

An ideal stirred fermenter is assumed to be well mixed so that the contents are uniform in composition at all times. Another ideal fermenter is the plug-flow fermenter, the analysis of which is analogous to the ideal batch fermenter.

In a *tubular-flow fermenter*, nutrients and microorganisms enter one end of a cylindrical tube and the cells grow while they pass through. Since the long tube and lack of stirring device prevents complete mixing of the fluid, the properties of the flowing stream will vary in both longitudinal and radial direction. However, the variation in the radial direction is small compared to that in the longitudinal direction. The ideal tubular-flow fermenter without radial variations is called a *plug-flow fermenter* (PFF).

In reality, the PFF fermenter is hard to be found. However, the packed-bed fermenter and multi-staged fermenter can be approximated as PFF. Even though the steady-state PFF is operated in a continuous mode, the cell concentration of an ideal batch fermenter after time t will be the same as that of a steady-state PFF at the longitudinal location where the residence time i is equal to t (Figure 6.4). Therefore, the following analysis applies for both the ideal batch fermenter and steady-state PFF.

If liquid medium is inoculated with a seed culture, the cell will start to grow exponentially after the lag phase. The change of the cell concentration in a batch fermenter is equal to the growth rate of cells in it:

$$\frac{dC_X}{dt} = r_X = \mu\, C_X \tag{6.20}$$

To derive the performance equation of a batch fermentation, we need to integrate Eq. (6.20) to obtain:

$$\int_{C_{X_0}}^{C_X} \frac{dC_X}{r_X} = \int_{C_{X_0}}^{C_X} \frac{dC_X}{\mu\, C_X} = \int_{t_0}^{t} dt = t - t_0 \tag{6.21}$$

It should be noted that Eq. (6.21) only applies when r_X is larger than zero. Therefore, t_1 in Eq. (6.21) is not the time that the culture was inoculated, but the time that the cells start to grow, which is the beginning of the accelerated growth phase.

According to Eq. (6.21), the batch growth time $t - t_0$ is the area under the $1/r_X$ versus C_X curve between C_{X_0} and C_X as shown in Figure 6.5. The solid line in Figure 6.5 was calculated with the Monod equation and the shaded area is equal to $t - t_0$. The batch growth time is rarely estimated by this graphical method since the C_X versus t

curve is a more straightforward way to determine it. However, the graphical representation is useful in comparing the performance of the various fermenter configurations, which is discussed later. At this time just note that the curve is U shaped, which is characteristic of *auto catalytic reactions:*

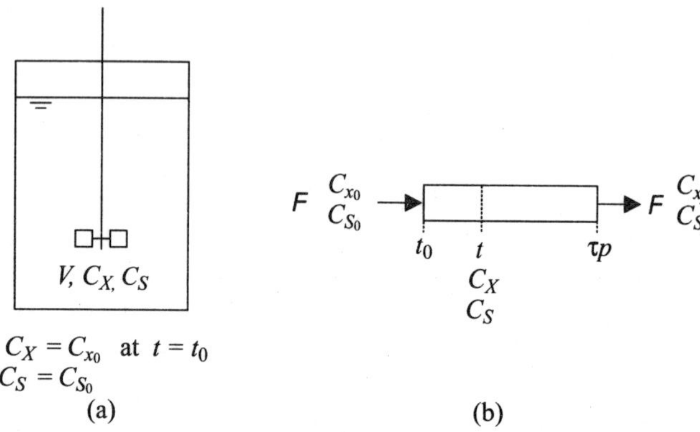

Fig. 6.4 Schematic diagram of (a) batch stirred-tank fermenter and plug-flow fermenter

$$S + X \longrightarrow X + X$$

The rate for an autocatalytic reaction is slow at the start because the concentration of X is low. It increases as cells multiply and reaches a maximum rate. As the substrate is depleted and the toxic products accumulate, the rate decreases to a low value.

If Monod kinetics adequately represents the growth rate during the exponential period, we can substitute Eq. (6.11) into Eq. (6.21) to obtain

$$\int_{C_{x_0}}^{C_x} \frac{(K_S + C_S)dC_X}{\mu_{max} C_S C_X} = \int_{t_0}^{t} dt \tag{6.22}$$

Eq. (6.22) can be integrated if we know the relationship between C_S and C_X. It has been observed frequently that the amount of cell mass produced is proportional to the amount of a limiting substrate consumed. The *growth yield* $(Y_{X/S})$ is defined as

$$Y_{X/S} = \frac{\Delta C_X}{-\Delta C_S} = \frac{C_X - C_{X_0}}{-(C_S - C_{S_0})} \tag{6.23}$$

Substitution of Eq. (6.23) into Eq. (6.22) and integration of the resultant equation gives a relationship which shows how the cell concentration changes with respect to time:

$$(t - t_0)\mu_{max} = \left(\frac{K_S Y_{X/S}}{C_{X_0} + C_{S_0} Y_{X/S}} + 1 \right) \ln \frac{C_X}{C_{X_0}} + \frac{K_S Y_{X/S}}{C_{X_0} + C_{S_0} Y_{X/S}} \ln \frac{C_{S_0}}{C_S} \tag{6.24}$$

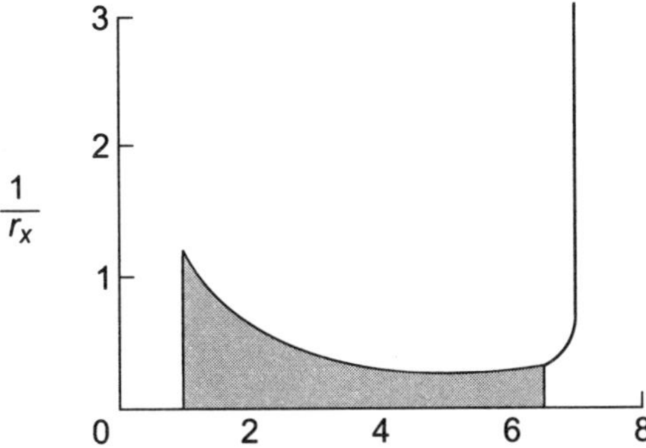

Fig. 6.5 A graphical representation of the batch growth time $t - t_1$ (shaded area). The solid line represents the Monod model with μ_{max} = 0.935 hr^{-1}, K_S =0.71 g/L, $Y_{X/S}$ = 0.6, C_{X_0} = 1 g/L, and C_{S_0} = 10 g/L.

The Monod kinetic parameters, μ_{max} and K_S, cannot be estimated with a series of batch runs as easily as the Michaelis-Menten parameters for an enzyme reaction. In the case of an enzyme reaction, the initial rate of reaction can be measured as a function of substrate concentration in batch runs. However, in the case of cell cultivation, the initial rate of reaction in a batch run is always zero due to the presence of a lag phase, during which Monod kinetics does not apply. It should be noted that even though the Monod equation has the same form as the Michaelis-Menten equation, the rate equation is different. In the Michaelis-Menten equation,

$$\frac{dC_P}{dt} = \frac{r_{max}C_S}{K_M + C_S} \tag{6.25}$$

while in the Monod equation,

$$\frac{dC_X}{dt} = \frac{\mu_{max}C_S C_X}{K_S + C_S} \tag{6.26}$$

There is a C_X term in the Monod equation which is not present in the Michaelis-Menten equation.

Example 6.1

The growth rate of *E. coli* be expressed by Monod kinetics with the parameters of μ_{max} = 0.935 hr^{-1} and K_S = 0.71 g/L (Monod, 1949). Assume that the cell yield $Y_{X/S}$ is 0.6 g dry cells per g substrate. If C_{X_0} is 1 g/L and C_{S_1} = 10 g/L when the cells start to grow exponentially, at t_0 = 0, show how In C_X, C_X, C_S, d In C_X/dt, and dC_X/dt change with respect to time.

Solution:

You can solve Eqs. (6.23) and (6.24) to calculate the change of C_X and C_S with respect to time, which involves the solution of a nonlinear equation. Alternatively, the Advanced Continuous Simulation Language (ACSL, 1975) can be used to solve Eqs. (6.22) and (6.23).

Table 6.1   ACSL Program for Example 6.1

```
CONSTANT MUMAX..=0.935, KS=0.71, YXS=0.6, ...
   CX2=1., CS2=10, TSTOP=10
CINTERVAL CINT=0.1 $ 'COMMUNICAITON INTERVAL¹
VARIABLE T=0.0
   CX=INTEG(MUMAX*CS*CX/(KS+CS),CX2)
   CS=CS2-(CX-CX2)/YXS
   LNCX=ALOG(CX)
   DLNCX=MUMAX*CS/(KS+CS)
   DCXDT=MUMAX*CS*CX/(KS+CS)
   TERMT(CS.LE.O)  END
   END
```

Fig. 6.6   Solution to Example 6.1.

Table 6.1 shows the ACSL program and Figure 6.6 shows the results. Comparison of Figure 6.6 with Figure 6.1 shows that Monod kinetics can predict the cell growth from the start of the exponential growth phase to the stationary phase.

6.4.2   Ideal Con tin uous Stirred-tank Fermen ter

Microbial populations can be maintained in a state of exponential growth over a long period of time by using a system of continuous culture. Figure 6.7 shows the block diagram for a continuous stirred-tank fermenter (CSTF). The growth chamber is connected to a reservoir of sterile medium. Once growth has been initiated, fresh medium is continuously supplied from the reservoir.

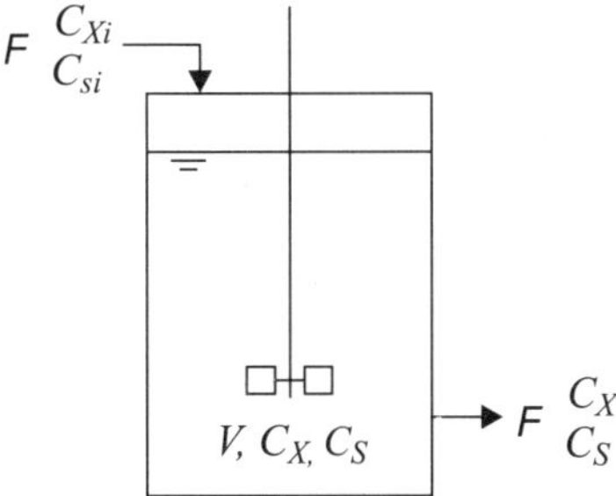

Fig. 6.7 Schematic diagram of continuous stirred-tank fermenter (CSTF)

Continuous culture systems can be operated as *chemostat* or as *turbidostat*. In a chemostat the flow rate is set at a particular value and the rate of growth of the culture adjusts to this flow rate. In a turbidostat the turbidity is set at a constant level by adjusting the flow rate. It is easier to operate chemostat than turbidostat, because the former can be done by setting the pump at a constant flow rate, whereas the latter requires an optical sensing device and a controller. However, the turbidostat is recommended when continuous fermentation needs to be carried out at high dilution rates near the washout point, since it can prevent washout by regulating the flow rate in case the cell loss through the output stream exceeds the cell growth in the fermenter.

The material balance for the microorganisms in a CSTF (Figure 6.7) can be written as

$$FC_{X_i} - FC_X + Vr_X = V\frac{dC_X}{dt} \tag{6.27}$$

where r_X is the rate of cell growth in the fermenter and dC_X/dt represents the change of cell concentration in the fermenter with time.

For the steady-state operation of a CSTF, the change of cell concentration with time is equal to zero $(dC_X/dt = 0)$ since the microorganisms in the vessel grow just fast enough to replace those lost through the outlet stream, and Eq. (6.27) becomes

$$\tau_m = \frac{V}{F} = \frac{C_X - C_{X_i}}{r_X} \tag{6.28}$$

Eq. (6.28) shows that the required residence time is equal to $C_X - C_{X_i}$, times $1/r_X$, which is equal to the area of the rectangle of width $C_X - C_{X_i}$ and height $1/r_X$ on the $1/r_X$ versus C_X curve.

Figure 6.8 shows the $1/r_X$ versus C_X curve. The shaded rectangular area in the figure is equal to the residence time in a CSTF when the inlet stream is sterile. This graphical illustration of the residence time can aid us in comparing the effectiveness of fermenter systems. The shorter the residence time in reaching a certain cell concentration, the more effective the fermenter. The optimum operation of fermenters based on this graphical illustration is discussed in the next section.

If the input stream is sterile ($C_{X_i} = 0$), and the cells in a CSTF are growing exponentially ($r_X = \mu C_X$), Eq. (6.28) becomes

$$\tau_m = \frac{1}{\mu} = \frac{1}{D} \tag{6.29}$$

where D is known as *dilution rate* and is equal to the reciprocal of the residence time (τ_m). Therefore, for the steady-state CSTF with sterile feed, the specific growth rate is equal to the dilution rate. In other words, the specific growth rate of a microorganism can be controlled by changing the medium flow rate. If the growth rate can be expressed by Monod equation, then

$$D = \mu = \frac{1}{\tau_m} = \frac{\mu_{max} C_S}{K_S + C_S} \tag{6.30}$$

From Eq. (6.30), C_S can be calculated with a known residence time and the Monod kinetic parameters as:

$$C_S = \frac{K_S}{\tau_m \mu_{max} - 1} \tag{6.31}$$

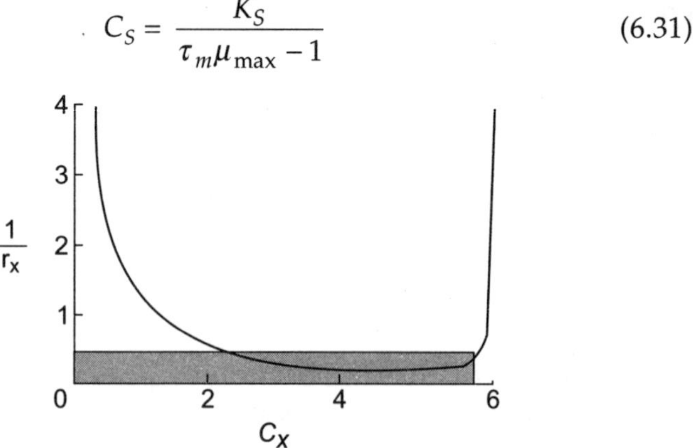

Fig. 6.8 A graphical representation of the estimation of residence time for the CSTF. The line represents the Monod model with $\mu_{max} = 0.935$ hr^{-1}, $K_S = 0.71$ g/L, $Y_{X/S} = 0.6$, $C_{S_i} = 10$ g/L, and $C_{X_i} = 0$.

It should be noted, however, that the preceding expression is only valid when $\tau_m \, \mu_{max} > 1$. If $\tau_m \, \mu_{max} < 1$, the growth rate of the cells is less than the rate of cells leaving with the outlet stream.

Consequently, all of the cells in the fermenter will be washed out and Eq. (6.31) is invalid.

If the growth yield $(Y_{X/S})$ is constant, then

$$C_X = Y_{X/S}\,(C_{S_i} - C_S) \tag{6.32}$$

Substituting Eq. (6.31) into Eq. (6.32) yields the correlation for C_X as

$$C_X = Y_{X/S}\left(C_{Si} - \frac{K_S}{\tau_m \mu_{\max} - 1}\right) \tag{6.33}$$

$$C_p = C_{Pi} + Y_{P/S}\left(C_{Si} - \frac{K_S}{\tau_m \mu_{\max} - 1}\right) \tag{6.34}$$

Again, Eqs. (6.33) and (6.34) are valid only when $\tau_m\,\mu_{\max} > 1$

In this section, we set a material balance for the microbial concentration and derive various equations for the CSTF. The same equations can be also obtained by setting material balances for the substrate concentration and product concentration.

6.4.3. Evaluation of Monod Kinetic Parameters

The equality of the specific growth rate and the dilution rate of the steady-state CSTF shown in Eq. (6.30) is helpful in studying the effects of various components of the medium on the specific growth rate. By measuring the steady-state substrate concentration at various flow rates, various kinetic models can be tested and the value of the kinetic parameters can be estimated. By rearranging Eq. (6.30), a linear relationship can be obtained as follows:

$$\frac{1}{\mu} = \frac{K_S}{\mu_{\max}} \frac{1}{C_S} + \frac{1}{\mu_{\max}} \tag{6.35}$$

where μ is equal to the dilution rate (D) for a chemostat. If a certain microorganism follows Monod kinetics, the plot of $1/\mu$ versus $1/C_S$ yields the values of $\mu_{\max}$ and K_S by reading the intercept and the slope of the straight line. This plot is the same as the Lineweaver-Burk plot for Michaelis-Menten kinetics. It has the advantage that it shows the relationship between the independent variable (C_S) and dependent variable (μ). However, $1/\mu$ approaches infinity as the substrate concentration decreases. This gives undue weight to measurements made at low substrate concentrations and insufficient weight to measurements at high substrate concentrations.

Eq. (6.30) can be rearranged to give the following linear relationships, which can be employed instead of Eq. (6.35) for a better estimation of the parameters in certain cases.

$$\frac{C_S}{\mu} = \frac{K_S}{\mu_{max}} + \frac{C_S}{\mu_{max}} \tag{636}$$

$$\mu = \mu_{max} - K_S \frac{\mu}{C_S} \tag{6.37}$$

However, the limitation of this approach to determine the kinetic parameters is in the difficulty of running a CSTF. For batch runs, we can even use shaker flasks to make multiple runs with many different conditions at the same time. The batch run in a stirred fermenter is not difficult to carry out, either. Since there is no input and output connections except the air supply and the length of a run is short, the danger of contamination of the fermenter is not serious.

For CSTF runs, we need to have nutrient and product reservoirs which are connected to the fermenter aseptically. The rate of input and output stream needs to be controlled precisely. Sometimes, the control of the outlet flowrate can be difficult due to the foaming or plugging by large cell aggregates. Since the length of the run should last several days or even weeks to reach a steady state and also to vary the dilution rates, there is always a high risk for the fermenter to be contaminated. Frequently, it is difficult to reach a steady state because of the cell's mutation and adaptation to new environment.

Furthermore, since most large-scale fermentations are carried out in batch mode, the kinetic parameters determined by the chemostat study should be able to predict the growth in a batch fermenter. However, due to the significantly different environments of batch and continuous fermenters, the kinetic model developed from the CSTF runs may fail to predict the growth behavior of a batch fermenter. Nevertheless, the verification of a kinetic model and the evaluation of kinetic parameters by running chemostat is the most reliable method because of its constant medium environment.

The data from batch runs can be used to determine the kinetic parameters, even though this is not a highly recommended procedure. The specific growth rate during a batch run can be estimated by measuring the slope of the cell concentration versus time curve at the various points. The substrate concentration needs to be measured at the same points where the slope is read. Then, the plots according to Eqs. (6.35), (6.36), (6.37) can be constructed to determine the kinetic parameters. However, the parameter values obtained in this method need to be checked carefully whether they are in the reasonable range for the cells tested.

Example 6.2

A chemostat study was performed with yeast. The medium flow rate was varied and the steady-state concentration of cells and glucose in the fermenter were measured and recorded. The inlet concentration of glucose was set at 100 g/L. The volume of the fermenter contents was 500 mL. The inlet stream was sterile.

Flow rate F, mL/hr	Cell Conc. C_X, g/L	Substrate Conc. C_S, g/L
31	5.97	0.5
50	5.94	1.0
71	5.88	2.0
91	5.76	4.0
200	0	100

a. Find the rate equation for cell growth.
b. What should be the range of the flow rate to prevent washout of the cells?

Solution:

a. Let's assume that the growth rate can be expressed by Monod kinetics. If this assumption is reasonable, the plot of $1/\mu$ versus $1/C_S$ will result in a straight line according to Eq. (6.35). The dilution rate for the chemostat is

$$D=\frac{F}{V}$$

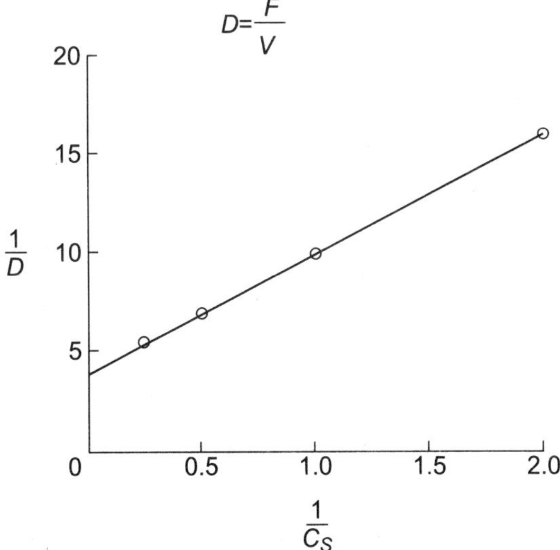

Fig. 6.9　The plot of $1/D$ versus $1/C_S$ for Example 6.2.

The plot of $1/D$ versus $1/C_S$ is shown in Figure 6.9 which shows a straight line with intercept

$$\frac{1}{\mu_{max}} = 3.8$$

and slope

$$\frac{K_S}{\mu_{max}} = 5.2$$

Therefore, $\mu_{max} = 0.26 \text{ hr}^{-1}$, and $K_S = 1.37 \text{ g/L}$. The rate equation of cell growth is

$$r_X = \frac{0.26 C_S C_X}{1.37 + C_S}$$

b. To prevent washout of the cells, the cell concentration should be maintained so that it will be greater than zero. Therefore, from Eq. (6.33).

$$C_X = Y_{X/S}\left(C_{S_i} - \frac{K_S}{\tau_m \mu_{max} - 1}\right) > 0$$

Solving the preceding equation for τ_m yields

$$\tau_m = \frac{V}{F} = \frac{K_S + C_{S_i}}{C_{Si} \, \mu_{max}}$$

Therefore,

$$F = \frac{V C_{S_i} \mu_{max}}{K_S + C_{S_i}} = \frac{0.5(100)(0.26)}{1.37 + 100} = 1.128 \text{ L/hr}$$

6.4.4 Productivity of CSTF

Normally, the productivity of the fermenter is expressed as the amount of a product produced per unit time and volume. If the inlet stream is sterile ($C_{X_i} = 0$), the productivity of cell mass is equal to C_X/τ_m, which is equal to the slope of the straight line $\overline{OAB}$ of the C_X vs. τ_m curve, as shown in Figure 6.10. The productivity at point A is equal to that at point B. At point A, the cell concentration of the outlet stream is low but the residence time is short, therefore, more medium can pass through. On the other hand, at point B the cell concentration of the outlet stream is high, but the residence time is long, so a smaller amount of medium passes through. Point A is an unstable region because it is very close to the washout point D, and because a small fluctuation in the residence time can bring about a large change in the cell concentration. As the slope of the line increases, the productivity increases, and the length of $\overline{AB}$ decreases. The slope of the line will have its maximum value when it is tangent to the C_X curve. Therefore, the value of the maximum productivity is equal to the slope of line $\overline{OC}$. The maximum productivity will be attained at the point D.

The operating condition for the maximum productivity of the CSTF can be estimated graphically by using $1/r_X$ versus C_X curve. The maximum productivity can be attained when the residence time is the minimum. Since the residence time is equal to the area of the rectangle of width C_X and height $1/r_X$ on the $1/r_X$ versus C_X curve, it is the minimum when the $1/r_X$ is the minimum, as shown in Figure 6.11.

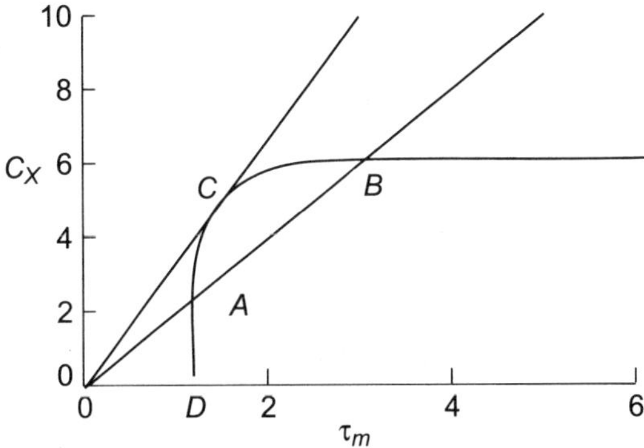

Fig. 6.10 The change of the concentrations of cells and substrate as a function of the residence time. Productivity is equal to the slope of the straight line $\overline{OAB}$. The curve is drawn by using the Monod model with $,\mu_{max}$ = 0.935 hr^{-1}, K_S = 0.71 g/L, $Y_{X/S}$ = 0.6, and C_{S_i} = 10 g/L.

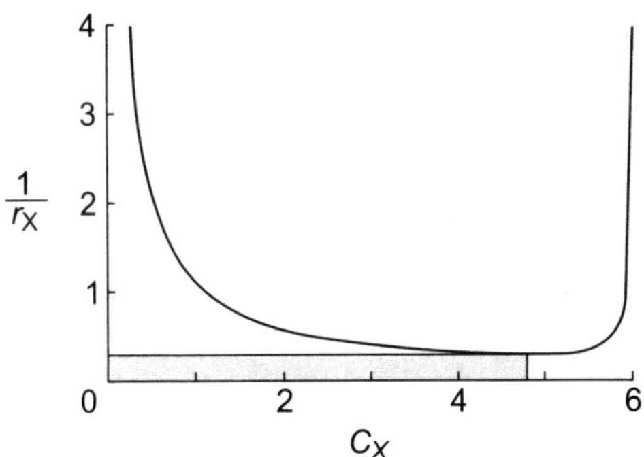

Fig. 6.11 A graphical illustration of the CSTF with maximum productivity. The solid line represents the Monod model with μ_{max} = 0.935 hr^{-1}, K_S = 0.71 g/L, $Y_{X/S}$ = 0.6, C_{S_i} = 10 g/L, and C_{X_i} = 0.

It would be interesting to derive the equations for the cell concentration and residence time at this maximum cell productivity. The cell productivity for a steady-state CSTF with sterile feed is

$$\frac{C_X}{\tau_m} = r_X = \frac{\mu_{max}C_SC_X}{K_S + C_S} \tag{6.38}$$

The productivity is maximum when $dr_X/dC_X = 0$. After substituting $C_S = C_{S_i} - C_X Y_{X/S}$ into the preceding equation, differentiating with respect to C_X, and setting the resultant equation to zero, we obtain the optimum cell concentration for the maximum productivity as

$$C_{X,opt} = Y_{X/S} C_{Si} \frac{\alpha}{\alpha + 1} \tag{6.39}$$

where

$$\alpha = \sqrt{\frac{K_S + C_{S_i}}{K_S}} \tag{6.40}$$

Since

$$C_S - C_{Si}/Y_X/S$$

$$C_{S.opt} = \frac{C_{S_i}}{\alpha + 1} \tag{6.41}$$

Substituting Eq. (6.41) into Eq. (6.38) for C_S yields the optimum residence time:

$$\tau_{m,opt} = \frac{\alpha}{\mu_{max}(\alpha - 1)} \tag{6.42}$$

6.4.5   Comparison of Batch and CSTF

As discussed earlier, the residence time required for a batch or steady-state PFF to reach a certain level of cell concentration is

$$\tau_b = t_0 + \int_{C_{X_0}}^{C_X} \frac{dC_X}{r_X} \tag{6-43}$$

where t_0 is the time required to reach an exponential growth phase. The area under the $1/r_X$ versus C_X curve between C_X and C_X is equal to $t_b - t_1$., as shown in Figure 6.5.

On the other hand, the residence time for the CSTF is expressed as Eq. (6.28) which is equal to the area of the rectangle of width $C_X - C_{X_i}$ and height $1/r_X$.

Since the $1/r_X$ versus C_X curve is U shaped, we can make the following conclusions for *single fermenter*.

1. The most productive fermenter system is a CSTF operated at the cell concentration at which value of $1/r_X$ is minimum, as shown in Figure 6.12(a), because it requires the smallest residence time.

2. If the final cell concentration to be reached is in the stationary phase, the batch fermenter is a better choice than the CSTF because the residence time required for the batch as shown in Figure 6.12(b) is smaller than that for the CSTF.

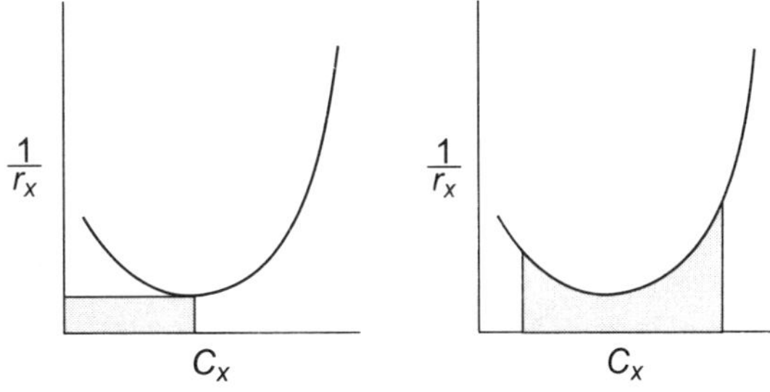

Fig. 6.12 Graphical illustration of the residence time required (shaded area) for the (a) CSTF and (b) Batch fermenter.

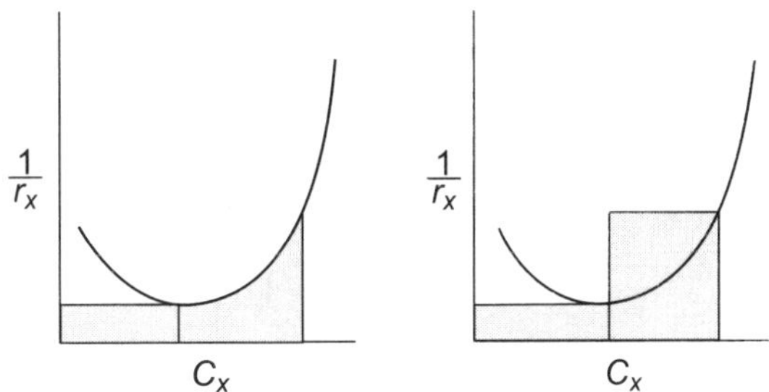

Fig. 6.13 Graphical illustration of the total residence time required (shaded area) when two fermenters are connected in series: (a) CSTF and PFF, and (b) two CSTFs.

6.5   MULTIPLE FERMENTERS CONNECTED IN SERIES

A question arises frequently whether it may be more efficient to use multiple fermenters connected in series instead of one large fermenter. Choosing the optimum fermenter system for maximum productivity depends on the shape of the $1/r_X$ versus C_X curve and the process requirement, such as the final conversion.

In the $1/r_X$ versus C_X curve, if the final cell concentration is less than $C_{X,opt}$ one fermenter is better than two fermenters connected in series, because two CSTFs connected in series require more residence time than one CSTF does. However, if the final cell concentration is much larger than $C_{X,opt}$ the best combination of two fermenters for a minimum total residence time is a CSTF operated at $C_{X,opt}$ followed by a PFF, as shown in (Figure 6.13a). A CSTF operated at $C_{X,opt}$ followed by another CSTF connected in series is also better than one CSTF (Figure 6.13b).

6.5.1   CSTF and PFF in Series

Figure 6.14 shows the schematic diagram of the two fermenters connected in series, CSTF followed by PFF. The result of the material balance for the first fermenter is the same as the single CSTF which we have already developed. If the input stream is sterile ($C_{Xi} = 0$), the concentrations of the substrate, cell and product can be calculated from Eqs. (6.31), 6.33), and 6.34), as follows:

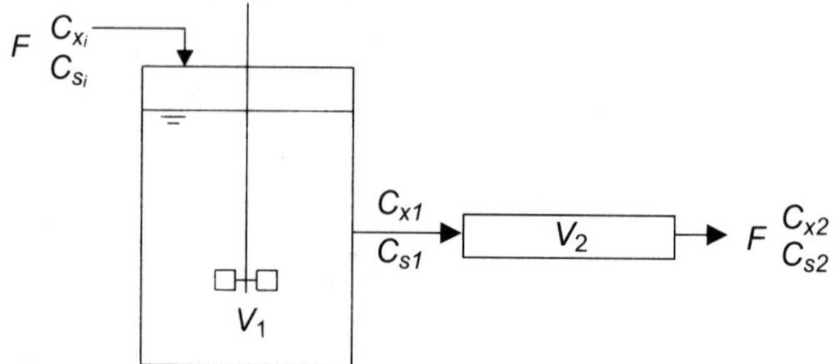

Fig. 6.14 Schematic diagram of the two fermenters, CSTF and PFF, connected in series.

$$C_{S_1} = \frac{K_S}{\tau_{m_1}\mu_{\max} - 1} \tag{6.44}$$

$$C_{X_1} = Y_{X/S}\left(C_{S_i} - \frac{K_S}{\tau_{m_1}\mu_{\max} - 1}\right) \tag{6.45}$$

$$C_{P_1} = C_{P_i} + Y_{P/S}\left(C_{S_i} - \frac{K_S}{\tau_{m_1}\mu_{\max} - 1}\right) \tag{6.46}$$

For the second PFF, the residence time can be estimated by

$$\tau_{p2} = \int_{C_{X_1}}^{C_{X_2}} \frac{dC_X}{r_X} = \int_{C_{X_1}}^{C_{X_2}} \frac{(K_S + C_S)dC_X}{\mu_{\max}C_S C_X} \tag{6.47}$$

Since growth yield can be expressed as

$$Y_{X/S} = \frac{C_{X_2} - C_{X_1}}{C_{S_1} - C_{S_2}} \tag{6.48}$$

Integration of Eq. (6.47) after the substitution of Eq. (6.48) will result,

$$\tau_{p2}\,\mu_{\max} = \left(\frac{K_S Y_{X/S}}{C_{X_1} + C_{S_1}Y_{X/S}} + 1\right)\ln\frac{C_{X_2}}{C_{X_1}} + \frac{K_S Y_{X/S}}{C_{X_1} + C_{S_1}Y_{X/S}}\ln\frac{C_{S_1}}{C_{S_2}} \tag{6.49}$$

If the final cell concentration (C_{X_2}) is known, the final substrate concentration (C_{S_2}) can be calculated from Eq. (6.48). The residence time ofthe second fermenter can then be calculated using Eq. (6.49). If the residence time of the second fermenter is known,

Eqs. (6.48) and (6.49) have to be solved simultaneously to estimate the cell and substrate concentrations. Another approach is to integrate Eq. (6.47) numerically until the given τ_{p2} value is reached.

6.5.2 Multiple CSTFs in Series

If the final cell concentration is larger than $C_{X,opt}$, the best combination of two fermenters for a minimum total residence time is a CSTF operated at $C_{X,opt}$ followed by a PFF, as explained already. However, the cultivation of microorganisms in the PFF is limited to several experimental cases, such as the tubular loop batch fermenter (Russell et al., 1974) and scraped tubular fermenter (Moo-Young et al., 1979). Furthermore, the growth kinetics in a PFF can be significantly different from that in a CSTF.

Another more practical approach is to use multiple CSTFs in series, since a CSTF operated at $C_{X,opt}$ followed another CSTF connected in series is also better than one CSTF.

Hill and Robinson (1989) derived an equation to predict the minimum possible total residence time to achieve any desired substrate conversion. They found that three optimally designed CSTFs connected in series provided close to the minimum possible residence time for any desired substrate conversion.

Figure 6.15 shows the schematic diagram of the multiple CSTFs connected in series. For the nth steady-state CSTF, the material balance for the microorganisms can be written as

$$F(C_{X_{n-1}} - C_{X_n}) + V_n R_{X_n} = 0 \tag{6.50}$$

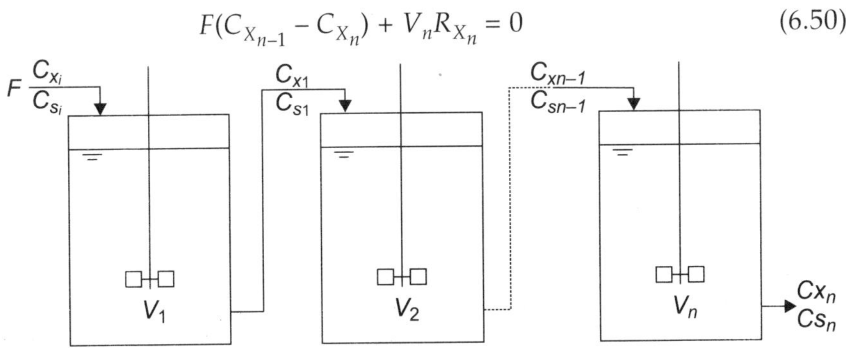

Fig. 6.15 Schematic diagram of the multiple CSTFs connected in series.

$$r_{X_n} = \frac{\mu_{max} C_{Sn} C_{X_n}}{K_S + C_{Sn}} \tag{6.51}$$

Growth yield can be expressed as

$$Y_{X/S} = \frac{C_{X_n} - C_{X_{n-1}}}{C_{S_{n-1}} - C_{S_n}} \tag{6.52}$$

By solving Eqs. (6.50), (6.51), and (6.52) simultaneously, we can calculate either dilution rate with the known cell concentration, or vice versa.

The estimation of the cell or substrate concentration with the known dilution rate can be done easily by using graphical technique as shown in Figure 6.1 (Luedeking, 1967). From Eq. (6.50), the dilution rate of the first reactor when the inlet stream is sterile is

$$D_1 = \frac{F}{V_1} = \frac{r_{X_1}}{C_{X_1}} \tag{6.53}$$

which can be represented by the slope of the straight line connecting the origin and $(C_{Xi}S_{Xi})$ in Figure 6.16. Similarly, for the second fermenter

$$D_2 = \frac{F}{V_2} = \frac{r_{X_2}}{C_{X_2} - C_{X1}} \tag{6.54}$$

which is the slope of the line connecting $(C_{X_1,} = 0)$ and $(C_{X_2,}r_{X2})$.

Therefore, by knowing the dilution rate of each fermenter, you can estimate the cell concentration of each fermenter, or vice versa.

Example 6.3

Suppose you have a microorganism that obeys the Monod equation:

$$\frac{dC_X}{dt} = \frac{\mu_{max}C_S C_X}{K_S + C_S}$$

where $\mu_{max} = 0.7 \text{ hr}^{-1}$ and $K_S = 5 \text{ g/L}$. The cell yield $(Y_{X/S})$ is 0.65. You want to cultivate this microorganism in either one fermenter or two in series. The flow rate and the substrate concentration of the inlet stream should be 500 L/hr and 85 g/L, respectively. The substrate concentration of the outlet stream must be 5 g/L

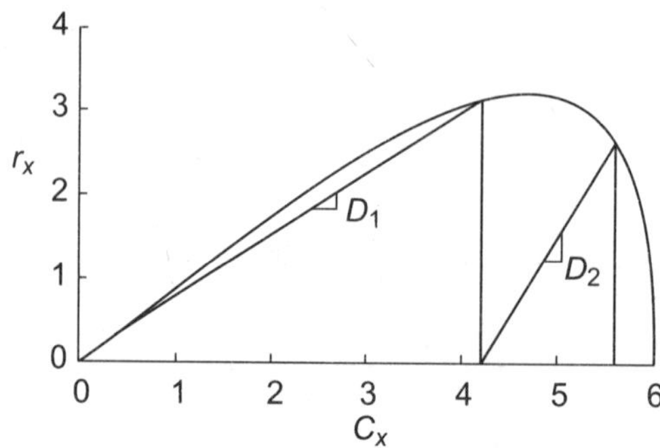

Fig. 6.16 Graphical solution for a two-stage continuous fermentation. The line represents the Monod model with $\mu_{max} = 0.935 \text{ hr}^{-1}$, $K_s = 0.71 \text{ g/L}$, $Y_{X/S}=0.6$, $C_{Si} = 10\text{g/L}, C_{Xi} = 0.$}

a. If you use one CSTF, what should be the size of the fermenter? What is the cell concentration of the outlet stream?

b. If you use two CSTFs in series, what sizes of the two fermenters will be most productive? What are the concentration of cells and substrate in the outlet stream of the first fermenter?

c. What is the best combination of fermenter types and volumes if you use two fermenters in series?

Solution:

a. For a single steady-state CSTF with a sterile feed, the dilution rate is equal to specific growth rate:

$$D = \frac{F}{V} = \frac{\mu_{max}C_S}{K_S + C_S} = \frac{0.7(5)}{5+5} = 0.35$$

$$V = \frac{F}{D} = \frac{500}{0.35} = 1{,}429 \text{ L}$$

The cell concentration of the outlet stream is

$$C_X = Y_{X/S}(C_{Si} - C_s) = 0.65\,(85 - 5) = 52 \text{ g/L}$$

b. For two CSTFs in series, the first fermenter must be operated at $C_{X,opt}$ and $C_{S,opt}$

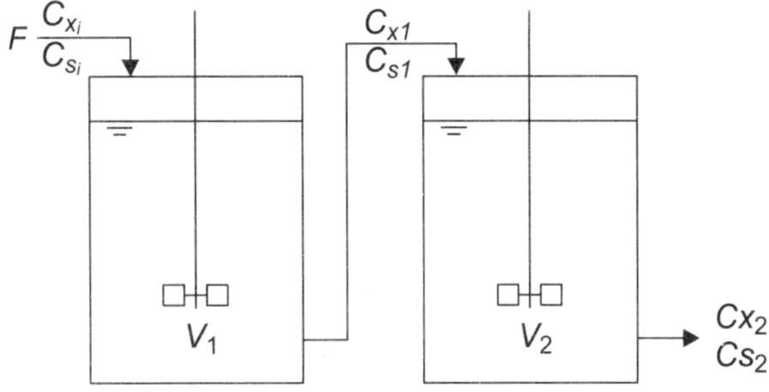

Therefore, from Eq. (6.39) through Eq. (6. 42),

$$\alpha = \sqrt{\frac{K_S + C_{S_i}}{K_S}} = \sqrt{\frac{5+85}{5}} = 4.2$$

$$C_{X_1} = C_{X_{opt}} = Y_{X/S}\,C_{S_i}\,\frac{\alpha}{\alpha+1} = 0.65(85)\frac{4.2}{4.2+1} = 45 \text{g/L}$$

$$C_{S_1} = C_{S_{opt}} = \frac{C_{S_i}}{\alpha+1} = \frac{85}{4.2+1} = 1.6 \text{ g/L}$$

$$\tau_{m_1} = \tau_{m_{opt}} = \frac{\alpha}{\mu_{max}(\alpha-1)} = \frac{4.2}{0.7(4.2-1)} = 1.9 \text{ hr}$$

$$V_1 \, \tau_{m1} F = 1.9(500) = 950 \text{ L}$$

For the second fermenter, from Eq. (6.50),

$$F(C_{X_1} - C_{X_2}) + \frac{V_2 \mu_{max} C_{S_2} C_{X_2}}{K_S + C_{S_2}} = 0$$

By rearranging the preceding equation for V_2,

$$V_2 = \frac{F(C_{X_2} - C_{X_1})}{\mu_{max} C_{S_2} C_{X_2} /(K_S + C_{S_2})} = \frac{500(52 - 45)}{0.7(5)(52)/(5 + 5)} = 192 \text{ L}$$

The total volume of the two CSTFs is

$$V = V_1 + V_2 = 950 + 192 = 1{,}142 \text{ L}$$

which is 20 percent smaller than the volume required when we use a single CSTF.

c. The best combination is a CSTF operated at the maximum rate followed by a PFF. The volume of the first fermenter is 950 L as calculated in part (b).

For the second PFF, from Eq. (6.54)

$$\tau_{p2} = \frac{1}{\mu_{max}} \left(\frac{K_S Y_{X/S}}{C_{X_1} + C_{S_1} Y_{X/S}} + 1 \right) \ln \frac{C_{X_2}}{C_{X_1}} + \frac{K_S Y_{X/S}}{C_{X_1} + C_{S_1} + C_{S_1} Y_{X/S}} \ln \frac{C_{S_1}}{C_{S_2}}$$

$$= \frac{1}{0.7} \left[\left(\frac{5(0.65)}{45 + 16(0.65)} + 1 \right) \ln \frac{52}{45} + \frac{5(0.65)}{45 + 16(0.65)} \ln \frac{16}{5} \right] = 0.32$$

Therefore,

$$V_2 \, \tau_{p_2} F = 0.32 \, (500) = 160 \text{L}$$

The total volume of the CSTF and PFF is

$$V = V_1 + V_2 = 950 + 160 = 1{,}110 \text{ L}$$

which is 22 percent smaller than the volume required when we use a single CSTF. The additional saving by employing the second PFF instead of a second CSTF is not significant in this case.

6.6 CELL RECYCLING

For the continuous operation of a PFF or CSTF, cells are discharged with the outlet stream which limits the productivity of fermenters. The productivity can be improved by recycling the cells from the outlet stream to the fermenter.

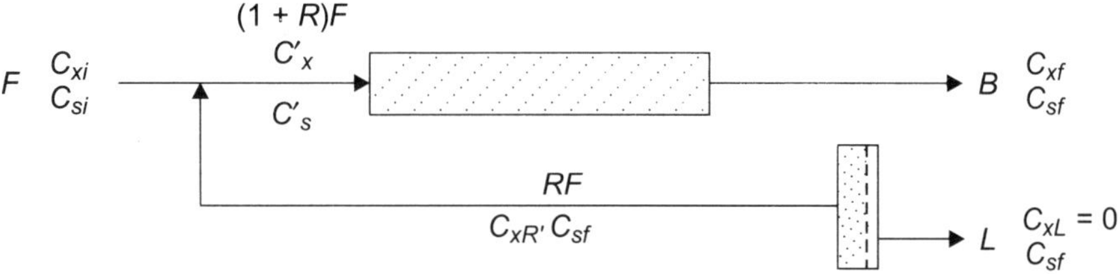

Fig. 6.17 Schematic diagram of PFF with cell recycling.

6.6.1 PFF with Cell Recycling

A PFF requires the initial presence of microorganisms in the inlet stream as a batch fermenter requires initial inoculum. The most economical way to provide cells in the inlet stream is to recycle the part of the outlet stream back to the inlet with or without a cell separation device. Figure 6.17 shows the schematic diagram of a PFF with cell recycling. Unlike the CSTF, the PFF does not require the cell separator in order to recycle, though its presence increases the productivity of the fermenter slightly as will be shown later. The performance equation of the PFF with Monod kintics can be written as:

$$\frac{V}{(1+R)F} = \frac{\tau_p}{1+R} = \int_{C_X}^{C_{X_f}} \frac{dC_X}{r_X} = \int_{C_X}^{C_{X_f}} \frac{(K_S + C_S)dC_X}{\mu_{max}C_S C_X} \tag{6.55}$$

where τ_p is the residence time based on the flow rate of the overal system. The actual residence time in the fermenter is larger than τ_p due to the increase of the flow rate by the recycling.

If the growth yield is constant,

$$C_S = C_S' - \frac{1}{Y_{X/S}}(C_X - C_X') \tag{6.56}$$

Substituting Eq. (6.56) into Eq. (6.55) for Cs and integrating it will result:

$$\frac{\tau_p \mu_{max}}{1+R} = \left(\frac{K_S Y_{X/S}}{C_X + C_S Y_{X/S}} + 1\right)\ln\frac{C_{X_f}}{C_X'} + \frac{K_S Y_{X/S}}{C_X' + C_S Y_{X/S}}\ln\frac{C_S'}{C_{S_f}} \tag{6.57}$$

where C_X' and C_S' can be estimated from the cell and substrate balance at the mixing point of the inlet and the recycle stream as

$$C_X' = \frac{C_{X_i} + RC_{X_R}}{1+R} \tag{6.58}$$

$$C_S' = \frac{C_{S_i} + RC_{S_R}}{1+R} \tag{6.59}$$

The cell concentrations of the outlet stream, can be estimated from the overall cell balance

$$C_{X_f} = \frac{1}{\beta}[C_{X_i} + Y_{X/S}(C_{S_i} - C_{S_f})] \tag{6.60}$$

The cell concentrations of the recycle stream, can be estimated from the cell balance over the filter as

$$C_{X_R} = \frac{1 + R + \beta}{R} C_{X_f} \tag{6.61}$$

where β is the bleeding rate which is defined as

$$\beta = \frac{B}{F} \tag{6.62}$$

Figure 6.18 shows the effect of the recycle rate (R) on the residence time of a PFF system with recycling. Note that τ is calculated based on the inlet flowrate that is the true residence time for the fermenter system. The actual τ in the PFF is unimportant because it decreases with the increase of the recycle rate. When $\beta = 1$, the bleeding rate is equal to the flow rate and the flow rate of filtrate stream L is zero, therefore, the recycle stream is not filtered. The residence time will be infinite if R is zero and decreases sharply as R is increased until it reaches the point the decrease is gradual. In this specific case, the optimum recycle ratio may be somewhere around 0.2.

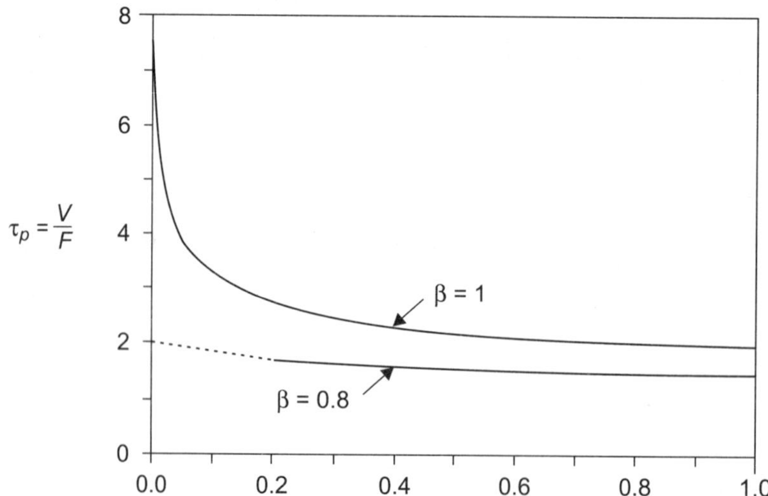

Fig. 6.18 The effect of the recycle rate (R) and the bleeding rate (β = B/F) on the residence time (τ = V/F hr) of a PFF system with recycling. The curve is drawn by using the Monod model with, μ_{max} = 0.935 hr^{-1}, K_S = 0.71 g/L, $Y_{X/S}$ = 0.6, C_{Si} =10 g/L, C_{Sf} = 1.3 g/L, and C_{Xi} = 0.

Another curve in Figure 6.18 is for $\beta = B/F = 1.8$. The required residence time can be reduced by concentrating the recycle stream 25 to 40% when R is between 0.2 and 1.0. When $R \leq 1.2$, the part of curve is noted as a dotted line because it may be difficult to reduce the recycle ratio below 0.2 when $\beta = 0.8$. For example, in order to maintain $R = 0.1$, The amount of $1.3F$ needs to be recycled and concentrated to $0.1F$, which may be difficult depending on the cell concentration of the outlet. The higher the concentration factor in a filter unit is, the higher the danger of the filter failure can be expected.

The analysis in this section and the next can be also applied for a cell settler as a cell separator. The outlet stream of the cell settler will be equal to $F = B + L$ and its concentration will be $(B/F)C_{Xf} = \beta C_{Sf}$.

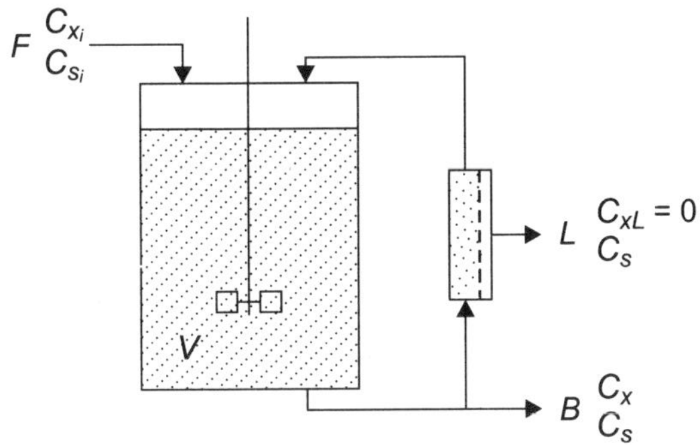

Fig. 6.19 Schematic diagram of CSTF with cell recycling.

6.6.2 CSTF with Cell Recycling

The cellular productivity in a CSTF increases with an increase in the dilution rate and reaches a maximum value. If the dilution rate is increased beyond the maximum point, the productivity will be decreased abruptly and the cells will start to be washed out because the rate of cell generation is less than that of cell loss from the outlet stream. Therefore, the productivity of the fermenter is limited due to the loss of cells with the outlet stream. One way to improve the reactor productivity is to recycle the cell by separating the cells from the product stream using a cross-flow filter unit (Figure 6.19).

The high cell concentration maintained using cell recycling will increase the cellular productivity since the growth rate is proportional to the cell concentration. However, there must be a limit in the increase of the cellular productivity with increased cell concentration because in a high cell concentration environment, the nutrient-transfer rate will be decreased due to overcrowding and aggregation of cells. The maintenance of the extremely high cell concentration is also not practical because the filter unit will fail more frequently at the higher cell concentrations.

If all cells are recycled back into the fermenter, the cell concentration will increase continuously with time and a steady state will never be reached. Therefore, to operate a CSTF with recycling in a steady-state mode, we need to have a bleeding stream, as shown in Figure 6.19. The material balance for cells in the fermenter with a cell recycling unit is

$$FC_{X_i} - BC_X + V \mu C_X = V \frac{dC_X}{dt} \tag{6.63}$$

It should be noted that actual flow rates of the streams going in and out of the filter unit do not matter as far as overall material balance is concerned. For a steady-state CSTF with cell recycling and a sterile feed,

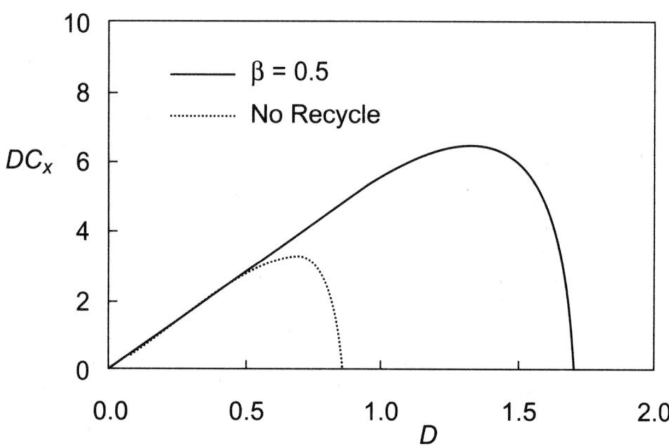

Fig. 6.20 The effect of bleeding ratio on the cellular productivity (βDC_X).

$$\beta D = \frac{\beta}{\tau_m} = \mu \tag{6.64}$$

Now, βD instead *of* D is equal to the specific growth rate. When $\beta = 1$, cells are not recycled, therefore, $D = \mu$.

If the growth rate can be expressed by Monod kinetics, substitution of Eq. (6.11) into Eq. (6.64) and rearrangement for C_S yields

$$C_S = \frac{\beta K_S}{\tau_m \mu_{max} - \beta} \tag{6.65}$$

which is valid when $\tau_m \mu_{max} > \beta$. The cell concentration in the fermenter can be calculated from the value of C_S as

$$C_X = \frac{Y_{X/S}}{\beta}(C_{S_i} - C_S) \tag{6.66}$$

Figure 6.20 shows the effect of bleeding ratio on the cell productivity for the Monod model. As β is reduced from 1 (no recycling) to 0.5, the cell productivity is doubled.

6.7   ALTERNATIVE FERMENTERS

Many alternative fermenters have been proposed and tested. These fermenters were designed to improve either the disadvantages of the stirred tank fermenter-high power consumption and shear damage, or to meet a specific requirement of a certain fermentation process, such as better aeration, effective heat removal, cell separation or retention, immobilization of cells, the reduction of equipment and operating costs for inexpensive bulk products, and unusually large designs.

Fermenters are usually classified based on their vessel type such as tank, column, or loop fermenters. The tank and column fermenters are both constructed as cylindrical vessels. They can be distinguished based on their height-to-diameter ratio *(H/D)* as (Schiigerl, 1982):

$$H/D < 3 \text{ for the tank and}$$

$$H/D > 3 \text{ for the column fermenter}$$

A loop fermenter is a tank or column fermenter with a liquid circulation loop, which can be a central draft tube or an external loop.

Table 6.2   Classifications of Fermenters

	Primary Source of Mixing		
Vessel Type	*Compressed Air*	*Internal Moving Parts*	*External Pumping*
Tank	—	stirred-tank	—
Column	bubble column	multistage	sieve tray
	tapered column	(or cascade)	packed-bed
Loop	air-lift pressure cycle	propeller loop	jet loop

Another way to classify fermenters is based on how the fermenter contents are mixed: by compressed air, by a mechanical internal moving part, or by external liquid pumping. Representative fermenters in each category are listed in Table 6.2 and the advantages and disadvantages of three basic fermenter types are listed in Table 6.3.

6.7.1   Column Fermenter

The most simple fermenter is the bubble column fermenter (or tower fermenter), which is usually composed of a long cylindrical vessel with a sparging device at the bottom [Figure 6.21 (a)–(c)]. The fermenter contents are mixed by the rising bubbles which also provide the oxygen needs of the cells. Since it does not have any moving parts, it is energy efficient with respect to the amount of oxygen transfer per unit energy input. As the cells settle, high cell concentrations can be maintained in the lower portion of the column without any separation device.

Table 6.3 Advantages and Disadvantages of Three Basic Fermenter Configurations

Type	Advantages	Disadvantages
Stirred-tank	1. Flexible and adaptable 2. Wide range of mixing intensity 3. Can handle high viscosity media	1. High power consumption 2. Damage shear sensitive cells 3. High equipment costs
Bubble Column	1. No moving parts 2. Simple 3. Low equipment costs 4. High cell concentration	1. Poor mixing 2. Excessive foaming. 3. Limited to low viscosity system
Air-lift	1. No moving parts 2. Simple 3. High gas absorption efficiency 4. Good heat transfer	1. Poor mixing 2. Excessive foaming 3. Limited to low viscosity system

However, the bubble column fermenter is usually limited to aerobic fermentations and the rising bubbles may not provide adequate mixing for optimal growth. Only the lower part of the column can be maintained with high cell concentrations, which leads to the rapid initial fermentation followed by a slower one involving less desirable substrates. As the cell concentration increases in a fermenter, high air-flow rates are required to maintain the cell suspension and mixing. However, the increased air-flow rate can cause excessive foaming and high retention of air bubbles in the column which decreases the productivity of the fermenter. As bubbles rise in the column, they can coalesce rapidly leading to a decrease in the oxygen-transfer rate. Therefore, column fermenters are inflexible and limited to a relatively narrow range of operating conditions.

To overcome the weaknesses of the column fermenter, several alternative designs have been proposed. A tapered column fermenter [Figure 6.21 (b)] can maintain a high air-flow rate per unit area at the lower section of the fermenter where the cell concentration is high.

Several sieve plates can be installed in the column [Figure 6.21 (c)] for the effective gas-liquid contact and the breakup of the coalesced bubbles. To enhance the mixing without internal moving parts, the fermentation broth can be pumped out and recirculated by using an external liquid pump [Figure 6.21 (d)(e)].

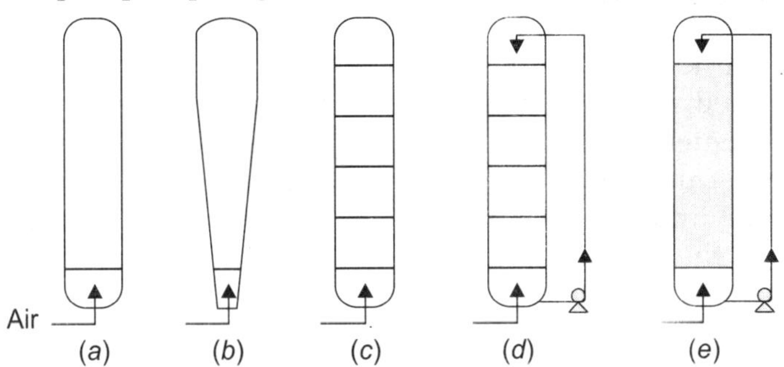

Fig. 6.21 Column fermenters: (a) bubble column, (b) tapered bubble column, (c) sieve-tray bubble column, (d) sieve-tray column with external pumping, and (e) packed-bed column with external pumping.

6.7.2 Loop Fermenter

A loop fermenter is a tank or column fermenter with a liquid circulation loop, which can be a central draft tube or external loop. Depending on how the liquid circulation is induced, it can be classified into three different types: air-lift, stirred loop, and jet loop (Figure 6.22).

The liquid circulation of the air-lift fermenter is induced by sparged air which creates a density difference between the bubble-rich part of the liquid in the riser and the denser bubble-depleted part of the liquid in the downcomer as shown in Figure 6.22 (a). The liquid circulation and mixing can be enhanced by by circulating liquid externally using a pump as shown in Figure 6.22 (b). However, adding the pump diminishes the real advantages of an air-lift fermenter for being simple and energy efficient.

The ICI pressure cycle fermenter (Imperial Chemical Industries Ltd., England) is an air-lift fermenter with an outer loop, which was developed for the aerobic fermentation requiring heat removal such as the single-cell protein production from methanol. Medium and air are introduced into the upper and lower parts of the loop as shown in Figure 6.22 (c). The air serves two purposes: It provides the oxygen needed for the growth of the microorganisms and the rising air creates natural circulation of the liquid in the fermenter through the loop. A heat exchanger to cool the liquid medium is installed in the loop. It was claimed that the fermenter gives a high rate of oxygen absorption per unit of volume, that it uses a high proportion of oxygen in the air passed through the fermenter, and that the high circulation of the fermentation liquor provides good mixing (Technical Brochure, ICI Ltd.).

6.8 STRUCTURED MODEL

So far, the kinetic models described in this chapter have been unstructured, distributed models based on assumptions that cells can be represented by a single component, such as cell mass or cell number per unit volume and the population of the cellular mass is distributed uniformly throughout the culture. These models do not recognize the change in the composition of cells during growth. Unstructured, distributed models such as Monod's equation can satisfactorily predict growth behavior of many situations. However, they cannot account for lag phases, sequential uptake of substrates, or changes in mean cell size during the growth cycle of batch culture.

Structured models recognize the multiplicity of cell components and their interactions. Many different models have been proposed based on the assumptions made for cell components and their interactions. More comprehensive review of these models can be found in Tsuchiya et al. (1966) and Harder and Roels (1982). In this section, general equations describing structured models are derived and a simple structured model is introduced.

6.8.1 General Structured Model

Let's define a system as an individual cell or multiple cells. The system does not contain any of the abiotic phase of culture. Instead, it is the biotic[2] phase only, which possesses mass m on a dry basis and specific volume $\hat{v}$. Let's assume that there are c components in the cell and the mass of the jth component per unit volume of system is $(\hat{C}_{X_i})$. It is also assumed that there exist kinetic rate expressions for p reactions occurring in the system and the rate of jth component formed from the ith reaction per unit volume of system is $\hat{r}_{X_{i,j}}$

Then, during batch cultivation, the change of the jth component in the system with respect to time can be expressed as (Fredrickson, 1976):

$$\frac{d\left(m\hat{v}\hat{C}_{X_j}\right)}{dt} = m\hat{v}\sum_{i=1}^{p} r_{X_{i,j}} \tag{6.67}$$

If we assume that the specific volume $\hat{v}$ is constant with time, Eq. (6.67) can be rearranged to

---

[2] Caused or produced by living beings.

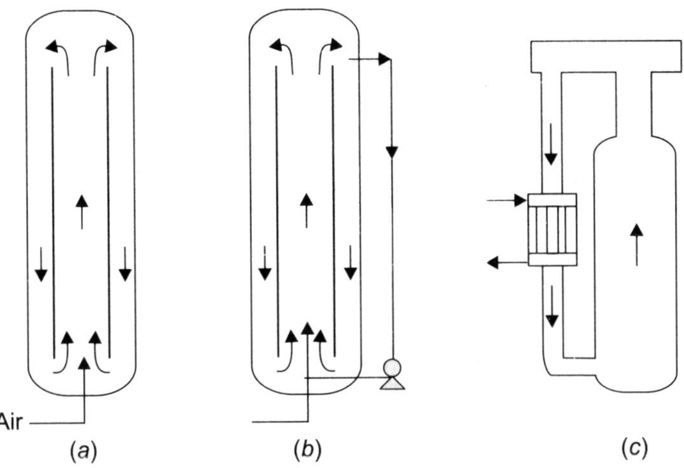

Fig. 6.22 Loop fermenters: (a) air-lift, (b) air-lift with external pumping (c) ICI pressure cycle.

$$\frac{d\hat{C}_{X_j}}{dt} = \sum_{i=1}^{p} r_{X_{i,j}} - \frac{1}{m}\frac{dm}{dt}\hat{C}_{X_j} \tag{6.68}$$

where the second term of the right-hand side of Eq. (6.68) represents the dilution of intracellular components by the growth of biomaterial. It should be noted that all variables denoted by circumflexes in the preceding equations are *intracellular properties*. Since the structural models recognize the multiplicity of cell components and their interactions in the cell, it makes more sense to express the model with intrinsic variables.

Since $\hat{C}_X = \sum_{j=1}^{n}\hat{C}_{Xj}$ the summation of Eq. (6.68) for all c components yields

$$\frac{d\hat{C}_X}{dt} = \sum_{j=1}^{c}\sum_{i=1}^{p}\hat{r}_{X_{i,j}} - \frac{1}{m}\frac{dm}{dt}\hat{C}_X \tag{6.69}$$

where $\hat{C}_X$ is the mass of biomaterial per unit volume of the biotic system that is equal to $1/\hat{v}$, which was assumed to be constant. Therefore, Eq. (6.69) is simplified to

$$\frac{1}{m}\frac{dm}{dt} = \frac{1}{\hat{C}_X}\sum_{j=1}^{c}\sum_{i=1}^{p}\hat{r}_{X_{i,j}} \tag{6.70}$$

which gives the relationship between the rate of cell growth and the kinetic rate expressions of all reactions occurring in cells.

The concentration terms m the preceding equations can be expressed as mass per unit culture volume V instead of that per biotic system volume $m\,v$. The two different definitions of concentration are related as

$$\hat{C}_{X_j} = \frac{V}{m\hat{v}}C_{X_j} \tag{6.71}$$

Note that the concentration based on the culture volume is no longer denoted with a circumflex. Substituting Eq. (6.71) into Eq. (6.68) and simplifying for the constant V yields

$$\frac{dC_{X_j}}{dt} = \frac{m\hat{v}}{V}\sum_{i=1}^{p}\hat{r}_{X_{i,j}} \tag{6.72}$$

It should be noted that even though concentration terms are expressed based on the total culture volume, kinetic parameters still remain on a biotic phase basis in the formulation.

6.8.2 Two-Compartment Model

One of the simplest structured models is the two-compartment model proposed by Williams (1967) and is based on the following assumptions:[3]

1. The cell comprises two basic compartments: a synthetic portion A such as precursor molecules and RNA, and a structural portion B such as protein and DNA.

$$\hat{C}_X = \hat{C}_{X_A} + \hat{C}_{X_B} \tag{6.73}$$

where ($\hat{C}_X$ is equal to 1 v, and assumed to be constant,

$$\frac{d\hat{C}_X}{dt} = 0 \tag{6.74}$$

2. The synthetic portion A is fed by uptake from a substrate S and the structural portion B is in turn fed from the synthetic portion as:

$$S \longrightarrow A \longrightarrow B \tag{6.75}$$

3. The rate of the first reaction in Eq. (6.75) is proportional to the product of substrate and cell concentrations as:

$$\hat{r}_{X_{1,A}} = k_1\overline{C}_S\hat{C}_X \tag{6.76}$$

where $\overline{C}_S$ is the mass of substrate per unit abiotic volume, $V\text{-}m\hat{v}$. The rate of the second reaction in Eq. (6.75) is proportional to the product of the concentrations of the synthetic portion and the structural portion as:

$$\hat{r}_{X_{2,B}} = -r_{X_{2,A}} = k_2\overline{C}_{X_A}\hat{C}_{X_B} \tag{6.77}$$

---

[3] The original Williams model (1967) was modified according to the suggestion made *by* Fredrickson (1976).

If we write Eq. (6.68) for each component, and substitute Eqs. (6.76) and (6. 77) for the reaction rates, we obtain

$$\frac{d\hat{C}_{X_A}}{dt} = k_1 \overline{C}_S \hat{C}_X - k_2 \hat{C}_{X_A} \hat{C}_{X_B} - \frac{1}{m}\frac{dm}{dt}\hat{C}_{X_A} \qquad (6.78)$$

$$\frac{d\hat{C}_{X_B}}{dt} = k_2 \hat{C}_{X_A} \hat{C}_{X_B} - \frac{1}{m}\frac{dm}{dt}\hat{C}_{X_A} \qquad (6.79)$$

The change of the substrate concentration is

$$\frac{d\overline{C}_S}{dt} = -\frac{k_1}{Y_{X/S}}\overline{C}_S \hat{C}_X \qquad (6.80)$$

where $Y_{X/S}$ is the yield constant.

Following, Eqs. (6.73), (6.74), (6.78), (6.79), and (6.80) can be solved

simultaneously to obtain the change of $\hat{C}_X$, $\hat{C}_{XA}$, $\hat{C}_{XB}$, $\overline{C}_S$, and m.

1. Cell division occurs if and only if the structural portion $\left(m\hat{v}\hat{C}_{X_B}\right)$ has doubled its initial value, and a dividing cell apportions each component equally to its two daughter cells. Therefore, the total number of cells in the system will be proportional to the structural portion as

$$n \propto m\hat{v}\hat{C}_{X_B} \qquad (6.81)$$

The average mass of a cell is equal to the total cell mass divided by total cell number. Therefore,

$$\text{Average mass of cell} = \frac{m}{n} \propto \frac{m\hat{v}\hat{C}_X}{m\hat{v}\hat{C}_{X_B}} = \frac{\hat{C}_X}{\hat{C}_{X_B}} \qquad (6.82)$$

As a result, this model can also predict the change of the average cell size with respect to time, which is not possible with the Monod model. Eqs. (6.73), (6.78), (6.79), and (6.80) can be expressed with the concentrations in terms of mass per unit culture volume as

$$C_X = C_{XA} + C_{XB} \qquad (6.83)$$

$$\frac{dC_{X_A}}{dt} = \left(\frac{V}{V - m\hat{v}}\right)k_1 C_S C_X - \left(\frac{V}{m\hat{v}}\right)k_2 C_{X_A} C_{X_B} \qquad (6.84)$$

$$\frac{dC_{X_B}}{dt} = \left(\frac{V}{m\hat{v}}\right)k_2 C_{X_A} C_{X_B} \qquad (6.85)$$

$$\frac{dC_S}{dt} = -\frac{1}{Y_{X/S}}\left(\frac{V}{V - m\hat{v}}\right)k_2 C_S C_X \qquad (6.86)$$

where $C_X = (m\hat{v}/V)\hat{C}_X$, $C_{X_A} = (m\hat{v}/V)\hat{C}_{X_A}$, $C_{X_B} = (m\hat{v}/V)\hat{C}_{X_B}$,

and $$C_S = [(V - m\hat{v})/V]\overline{C}_S$$

The mass m is related to C_X as

$$m = C_X V \tag{6.87}$$

Now, the total number of cells is proportional to C_{X_B} and the average mass of a cell to C_X/C_{X_B}.

Figure 6.23 illustrates simulation curves of a batch culture, which show the changes of mass $(C_X/C_{X_{max}})$, number $(C_X/C_{X_{Bmax}})$, and size of cells (C_X/C_{X_B}), and the substrate concentration (C_S/C_{S_0}) in dimensionless form. This curve shows the following features:

1. During a lag phase, cells grows in size but not in number.

2. Cells are largest during the exponential phase.

3. The total cell mass reaches its asymptote before cell number.

4. Cells no longer grow or divide during a stationary phase

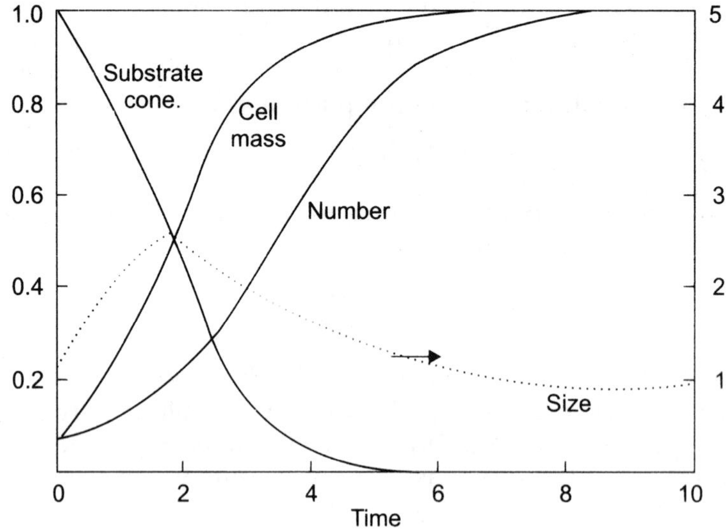

Fig. 6.23 Simulation results of a batch culture using the two compartment model proposed by Williams (1967), which show the changes of mass, number, and size of cells, and the substrate concentration in dimensionless form. Parameters: C_{S_0} = 1 g/L, $C_{X_{A0}}$ = 0, $C_{X_{B0}}$ = 0.05 g/L, k_1 = 1.25 L/g hr, k_2 = 0.0002 L/g hr, $\hat{v}$ = 0.0002 L/g

Even though Williams's model provides many features that unstructured models are unable to predict, it requires only two parameters, which is the same number of parameters required for Monod kinetics.

6.9   NOMENCLATURE

B	flow rate of bleeding stream, m^3/s
C	concentration, mass per unit volume of culture, kg/m^3
$\hat{C}$	mtracellular concentration, mass per unit volume of biotic phase, kg/m^3
$\overline{C}$	extracellular concentration, mass per unit volume of abiotic phase, kg/m
C_1, C_2	constants
C_N	cell number density, number of cells/m
D	dilution rate, s^{-1}
F	flow rate, m^3/s
K_S	system coefficient for the Monod kinetics, kg/m
L	flow rate of filtrate stream, m^3/s
$\overline{m}$	average mass of cell in a system, kg
m	mass of biotic phase, kg
n	number of cells
r	rate of growth per unit volume, kg/m^3s
$\overline{r}_{i,j}$	rate of jth component formed from ith reaction per unit volume of a system, intracellular property, $kg/m\ s$
$r_{i,j}$	rate of jth component formed from ith reaction per unit volume of a system, extracellular property, $kg/m^3\ s$
t	time, s
t_d	doubling time, s
$\hat{v}$	specific volume of a system containing only biotic phase, m^3/kg
V	working volume of fermenter, m^3
Y	yield constant
β	bleeding ratio, defined as B/F
$\overline{\delta}$	average cell division rate, s^{-1}
μ	specific growth rate, s^{-1} or kg/m^3s
ρ	density, kg/m^3
τ	residence time, s

SUBSCRIPT

b	batch fermenter
i	input stream

m	mixed fermenter
n	cell in number basis
X	cell in dry weight basis
P	product
p	plug-flow fermenter
S	substrate

6.10 PROBLEMS

6.1 Derive the relationship giving the change with respect to time of the cell concentration in a batch fermenter, Eq. (6.24).

6.2 Aiba et al. (1968) reported the results of a chemostat study on the growth of a specific strain of baker's yeast as shown in the following table. The inlet stream of the chemostat did not contain any cells or products.

Dilution Rate D, hr^{-1}	Inlet cone. (g/L) glucose C_{Si}	Steady-state concentration (g/L)		
		Glucose C_S	Ehanol C_P	Cells C_X
0.084	21.5	0.054	7.97	2.00
0.100	10.9	0.079	4.70	1.20
0.160	21.2	0.138	8.57	2.40
0.198	20.7	0.186	8.44	2.33
0.242	10.8	0.226	4.51	1.25

    a. Find the rate equation for cell growth.

    b. Find the rate equation for product (ethanol) formation.

6.3 Andrews (1968) proposed the following model for the growth of microorganisms utilizing inhibitory substrates.

$$\mu = \frac{\mu_{max}}{1 + \dfrac{K_S}{C_S} + \dfrac{C_S}{K_I}}$$

Assume that a chemostat study was performed with a microorganism. The volume of the fermenter content was 1 L. The inlet stream was sterile. The flow rate and inlet substrate concentration were varied and the steady-state concentration of glucose in the fermenter was measured and recorded as shown in the table (the data are arbitrary).

    a. Determine the kinetic parameters (μ_{max}, K_S, and K_i) of this microorganism.

    b. If the cell yield, $Y_{X/S}$, is 0.46 g/g, what is the steady-state cell concentration when the flow rate is 0.20 L/h?

c. Andrews concluded in his paper that the primary result of substrate inhibition in a continuous culture may be process instability. Explain what might happen if you suddenly increase the substrate concentration from 30 to 60 g/L and why.

Flow rate L/hr	Inlet glucose conc. G_{Si} g/L	Steady-state glucose conc. Cs, g/L
0.20	30	0.5
0.25	30	0.7
0.35	30	1.1
0.50	30	1.6
0.70	30	3.3
0.80	30	10
0.50	60	30
0.60	60	22
0.70	60	15

6.4 Rate equations for the cells (yeast), substrate in the ethanol fermentation process are given (glucose),   and   product   as follows:

$$r_X = \frac{dC_X}{dt} = \mu_{max}\left(\frac{C_S}{K_S + C_S}\right)\left(1 - \frac{C_P}{C_{Pm}}\right)^n C_X$$

$$r_S = \frac{dC_S}{dt} = -\frac{1}{Y_{X/P}}\frac{dC_X}{dt}$$

$$r_P = \frac{dC_P}{dt} = -\frac{1}{Y_{X/P}}\frac{dC_X}{dt}$$

where $K_S = 1.6$ g/L, $\mu_{max} = 0.24$ hr^{-1}, $Y_{X/P} = 0.06$, $Y_{X/S} = 0.06$, $C_{Pm} = 100$ g/L, $C_{P_0} = 0$, $C_{xo} = 0.1$ g/L, and $n = 2$.

a. Calculate the change of C_X, C_p, and C_S as a function of time when $C_S = 001$ g/L.

b. Show the effect of the initial substrate concentration on the C_X versus t curve.

c. Show the effect of the maximum growth rate (μ_{max}) on the C_X versus t curve ($C_{S_1} = 100$ g/L).

6.5 The growth rate of *E. coil* in synthetic medium can be expressed by Monod kinetics as

$$r_X = \frac{0.935 C_S C_X}{0.71 + C_S} \qquad [\text{g/L hr}]$$

where C_S is the concentration of a limiting substrate, glucose. You are going to cultivate *E. coli* in a steady-state CSTF (working volume: 10 L) with a flow rate of 7 L/hr. The initial substrate concentration is 10 g/L and the cell yield constant $(Y_{X/S})$ is 0.6. The feed stream is sterile.

   a. What will be the doubling time and the division rate of the cells in the CSTF?

   b. What will be the cell and substrate concentrations of the outlet stream?

   c. If you connect one more 10/L CSTF to the first one, what will be the cell and substrate concentrations in the second fermenter?

   d. If you increase the flow rate from 7 to 10 L/hr for these two fermenters connected in series, what will happen and why? Make a recommendation to avoid the problem if there is any.

6.6 Suppose that the growth rate of a microorganism can be expressed as the following equation:

$$r_X = \mu_{max} \left[1 - \exp(-C_S/K_S)\right] C_X$$

where $\mu_{max} = 1.365 \text{ hr}^{-1}$ and $K_S = 6.8$ g/L. The cell yield $Y_{X/S}$ is found to be 0.45.

   a. If you cultivate this microorganism in a 10 L CSTF with the flow rate of 2.8 L/hr, what will be the steady-state cell concentration of the outlet stream? The substrate concentration of the inlet stream is 13g/L. The inlet stream is sterile.

   b. Explain the difference between this model and Monod model by using μ versus C_S graph.

6.7 Herbert et al.(1956) reported that the growth kinetics of *Aerobacter cloacae* in a chemically defined medium (glycerol as a limiting substrate) could be expressed by Monod kinetics as follows:

$$r_X = \frac{dC_X}{dt} = \frac{\mu_{max} C_S C_X}{K_S + C_S}$$

where $\mu_{max} = 0.85 \text{ hr}^{-1}$ and $K_S = 1.23 \times 10^{-2}$ g/L. The yield was found to be 0.53 g dry weight of organism/g glycerol used.

You are a biochemical engineer who has been assigned the task of designing the most effective continuous fermentation system to grow the microorganism (*Aerobacter cloacae*] with glycerol as its limiting substrate. For the following three questions, the concentration of glycerol in the feed stream and that of glycerol in the outlet stream should be 3 g/L and 0.1 g/L, respectively.

a. Since you have learned that the $1/r_X$ versus C_S curve for Monod kinetics has a U shape, you have recommended that the most effective system would be the combination of a continuous stirred-tank fermenter (CSTF) and a plug-flow fermenter (PFF). You were quite sure of this because the substrate concentration in the outlet stream has to be so low. However, your boss is insisting that the use of second PFF in addition to the first CSTF will not improve the productivity very much. Who is right? Prove whether you are right or wrong by drawing the $1/r_X$ versus C_S curve for this microorganism. Does it have a U shape? Discuss why you are right or wrong. (If you are right, think about how you can nicely correct you boss's wrong idea. If you are wrong, it will teach you that you have to be careful not to make a quick conclusion without adequate analysis.)

b. Recommend the best fermenter system (fermenter type and volume) which can handle 100 L/hr of feed stream. The best fermenter system is defined as that which can produce the maximum amount of cells per unit time and volume.

c. If $K_S = 1.23$ g/L instead of 1.23×10^{-2} g/L, what is the best fermenter system (fermenter type, volume) which can handle 100 L/hr in the feed stream. Draw the block diagram of the fermenter system with the concentrations of the substrate and the cells in the inlet and outlet streams of each fermenter. How is this case different from the case of part (a) and why?

6.8 Suppose you have an organism that obeys the Monod equation:

$$\frac{dC_X}{dt} = \frac{\mu_{max} C_S C_X}{K_S + C_S}$$

where $\mu_{max} = 0.5$ hr^{-1} and $K_S = 1$ g/L.
The organism is being cultivated in a steady-state CSTF, where $F = 100$ L/hr, $C_{S_i} = 50$ g/L, and $Y_{X/S} = 0.5$.

a. What size vessel will give the maximum total rate of cell production?

b. What are the substrate and cell concentrations of the optimum fermenter in part (a)?

c. If the exiting flow from the fermenter in part (a) is fed to a second fermenter (CSTF), what should be the size of the second fermenter to reduce the substrate concentration to 1 g/L?

d. If the exiting flow from the first fermenter in part (a) is fed to a second fermenter whose size is the same as the first, what will be the cell and substrate concentrations leaving the second fermenter?

6.9 You are going to cultivate yeast, *Saccharomyces cerevisiae*, by using a 10 m -fermenter your company already owns. You want to find out the amount of ethanol the fermenter can produce. Therefore, a chemostat study was carried out and the Monod kinetic parameters for the microorganism grown in the glucose medium at 30°C, pH 4.8, were found to be: $K_S = 0.0025$ g/L and $\mu_{max} = 0.25$ h^{-1}. The ethanol yield ($Y_{P/S}$) is 0.44 (g/g) and cell yield ($Y_{X/S}$) is 0.019 (g/g). The inlet substrate concentration is 50 g/L.

    a. What flow rate will give the maximum total ethanol production in the continuous fermenter and what is the maximum ethanol production rate?

    b. If you want to convert 95 percent of the incoming substrate, what must the ethanol production rate be for the continuous fermenter?

    c. If you have two 5 m^3-fermenters instead of one 10 m^3-fermenter, what is your recommendation for the use of these fermenters to convert 95 percent of the incoming substrate? Would you recommend connecting two fermenters in series to improve the productivity? Why or why not?

6.10 You are a biochemical engineer in a pharmaceutical company. Your company is a major producer of penicillin. Currently, what kind of fermenter is your company using for penicillin production? Why?

    Your boss asked you to study the possibility of using an air-lift fermenter as a replacement since it has many advantages. What is your recommendation?

6.11 Consider an organism with the following data for a 1-L chemostat, using an inlet substrate concentration Csi of 30 g/L:

Flow rate	Concentration (g/L)		
(mL/hr)	Substrate	Cell	Products
F	C_S	C_X	C_P
27.5	10.0	12.0	1.04
24.2	5.56	14.7	1.27
22.1	3.70	15.8	1.37
18.8	2.32	16.7	1.44

    a. In a continuous, perfectly mixed vessel at steady state with no cell death, if inlet substrate concentration C_{Si} now equals 25.0 g/L and cell concentration C_{X_i} is 0.0 g/L, what dilution rate D will give the maximum total rate of cell production? What are the outlet cell and substrate concentrations at this dilution rate?

b. Using the preceding organism, your boss would like you to design a continuous reactor system with an inlet flow and substrate concentration of 250 L/hr and 25 g/L, respectively, which will produce an overall yield of product of 2100 kg/yr, given an operating time of 300 day/yr at 24 hr/day. Assume no cells or product in the system inlet. Would you recommend a single fermenter of two fermenters in series? Design a system which will meet production constraints while minimizing *total fermenter volume*. Report reactor volumes and effluent cell, substrate, and product concentrations for the proposed fermenter(s). [*Contributed by Brian S. Hooker, TriState University.*]

6.12 A strain of yeast is being cultivated in a 30-L CSTF with a cell recycling system (cell settler) as shown in the following figure. The cell settler was designed so that the cell concentration of its outlet stream is 30 percent of that of its inlet stream, whereas the substrate concentrations of the two streams are the same. The growth rate of the cells can be represented by the Monod kinetics with the parameters: $K_S = 0.05$ g/L, $\mu_{max} = 0.3$ h^{-1} , and $Y_{X/S} = 0.025$. Calculate the steady-state substrate and cell concentrations in the fermenter. The inlet substrate concentration is 100 g/L and the flow rate is 20 L/hr. The feed stream is sterile.

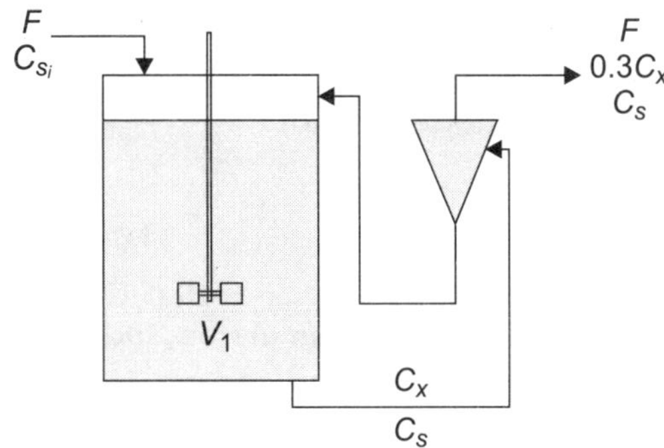

6.13 You need to cultivate hypothetical microbial cells with the Monod kinetic parameter values of $\mu_{max} = 5.0$ hr^{-1} and $K_S = 20$ g/L. The cell yield $(Y_{X/S})$ is 0.4 and the substrate concentration is 30 g/L. The required substrate conversion is 97%. Estimate the residence time required for the following fermenter configurations

a. Plug-flow fermenter with the inlet cell concentration of 0.5 g/L.

b. Plug-flow fermenter with a recycle stream (no cell separator) with the flow rate of 0.4F. The inlet cell concentration is 0.5 g/L. How your answer will be changed if you reduce the inlet cell concentration to zero?

    c. One steady-state CSTF with a recycle stream (no cell separator) with the flow rate of 0.4F. The inlet stream is sterile.

6.14 By using the structured model proposed by Ramkrishna *et al.* (1967), show the change of the concentrations [in g dry weight/L] of G-mass, H-mass, and inhibitor, and the fraction of G-mass with time during a batch cultivation of a microorganism which has the following parameters and initial conditions:

$$
\begin{aligned}
\mu &= 0.5 \text{ hr}^{-1} & aS' &= 2 \\
\mu' &= 2.5 \text{ hr}^{-1} & a_T &= 0 \\
K_S &= 0.2 \text{ g/L} & a_T{}' &= 0.2 \times 10^{-4} \\
K_S' &= 0.1 \text{ g/L} & a_{T1} &= 0.0267 \\
K &= 150 \text{ L/g hr} & a_{T1}{}' &= 0 \\
K' &= 70 \text{ L/g hr} & C_{S1} &= 10 \text{ g/L} \\
K_G &= 3.0 * 10^{-5} \text{ g/L} & C_{T1} &= 0 \text{ g/L} \\
K_G' &= 0.5 \times 10^{-5} \text{ g/L} & C_{XH1} &= 8.0 \times 10^{-6} \text{ g/L} \\
a_S &= 8 & C_{XG1} &= 1.0 \times 10^{-6} \text{ g/L}
\end{aligned}
$$

Compare your simulation result with Figure 14 of the paper by Ramkrishna et al. (1967). If you showed the change of the cell concentrations in g dry weight/liter, the shape of the curves are quite different from those in the paper. What are the differences? Explain. Do the parameter values predict realistic growth curves? What new features can this model predict which the Monod model cannot?

6.11 REFERENCES

Advanced Continuous Simulation Language: User Guide Reference Manual. Concord, MA: Mitchell and Gauthier, Assoc., Inc., 1975.

Aiba, S., A. E. Humphrey, and N. F. Millis, *Biochemical Engineering* (2nd ed.), p. 304. Tokyo, Japan: University of Tokyo Press, 1973.

Aiba, S., M. Shoda, and M. Nagatani, "Kinetics of Product Inhibition in Alcohol Fermentation," *Biotechnol. Bioeng.* **10** (1968):845-864.

Andrews, J. F., "A Mathematical Model for the Continuous Culture of Microorganisms Utilizing Inhibitory Substrates," *Biotechnol. Bioeng.* **10**(1968):707-723.

Bowen, R. L., "Unraveling the Mysteries of Shear-sensitive Mixing Systems," *Chem. Eng.* 9 (1986):55-63.

Carberry, J. J., *Chemical and Catalytic Reaction Engineering*, p. 52. New York, NY: McGraw-Hill Book Co., 1976.

Feder, J.and W. R.Tolbert, "The Large-Scale Cultivation of Mammalian Cells," *Scientific American* **248** (1983):36-43.

Fredrickson, A. G., "Formulation of Structured Growth Models," *Biotech. Bioeng.* **18** (1976):1481-1486.

Harder, A.and J. A. Roels, "Application of Simple Structured Models in Bioengineering," *Adv. Biochem. Eng.* **21** (1982):56-107.

Herbert, D., R. Elsworth, and R. C. Tellmg,"The Continuous Culture of Bacteria; a Theoretical and Experimental Study,".*j. Gen. Microbiol.* **14** (1956):601-622.

Hill, G. A. and C. W. Robinson, "Minimum Tank Volumes for CFST Bioreactors in Series," *Can. J. Chem. Eng.* **67** (1989):818-824.

Hooker, B. S., J. M. Lee, and G. An, "Cultivation of Plant Cells in a Stirred Vessel: Effect of Impeller Design," *Biotech. Bioeng.* **35** (1990):296-304.

Levenspiel, O., *Chemical Reaction Engineering* (2nd ed.), p. 150. New York, NY: John Wiley & Sons, 1972.

Luedeking, R., "Fermentation Process Kinetics," in *Biochemical and Biological Engineering Science,* ed.N.Blakebrough. London, England: Academic Press, Inc., 1967, pp. 181-243.

Monod,J., "The Growth of Bacterial Cultures," *Ann. Rev. Microbiol* **3** (1949):371-394.

Moo-Young, M., G.Van Dedem, and A. Bmder,"Design of Scraped Tubular Fermentors," *Biotech. Bioeng.* **21** (1979):593-607.

Ramkrishna, D., A. G. Fredrickson, and H. M. Tsuchiya,"Dynamics of Microbial Propagation: Models Considering Inhibitors and Variable Cell Composition," *Biotech. Bioeng.* **9** (1967):129-170.

Russell,!. W. F., I. J. Dunn, and H. W. Blanch, "The Tubular Loop Batch Fermentor: Basic Concepts," *Biotech. Bioeng.* **16** (1974): 1261-1272.

Schiigerl, K., "New Bioreactors for Aerobic Processes," *Int. Chem. Eng.* **22**(1982):591-610.

Tsuchiya, H. M., A. G. Fredrickson, and R. Aris, "Dynamics of Microbial Cell Populations," in *Adv. Chem. Eng.,* vol. 6, eds.T.E.Drew, J.W.Hoopes, Jr., and T.Vermeulen. New York, NY: Academic Press, 1966, pp. 125-206.

Williams, F. M., "A Model of Cell Growth Dynamics," *J. Theoret. Biol.* **15**(1967):190-207.

7

Genetic Engineering

The central tool for the new biotechnology is the recombinant DNA technique.[1] It allows direct manipulation of genetic material of individual cells. By inserting foreign genetic information into fast-growing microorganisms, we can produce foreign gene products (proteins) with higher rates and yields that have not been possible with any other cellular systems. This technology is also known as genetic engineering because it involves the manipulation of genetic materials.[2] In this chapter, basic principles involved in recombinant DNA technology and problems involved in cultivating the genetically engineered cells are briefly described.

7.1 *DNA* AND *RNA*

Deoxyribonucleic acid (DNA) is the most important molecule in living cells and contains all of the information that specifies the cell. DNA and ribonucleic acid (RNA) are macromolecules that are linear polymers built up from simple subunits, nucleotides.[3] The monomeric unit, *nucleotide,* has the following three components Figure 7.1:

1. A cyclic five-carbon (pentose) sugar: deoxynbose for DNA, and nbose for RNA.

2. A nitrogenous base of either purine or pyrimidine derivation, covalently attached to the 1'-carbon atom of the sugar by an N-glycosylic bond as shown in Figure 7.1

    a) The punnes: adenme (A) and guamne (G).

    b) The pyrimidines: cytosine (C), thymine (T) for DNA only, and uracil (U) for RNA only.

3. A phosphate attached to the 5' carbon of the sugar by phosphoester linkage.

---

[1] Read Chapter 1 for a general introduction to the new biotechnology.}

[2] The name *genetic engineering* should not mislead the readers that it is a field of engineering; it is a field of biological science.

[3] Macromolecules: a polymer, especially one composed of more than 100 repeated monomers.

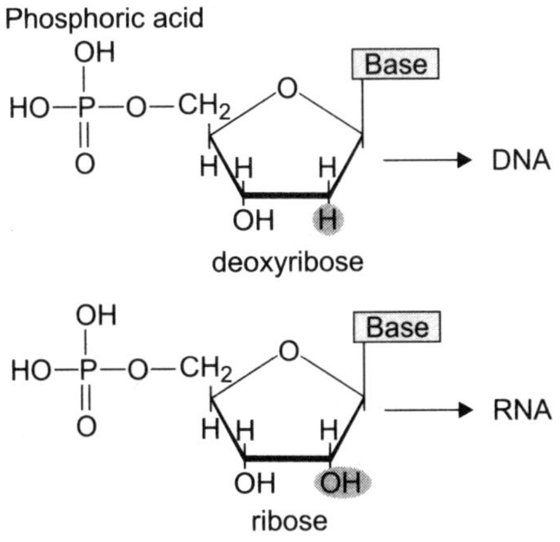

Fig. 7.1   Structure nucleotide

The nucleotides of DNA are called deoxyribonucleotides, since they contain the sugar deoxyribose, whereas, those of RNA are called nbonucleotides since they contain nbose instead. Each nucleotlde contains both a specific and a nonspecific region. The phosphate and sugar groups are the nonspecific portion of the nucleotide, while the purine and pyrimidine bases make up the specific portion.

Nucleotides are joined to one another linearly by a chemical bond between atoms in the nonspecific regions to form polynucleotides (Figure 7.2). The linkage (called *phosphodiester bonds]* is between a phosphate group and a hydroxyl group on the sugar component.

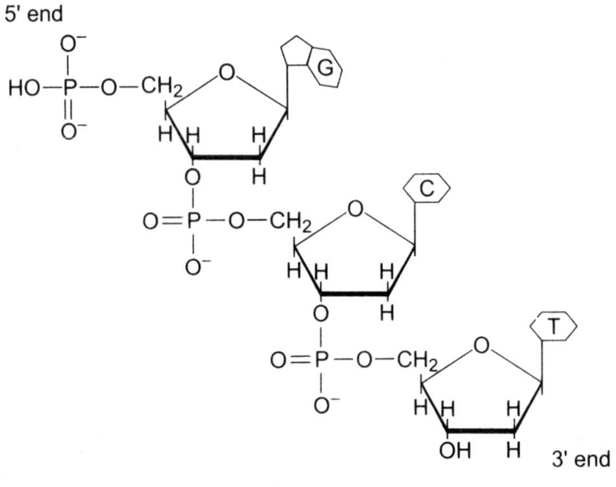

Fig. 7.2   Portion of DNA.

The most important feature of DNA is that it usually consists of two complementary strands coiled about one another to form a double helix (Figure 7.3). Each strand is the *poly nucleotide*. The diameter of the helix is about 20 Å and each chain makes a complete turn every 34 Å. There are ten nucleotides on each chain every turn of the helix.

The two chains are joined together by hydrogen bonds between the purme-to-pynmidme base pairs.[4] Ademne (purme) is always paired with thymme (pynmidme) and guanme (purme) with cytosme (pynmidme). As a result, chemical analysis of the molar content of the bases in DNA provided that the amount of adenine always equals that of thymine and that the amount of guanme always equals that of cytosme. This base pairing is so specific that adenine bonds only to thymine and guanine only to cytosine. This base pairing provides stabilization by hydrogen bonding between complementary bases.

Furthermore, this specificity of base pairing is what permits the transmission of genetic information from one generation to another. When cell duplication occurs, the DNA double helix unwinds, and two new DNA strands are formed that are complementary to the original strands. Thus, each of the new cells contains one of the original DNA strands and one newly synthesized strand in its double helix.

The sequence of bases (A, G, T, and C) in a strand of DNA specifies the order in which amino acids are assembled to form proteins. The *genetic code* is the collection of base sequences that correspond to each amino acid, *codon*. Since there are only four bases in DNA and twenty amino acids in protein, each codon must contain at least three bases.[5] Two bases cannot serve as codons because there are only 4^2 possible pairs of four bases, but three bases can serve because there are $4^3 = 64$ possible triplets. Since the number of possible triplets are more than enough, several codons designate the same ammo acid. In other words, the genetic code is highly redundant as shown in Table 7.1. For example, UCU, UCC, UCA, UCG, AGU, and AGC are all codes for serine.

7.2 CLONING OF A GENE

The cloning of a gene is the central operation of recombinant DNA technology. *Gene* is the basic unit of hereditary information located on a chromosome of the cell. *Chromosomes* are threadlike structures found in the nucleus, which are comprised primarily of DNA. The

---

[4] A hydrogen bond between the two chains is a weak attractive force between a covalently bounded hydrogen atom (H—N) and a negatively charged keto oxygen atoms (C=O).

[5] The twenty amino acids found in proteins are: Alanine (Ala), Arginine (Arg), Asparagine (Asn), Aspartic acid (Asp), Cysteine (Cys), Glutamic acid (Glu), Glutamine (Gin), Glycine (Gly), Histidine (His), Isoleucine (He), Leucine (Leu), Lysine (Lys), Methionine (Met), Phenylalanine (Phe), Proline (Pro), Serine (Ser), Threonine (Thr), Tryptophan(Trp), Tyrosine (Tyr), and Valine (Val).

genetic information is stored in an ordered sequence of nucleotide bases, comprising a segment of DNA. Each gene specifies the structure of a particular gene product, usually a protein.

7.2.1 Understanding of DNA Sequences

In order to manipulate the gene, it is necessary to learn about the overall organization of DNA sequences and how the functional units of DNA interact with one another in the total genetic repertoire of an organism, *genome*. Since the genome of mammalian cell is too large to learn about its organization, we should find the simplest genome so that the task can be accomplished.[6]

Table 7.1 The Universal Genetic Code

| First | Second Position | | | | Third |
Position	U	C	A	G	Position
U	Phe	Ser	Tyr	Cys	U
U	Phe	Ser	Tyr	Cys	C
U	Leu	Ser	Stop	Stop	A
U	Leu	Ser	Stop	Tip	G
C	Leu	Pro	His	Arg	U
C	Leu	Pro	His	Arg	C
C	Leu	Pro	Gin	Arg	A
C	Leu	Pro	Gin	Arg	G
A	He	Thr	Asn	Ser.	U
A	He	Thr	Asn	Ser	C
A	He	Thr	Lys	Arg	A
A	Met	Thr	Lys	Arg	G
G	Val	Ala	Asp	Gly	U
G	Val	Ala	Asp	Gly	C
G	Val	Ala	Glu	Gly	A
G	Val	Ala	Glu	Gly	G

The study of viruses revealed that their genomes are very small. SV40 virus of monkey has 5,243 base pairs in which there are five genes. Now, it is possible to study the gene structure with this small genome. However, still 5000-odd base pairs can be a great challenge

---

[6] A genome of mammalian cell can carry as much as 2.5 billion base pairs of information arrayed along its chromosomal DNA. The base sequences are arranged in discrete compartments of information: individual genes. There are between 50,000 and 100,000 genes in the genome of a mammal, which are too big to study the organization of DNA sequences.

to be sorted out. The discovery of *restriction endonucl eases*, bacterial enzymes that cut DNA at specific sequences, made the job easier, since it could reduce the long DNA molecules into a set of discreet fragments.

The restriction enzymes are present in most bacterial cells to restrict the ability of foreign DNA to take over the protein-synthetic machinery of cells. Their own DNA is protected by the restriction enzyme due to the presence of cellular enzymes that methylate specific nucleotides, so that the nucleotides cannot be recognized by the restriction enzymes.

Every restriction enzyme recognizes and cuts a specific DNA sequence which usually consists of four or six nucleotide pairs. For example, a restriction endonuclease *Eco* RI obtained from *Escherichia coli* cuts wherever it encounters the nucleotide sequence GAATTC, whereas *Bal* 1 from

Brevibacterium albidum cuts the sequence TGGCCA (Figure 7.4). There are over 400 different restriction enzymes purified from about 250 different microorganisms. The restriction enzymes break the double-stranded DNA molecules in two different ways as shown in Figure 7.4:

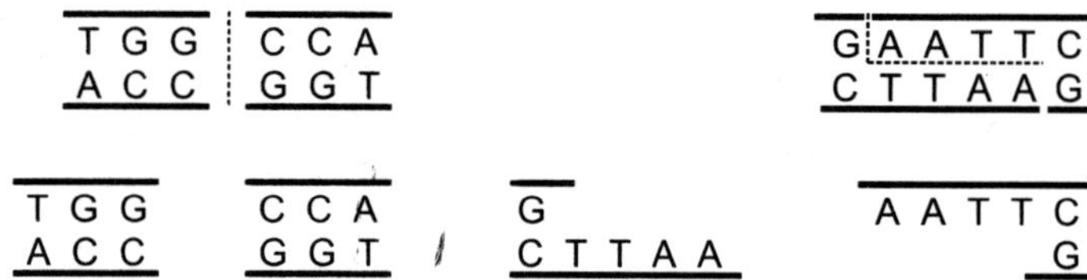

Fig. 7.4   Two types of cuts made by restriction enzymes: (a) flush ends and (b) cohesive ends.

1. Cuts on the line of symmetry to generate flush-end molecules.
2. Cuts on the positions symmetrically placed around the line of symmetry to generate cohesive-end (or sticky-end) molecules.

Since one restriction enzyme recognizes a unique sequence, the number of cuts made in a typical DNA molecule is generally small. The DNA fragments cut by the restriction endonucleases can be separated by their size by gel electrophoresis and be studied. These methods made it possible to establish the entire nucleotide sequence of the SV40 genome by 1978 (Weinberg, 1985). Because of the similarity of the molecular organization in all organisms, bacterial DNA and mammalian DNA are structurally compatible. Therefore, DNA segments from one life can readily be blended with the DNA of another form. This similarity extends to *plasmids* which are small extrachromosomal genetic elements found in a variety of bacterial species. They are double-stranded, closed circular DNA molecules.

7.2.2   The Joining of DNA Molecules

The basic procedures of recombinant DNA techniques are: (1) to join a DNA segment to a DNA molecule (known as vector) that can replicate, and (2) to provide an environment that allows reproduction of the joined DNA molecule.[7]

Four types of vector can be used to clone fragments of foreign DNA and propagate them in *Escherichia coll* . They are plasmids, bacteriophage\gamma, cosmids, and bacteriophage Ml3. All of these have the following properties necessary to be qualified as vectors:

1. They have ability to replicate autonomously in *E. coll.*
2. They contain a selection marker for the easy separation and purification of the vector from bacterial nucleic acids.
3. They have regions of DNA that are not essential for propagation in bacteria, so that foreign DNA can be inserted into these regions.
4. They can be inserted into host cells easily.

Plasmid DNA can be isolated from a culture of plasmid containing bacteria by adding detergent (such as sodium dodecyl sulfate) and by centrifuging the lysate.[8] The bacterial chromosome complex, much larger than the plasmid, moves to the bottom of the centrifuge tube, and the supercoiled plasmid and the linear chromosomal fragments remain in the supernatant. The supercoiled plasmid is again separated by centrifugation after treating with CsCl and ethidium bromide.

Plasmids contain genes for enzymes that are usually advantageous to the bacterial host. The *phenotypes*[9] conferred by different plasmids are (Maniatis et al., 1982): resistance to antibiotics, production of antibiotics, degradation of complex organic compounds, production of restriction and modification enzymes, and so on.

Plasmids can be introduced into bacteria after the bacteria is treated so that cells are temporarily permeable to small DNA molecules. This process is known as *transformation*. The bacteria that have been successfully transformed can be selected based on the new phenotype they received from the plasmid, such as resistance to antibiotics.

Some plasmids are present in cells in low-copy number – one or a few per cell, since the plasmid DNA replicates only once or twice before cell division. However, other plasmids exist in large numbers – from

---

[7] Vector is the DNA molecule used to introduce foreign DNA into host cells.

[8] The detergent alters the cell surface causing a leakage of cell constituent.

[9] Phenotype is the observable properties of an organism; produced by the genotype in cooperation with the environment.

10 to 100 per cell, since plasmid DNA replicates repeatedly until the proper copy number is reached.

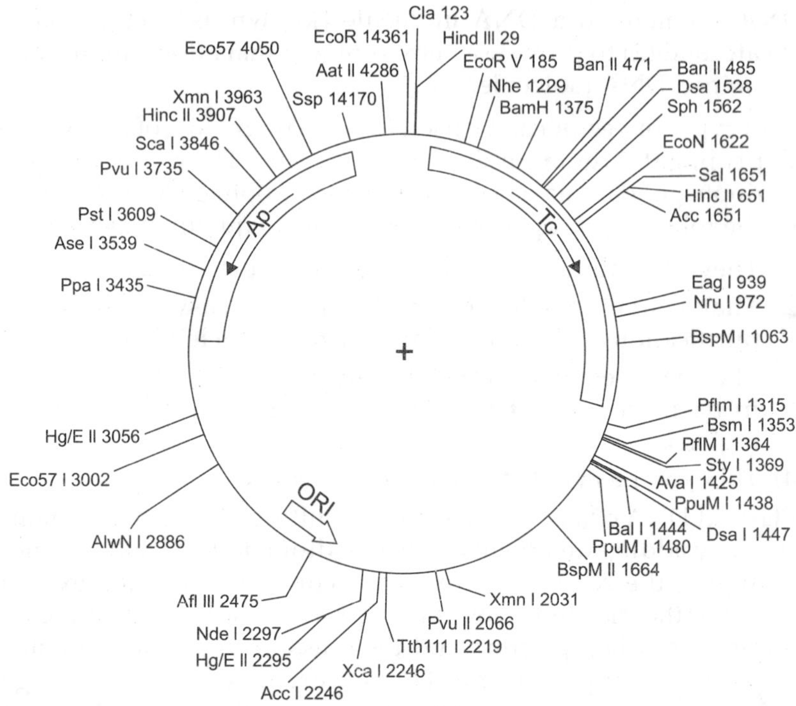

Fig. 7.5 Restriction map of pBR322 DNA [Courtesy of New England Biolabs, Inc., Beverly, MA].

The plasmids with large copy number are called relaxed plasmids, which is one of the useful properties as a cloning vector.

Figure 7.5 shows one of the most widely used vectors, pBR322, which is a relaxed plasmid with 4,363 base pairs in length. This vector contains both ampicillm (Ap) and tetracyclme (Tc) resistance genes. Numbering of the sequence begins within the unique *Eco* RI site: the first T in the sequence GAATTC is designated as nucleotide number 1. Numbering then continues around the molecule in the direction Tc to Ap.

The plasmid can be cut at a number of defined sites with restriction enzymes as explained in the previous section. The map in Figure 7.5 shows the restriction sites of those enzymes that cut the molecule once or twice; the unique sites are shown in bold type.

Therefore, the fragments generated by an enzyme can circularize by rejoining of complementary cohesive ends. Furthermore, the fragments generated by a particular enzyme acting on one DNA molecule have the same cohesive end with those generated by the same enzyme acting on another DNA molecule. Therefore, fragments from the two different molecules from two different organisms can be joined by hydrogen bonds reversibly when the fragments are mixed.

If the joint is ligated after basepairing, the fragments are joined permanently. The joining of these fragments is accomplished by *DNA ligase* (also called polynucleotide ligase), which bonds the recombmant molecules covalently by creating a phosphodiester bond between the 5′–PO$_4$ end of one polynucleotide and the 3′–OH end of another.

Figure 7.6 shows the overall procedure for the production of recombinant DNA. The plasmid is cut at a number of defined sites with a restriction enzyme. The DNA of a foreign genome is cleaved with the same restriction enzyme. Some of the resulting fragments may have the gene of interest. Plasmids and genome fragments are mixed and joined by DNA ligase. The recombinant plasmids are introduced into bacteria by cocultivation of plasmids and bacteria.

7.3 Stability of Recombinant Microorganisms

When recombinant microorganisms are cultivated to produce foreign proteins, the yield of the product may be decreased due to the plasmid loss from the organisms as they undergo the number of generations of growth. The instability is due to two characteristically different modes (Monbouquette and Ollis, 1986): (1) defective segregation of plasmid during cell division, or (2) structural instability resulting from mutations of plasmid DNA.

For segregational plasmid stability, cells need to be replicated so that the average number of plasmid copies per cell is doubled once per generation, and the plasmid copies need to be distributed equally to the daughter cells at cell division. It was reported that the problem of plasmid instability was encountered in systems employing *E. coll* (Kim and Ryu, 1984; Deretic et al., 1984; Caulcott et al., 1985), *B. subtilis* (Kadam et al., 1987), as well as yeast (Whiteway and Ahmed, 1984; Schwartz et al., 1988) as hosts.

The fraction of cells carrying plasmid in the total population was shown to be a function of the number of generations of growth (Imanaka and Aiba, 1981). Therefore, as the size of the fermenter increases, the number of generations of growth increases, and the fraction of cell carrying plasmid and the product yield decrease. The problem of the plasmid instability is more serious if the large-scale fermenter is operated continuously.

Example 7.1

Estimate the number of generations of growth needed for genetically modified microorganisms from 1 mL culture to a 33,000L production-scale fermenter. Assume that the inoculum size in each stage of the scale-up is 3 percent. The inoculum size is defined as the percentage of the fully grown cell culture volume in the total culture volume.

Solution:

For cells to grow from 3 percent to 100 percent of saturated cells, they have to multiply $100/3 = 33$ times, which requires the cells to double n times as,

$$2^n = 33$$

$$n = 5.04$$

Therefore, it takes about five generations. A typical inoculation steps with about 33 times volume increase can be as follows:

1. inoculum from 1 mL culture into a small Erlenmeyer flask (33 mL)
2. large flask (1L)
3. bench-scale fermenter (33L)
4. pilot-scale fermenter (1,000L)
5. production-scale fermenter (33,000L)

For the 3 percent inoculum, it takes about five generations of growth for the cultures to be fully grown in each step. Therefore, the total number of generations needed for 33,000L fermenter will be about 25.

7.3.1 Fermentation Kinetics of the Recombinant Cultures

Let's assume that the probability for the plasmid-carrying cells (X^+) to produce the plasmid-free cells (X) is p after one division. Then, N plasmid-carrying cells will produce $N(1-p)$ plasmid-carrying cells and Np plasmid-free cells after one division.

The total number of X^+ cells will be $N(2-p)$. During the exponential growth period, the growth rate of the plasmid-carrying cells will be expressed as (Imanaka and Aiba, 1981).

$$\frac{dC_{X^+}}{dt} = (1-p)\,\mu^+ C_{X^+} \tag{7.1}$$

where μ^+ is the specific growth rate of the plasma-carrying cells. C_{X^+} is the number of plasmid-carrying cells per unit volume. If the mass of the cells is approximately proportional to the number of cells, the preceding rate equation can be also applicable when C_{X^+} is mass of cells per unit volume.

On the other hand, the growth rate of the plasmid-free cells will be

$$\frac{dC_{X^-}}{dt} = p\mu^+ C_{X^+} + \mu^- C_{X^-} \tag{7.2}$$

If we assume that μ^+ *and* p are constant, the integration of Eq. (7.1) yields

$$C_{X^+} = C_{X_0^+} \exp[(1-p)\mu^+ t] \tag{7.3}$$

where $C_{X_0^+}$ is the initial concentration of the plasmid-carrying cells.

For the plasmid-free cells, substitution of Eq. (7.3) into Eq. (7.2) gives

$$\frac{dC_{X^-}}{dt} = \mu^- C_{X^-} = p\mu^+ C_{X_0^+} \exp[(1-p)\mu^+ t] \tag{7.4}$$

By solving the preceding linear first-order differential equation for the constant μ^+,

$$C_{X^-} = \frac{p\mu^+ C_{X_0^+}}{(1-p)\mu^+ - \mu^-} \{\exp[(1-p)\mu^+ t] - \exp(\mu^- t)\} + C_{X_0^-} \exp(\mu^+ t) \tag{7.5}$$

Therefore, Eqs. (7.3) and (7.5) predict how C_{X^+} and C_{X^-} change with time for given values of p, μ^+, and μ^-.

The fraction of the plasmid-carrying cells in the total population (f) can be defined as

$$f = \frac{C_{X^+}}{C_{X^+} + C_{X^-}} \tag{7.6}$$

Substitution of Eq. (7.3) and Eq. (7.5) to Eq. (7.6) yields

$$f = \frac{\exp\left[(1-p)\mu^+ t\right]}{\exp\left[(1-p)\mu^+ t\right] + \dfrac{p\mu^+ C_{X_0^+}}{(1-p)\mu^+ - \mu^-}\left\{\exp\left[(1-p)\mu^+ t\right] - \exp\left(\mu^- t\right)\right\}} \tag{7.7}$$

which shows the change of the fraction of the plasmid-carrying cells with respect to time during the exponential growth period of a batch fermentation. It would be interesting to see how f decreases with respect to the number of generations.

During the exponential growth period, the number of generations (n) of the plasmid-carrying cells can be calculated from the following relationship:

$$n = \frac{\mu^+ t}{\ln 2} \tag{7.8}$$

Combining Eqs. (7.7) and (7.8) will result in the equation of / for nth eneration as:

$$f_n = \frac{1 - \alpha - p}{1 - \alpha - p\left[2^{n(\alpha + p - 1)}\right]} \tag{7.9}$$

where *a* is the ratio of the specific growth rates as

$$\alpha = \frac{\mu^+}{\mu^-} \tag{7.10}$$

Eq. (7.9) can be used to predict the change of f with respect to the number of generations for the series of batch fermentations if we assume that cells multiplied exponentially during each batch cultivation. As it was illustrated earlier, it would take about 25 generations to scale up from a slant culture to a 33,000L production-scale fermenter. Figure 7.7 shows the effects *ofp* and a on f_{25}, which is decreased as the increase of α and p. When $p \leq 0.01$ and $\alpha < 1$, f_{25} is close to 1, that is, the plasmid-carrying cells are very stable. However, as α approaches 2.0, f_{25} also becomes zero. The typical values of α ranges from 1.0 to 2.0 (Imanaka and Aiba, 1981).

Figure 7.8 shows the change of f as a function of n and α. The p was set constant to be 0.01. The value of f decreases rapidly with the increase of n and α. When α is 1 .4, all of the plasmid-carrying cells lost their plasmids after about 33 generations.

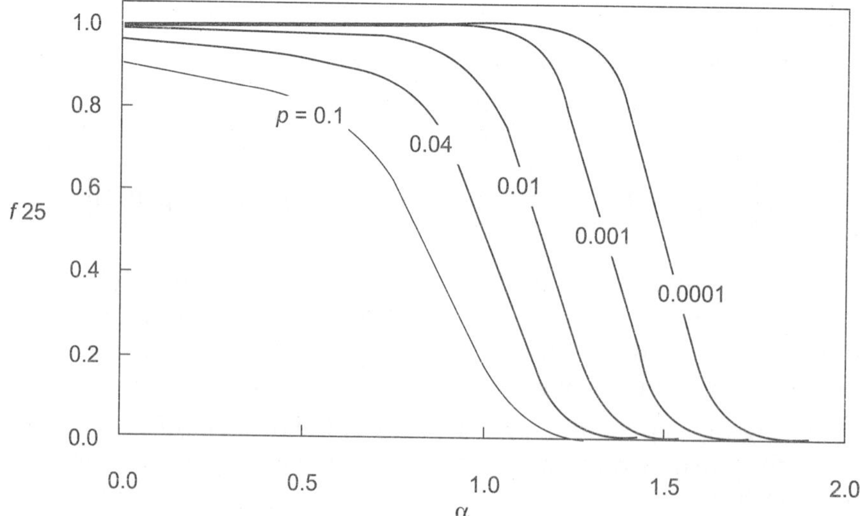

Fig. 7.7   The fraction of the plasmid-carrying cells after 25 genera-tions (f_{25}) as the function of α and p.

7.3.2 Continuous Stirred-tank Fermenter (CSTF)

Let's examine the stability of recombinant cells in the continuous stirred-tank fermenter. The material balance for the plasmid-carrying cells around a CSTF yields

$$-DC_{X^+} + (1 - p)\,\mu^+ C_{X^+} = \frac{dC_{X^+}}{dt} \tag{7.11}$$

Similarly, the material balance for the plasmid-free cells gives

$$-DC_{X^-} + p\mu^+ C_{X^+} + p\mu^- C_{X^-} = \frac{dC_X}{dt} \tag{7.12}$$

The addition of Eq. (7.11) and Eq. (7.12) gives the equation for the total cell concentration as

$$\left(\mu^+ C_{X^+} + \mu^- C_{X^-}\right) - D\left(C_{X^+} + C_{X^-}\right) = \frac{d\left(C_{X^+} + C_{X^-}\right)}{dt} \tag{7.13}$$

If the CSTF is operated so that the total concentration of cells is constant with time,

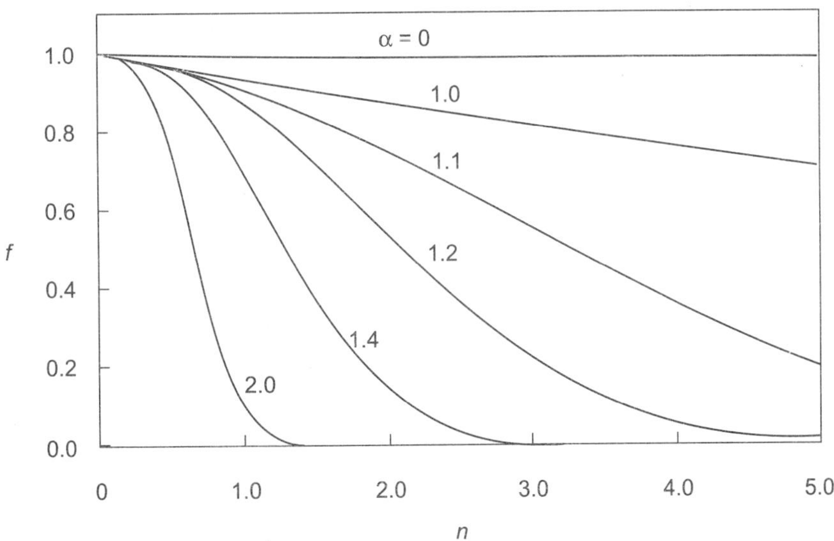

Fig. 7.8   The fraction of the plasmid-carrying cells (*f*) versus the number of generations n (p = 0.01).

$$\mu^+\left(C_{X^+} + \alpha\,C_{X^-}\right) = D\left(C_{X^+} + C_{X^-}\right) \tag{7.14}$$

If $\alpha = 1$, Eq. (7.14) is simplified to

$$\mu^+ = D \tag{7.15}$$

Therefore, the specific growth rate of the cells in the fermenter is constant and determined by the dilution rate. Eqs. (7.11) and (7.12) can be solved after the substitution of Eq. (7.15) to give

$$C_{X^+} = C_{X_0^+} \exp(-pDt) \tag{7.16}$$

$$C_{X^-} = C_{X_0^-} + C_{X_0^+}[1 - \exp(-pDt)] \tag{7.17}$$

Therefore, during the continuous fermentation, the concentration of plasmid-carrying cells will be reduced, whereas the plasmid-free cells will increase.

If $\alpha \neq 1$, μ^+ is no longer constant for a constant dilution rate during the steady-state operation of CSTF but depends on the cell concentrations (C_{X^+} and C_{X^-} and α according to Eq. (7.14). As a result, μ^+ also changes with time. By solving Eqs. (7.11), (7.12), and (7.14) simultaneously, we can estimate how C_{X^+} and C_{X^-} will change with time. It would be interesting to observe how the fraction of the plasmid-carrying cells will decrease with time. Substituting Eq. (7.14) into Eq. (7.11) and dividing with $C_{X^+} + C_{X^-}$ yields

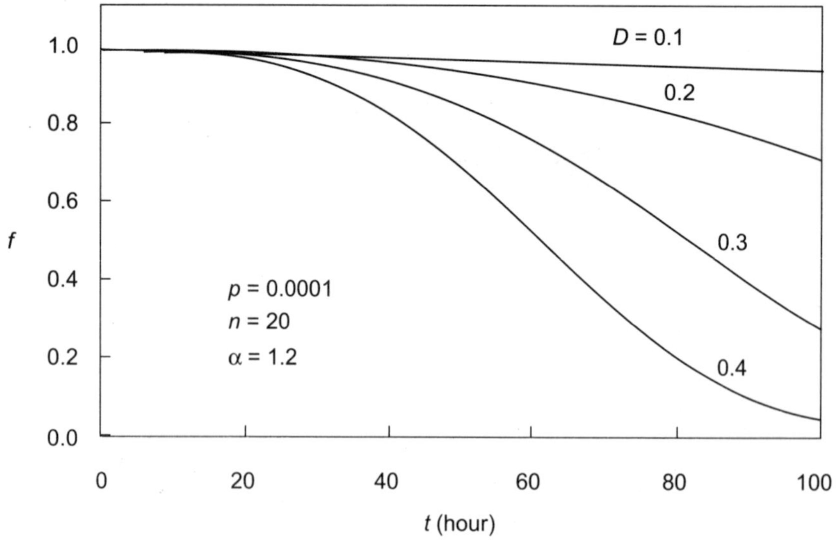

Fig. 7.9 The effect of dilution rate (*D*) on the fraction of plasmid-carrying cells during steady-state operation of CSTF. The initial value of *f* was calculated by assuming that the number of generations required for the step inoculation and initial batch and unsteady-state continuous fermentation was 20.

$$\frac{df}{dt} = -Df + \frac{(1-p)Df}{\alpha + (1+\alpha)f} \tag{7.18}$$

The solution of the preceding equation gives the change of f with time. The initial value of f for the solution of Eq. (7.18) can be estimated from Eq. (7.9).

Figure 7.9 shows the change of the fraction of plasmid-carrying cells with time during steady-state operation of CSTF. The initial value of f was calculated by assuming that the number of generations required for the step inoculation and initial batch and unsteady-state continuous fermentation was 20. The fraction of plasmid-carrying cells was decreased with time and with the increase of the dilution rate. When the p and D are sufficiently low, and α is close to 1, the CSTF can be operated economically for a long period of time.

However, if α increases to 1.4, CSTF will lose almost all plasmid-carrying cells within 100 hours of the steady-state CSTF operation, as shown in Figure 7.10.

7.3.3 Methods of Stabilization

Several methods have been suggested for the stabilization of recombmant systems that show the tendency to lose their plasmids.

They are as follows (Imanaka and Aiba, 1981):

1. Formulate media to favor the growth of plasmid-carrying cells over plasmid-free cells.
2. Put selective pressure against plasmid-free cells using *ctuxotrophic mutants* or antibiotic-resistant plasmids (Parker and DiBiasio, 1987). Auxotrophic mutant is the cell which is mutated so that it requires a specific growth substance beyond the minimum required for normal metabolism and reproduction.
3. Use temperature-dependent mutant plasmid or strain.
4. Use temperature-dependent control of gene expression. Plasmid is more likely to be kept unchanged when the gene expression is repressed. The higher the gene expressions, the more segregants tend to appear.
5. Use plasmid containing no transposable element. Transposable elements are DNA segments which can be inserted into several sites in a genome and can cause mutation.
6. Use recombination deficient strain.

7.4. GENETIC ENGINEERING OF PLANT CELLS

Although the utility of large-scale cultivation of plant cells} has been well recognized, plant tissue cultivation has not yet been employed industrially for the production of primary or secondary by-products except for a few test processes.[10] The undcrutilization of plant tissue

---

[10] First commercial process utilizing plant cells to manufacture shikonin was developed by a Japanese firm. Mitsue Petrochemical Industries Ltd. in 1983.

cultivation techniques is mainly due to the slow growth rate of plant cells compared to microbial cells and the low yield of products.

Recent developments in recombinant DNA technologies show great promise for solving this problem. Fast growing cultured cells can be selected and genetically modified to produce commercially valuable products in higher concentrations than normally produced by the cells. The potential products of recombinant DNA techniques are foreign proteins and secondary metabolites.

The production of secondary metabolites from genetically modified plant cells can increase productivity dramatically and bring in the rapid commercialization of large-scale plant cultivation. However, genes responsible for the biosynthesis of economically important secondary metabolites have not yet been isolated. Since secondary metabolites are usually biosynthesized via the joint action of many gene products, many genes are required for each biosynthetic pathway leading to the production of secondary metabolites.

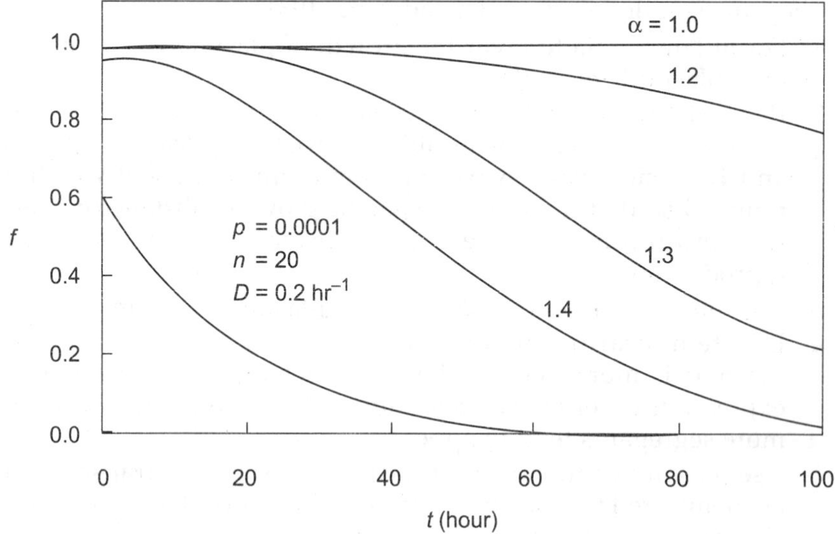

Fig. 7.10 The effect of a on the fraction of plasmid-carrying cells during steady-state operation of CSTF. The initial value of *f* was calculated by assuming that the number of generations required for the step inoculation and initial batch and unsteady-state continuous fermentation was 20.

On the other hand, genes responsible for the economically important proteins have been identified and inserted successfully into microorganisms, because proteins are immediate gene products. The technology to introduce the foreign gene into plant cells has been also developed for the agricultural applications of plant cell culture.

Therefore, the production of foreign proteins from genetically modified tobacco cells can be accomplished.

The potential advantages of employing plant cells as host cells for foreign products are as follows:

1. The protein products from recombinant plant cells may be more functional and potent as pharmaceuticals than those from microbial origin because a post-translational modification is likely to occur in plant cells.
2. The medium (mainly a carbohydrate source) to cultivate plant cells is well defined and inexpensive, while mammalian cells require very expensive serum as their medium.
3. The suspension plant cells can reach very high cell density (60 wet weight percent or 2.4 dry weight percent). This is higher than any other type of culture can attain.
4. Contaminations by bacteria or fungus are easily monitored in plant tissue cultures. Furthermore, these contaminants are usually not potent pathogens to human beings.

Recently, Hiatt et al. (1989) showed a great potential for the production of foreign proteins from plants by demonstrating the production of immunoglobulins and assembly of functional antibodies in genetically modified tobacco. They transformed tobacco leaf segments by using complementary DNAs from a mouse hybridoma messenger RNA and regenerated the segments to mature plants. Plants expressing single γ or κ immunoglobulin chains were crossed to yield progeny in which both chains were expressed simultaneously. A functional antibody accumulated to 1.3 percent of total leaf protein in plants expressing full-length cDNAs containing leader sequences.

With these future applications of genetic engineering techniques of plant cells in mind, the recombinant DNA technology for the plant cells is reviewed briefly (Lee and An, 1986).

7.4.1   Gene Manipulation

A foreign gene should be altered in order to be properly expressed in plant cells, because neither bacterial nor animal genes can be expressed in plants without modification. This indicates that the genetic information involved with gene expression such as production of *messenger-RNA* (m-RNA) from a gene and translation of the m-RNA into a protein is quite different in plants compared to other living organisms. For the best result, both upstream (promoter) and downstream (terminator) regions of the structural gene should be replaced with a plant sequence that contains appropriate information, as shown in Figure 7.11.

Plant promoters can be divided into two classes, constitutive and mducible. Constitutive promoters are functional in almost all tissues. Inducible promoters are normally silent until they are induced by environmental or developmental factors. For example, promoters for photosynthesis, defense mechanisms, and flower development are induced only under a certain condition. Selection of a plant promoter depends on the nature of the desired product. For most cases, a strong constitutive promoter is the best source for the production of a foreign product. However, if the foreign product is detrimental to plant growth, it may be necessary to use an mducible promoter so that the product will be synthesized only under desired conditions. Those promoters for photosynthesis are rapidly induced by white light and turned off during a dark period. The small subunit genes of the ribulose-1,5-bisphosphate carboxylase and the light-harvesting protein genes are most abundantly expressed in a green tissue. Therefore, the promoters from these genes are likely to be the best source for the light-induced expression. These promoters have been isolated from various plants and used for the expression of a bacterial gene.

Very little is known about plant terminators. It has not been determined whether an animal terminator can be functional in plants or not. Until such information is available, it is safer to use a plant terminator for the maximum expression of a foreign gene. *The nopaline synthase nos* terminator is most frequently used although several other plant terminators have also been isolated.

There should be no internal translation initiation signal (ATG) between a promoter sequence and a structural gene since an internal ATG sequence significantly reduces translation efficiency. The distance between the control regions (promoter and terminator) and the structural gene should not be too long. Otherwise a produced m-RNA would be less stable.

One of the important factors kept in mind when considering gene manipulation is the final location of the protein. The produced foreign protein can remain in the cytoplasm, be transported into an organelle, or be secreted to outside the cell. Most plant proteins are located in the cytoplasm. In order to be transported into cytoplasmic organelles or outside of the cell membrane, a specific signal (polypeptide) should be attached to the amino terminus of the desired protein. This signal peptide is cleaved off during the early stage of the transportation and only a mature protein is released to the final destination. It is not well understood yet whether a signal peptide sequence alone is enough for a proper transportation. Recent observations indicate that transportation of a foreign protein into a chloroplast requires only a signal peptide. However, the transfer mechanism into different organelles or into an extracellular environment would be significantly different from each other. If additional information within the main protein body is also necessary, it will be extremely difficult to engineer a foreign protein to be transported without altering the protein structure which may damage the function of the protein.

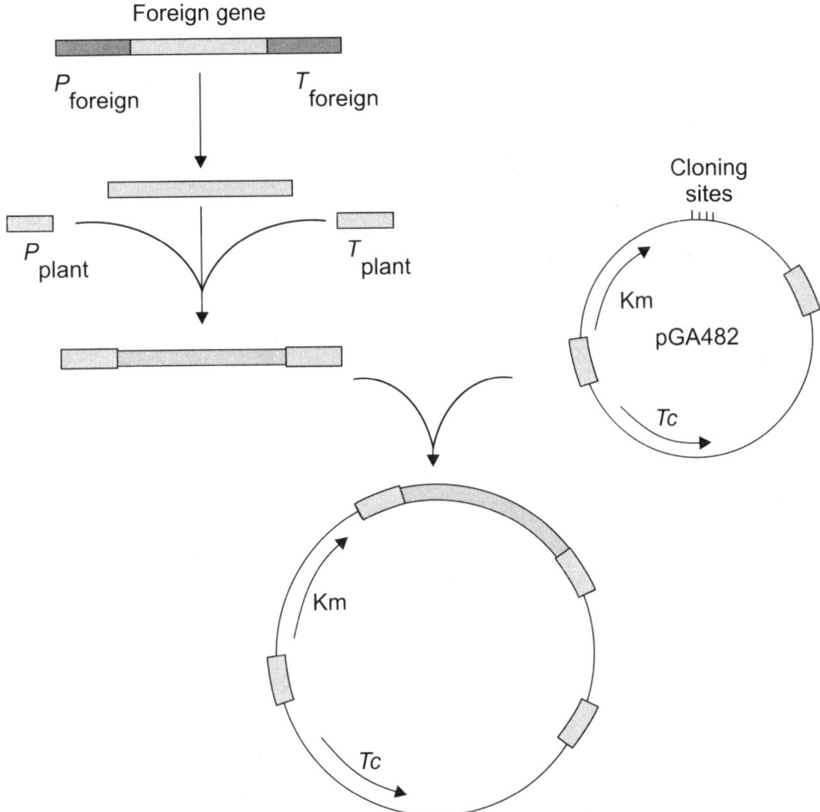

Fig. 7.11   Schematic diagram of foreign gene expression in plant. P: promoter, T: terminator, pGA482: plant cloning vector, Km: kanamycin resistance gene, Tc: tetracycline resistance gene.

Many proteins synthesized in a cell are modified. For example, carbohydrates are often attached to the proteins found in cell membrane. Some proteins become phosphorylated. Such modifications either activate or inactivate the function of the protein. Therefore, a foreign gene should be properly manipulated to obtain a correct *post-translational modification* . For example, if an attachment of a carbohydrate to a protein is desired, one may design the process so as to secrete the protein out of a cell membrane. During this procedure the protein can be properly modified. However, such modification mechanisms can be different in each living organism. Therefore, careful selection of a proper cell line is necessary.

7.4.2 Transformation

A technique widely used for the transformation of plants with novel DNA sequences is based on the tumor-inducing (Ti) plasmid of *Agrobacterium tumefaciens* . During the incubation of plant cells and the soil bacterium, a specific sequence called transfer DNA (T-DNA) of the Ti plasmid is transferred from the bacteria to plant cells. If an exogenous gene is inserted within the T-DNA, the gene can be transferred into a plant chromosome along with the T-DNA. Naturally occurring T-DNA carries phytohormone genes which lead to the tumor formation of the infected cells. In order to avoid the tumorigenic phenotype and facilitate a rapid selection of transformants, plant resistant markers have been developed by constructing chimeric gene fusions of bacterial antibiotic resistance gene and plant control regions as described in the previous section. In this way a kanamycin resistance marker and a chloramphenicol resistance marker have been constructed and used to replace the phytohormone genes on the T-DNA. Since Ti plasmids are difficult to manipulate directly *in vitro* by recombinant DNA methods due to their large size (about 200 kilobase pair), simplified systems have been developed.

The most advanced method utilizes a binary vector system. One of the binary vectors is shown in Figure 7.11. This system depends on a small vector which contains the minimum elements required *in cis* . Other functions necessary for the gene transfer mechanism are donated from a separate helper, Ti plasmid. These features permit an easy introduction and maintenance of an exogenous DNA containing a manipulated gene.

Plant cells can be transformed with a foreign gene by the cocultivation method. In the method, *A. tumefaciens* cells containing the binary vector and a helper Ti plasmid are mixed with actively growing cultured cells or freshly cut plant slices. After incubation of the cell mixture for two days, the transformed plant cells are selected on an antibiotic agar medium. Transformants can be normally visualized within two to three weeks after cocultivation. Several plants have been transformed with this technique.

7.5 NOMENCLATURE

C concentration, number of cells/m^3 or kg/m^3

D dilution rate, s^{-1}

f fraction of the plasmid-carrying cells in the total population, dimensionless

f_n fraction of the plasmid-carrying cells in the total population after n th generations, dimensionless

n number of generation, dimensionless

p probability per generation of producing a cell without plasmid, dimensionless

t time, s

α μ^-/μ^+, dimensionless

μ^+ specific growth rate of the plasmid-carrying cells, s^{-1}

μ^- specific growth rate of the plasmid-free cells, s^{-1}

Subscript

S substrate

X$^+$ plasmid-carrying cells

X$^-$ plasmid-free cells

7.6 PROBLEMS

7.1 Solve Eq. (7.4) with a suitable initial condition to obtain Eq. (7.5).

7.2 Estimate the number of generations of growth needed for genetically modified microorganisms from 1 mL culture to a 33,000L production-scale fermenter. Assume that the inoculum size in each stage of the scale-up is 5 percent except the first inoculation step.

7.3 For a recombinant microorganism *(E. coll)* containing plasmids, the probability per generation of producing a cell without plasmid *(p)* is 0.001 and the ratio of the specific growth rates μ^-/μ^+ is 1.2. Calculate the fraction of the plasmid-carrying cells when the cells are fully grown in a 64L fermenter. The inoculation was carried out by several steps from 1 mL stock culture. The inoculum size for each step is 2.5 percent.

7.4 The recombinant microorganism stated in Problem 7.3 ($p = 0.001$ and $\alpha = 1.2$) is being cultivated continuously in 64L fermenter with a flow rate of 15 L/hr. The total cell concentration in the reactor is maintained at a constant level. If the fraction of the plasmid-carrying cells (f) is 0.8 initially, how long can you operate this CSTF until the fraction f drops to 0.4?

7.5 You are maintaining a strain of recombinant microorganisms as a stock culture. When you checked the fraction of the plasmid-carrying cells in the total population (f) after three consecutive subculturmgs, the fraction was found to be 0.74. In each subculturing, you used 3 percent inoculum. The ratio of the specific growth rate of the plasmid-free and plasmid-carrying cells (μ^-/μ^+) is 1.38. Estimate the probability per generation of producing a cell without plasmid *(p)*.

7.7 REFERENCES

Caulcott, C. A., G. Lilley, E. M. Wright, M. K. Robinson, and G. T. Yarranton, "Investigation of the Instability of Plasmids Directing the Expression of Met-Prochymosin in *Escherichia coll* ," *J. General Microbiology* **131** (1985): 3355 – 3365.

Deretic, V., O. Francetic, and V. Glisin, "Instability of the plasmid carrying active penicillin acylase gene from *Escherichia coil* : conditions inducing insertional inactivation," *FEMS Microbiology Letters* **24** (1984): 173 – 177.

Hiatt, A., R. Cafferkey, and K. Bowdish, "Production of antibodies in transgemc plants," *Nature* **342** (1989): 76 – 78.

Imanaka, T. and S. Aiba, "A Perspective on the Application of Genetic Engineering: Stability of Recombinant Plasmid," *Ann. N.Y. Acad. Sci.* **369** (1981): 1 – 14.

Kadam, K. L., K. L. Wollweber, J. C. Grosch, and Y. C. Jao, "Investigation of Plasmid Instability in Amylase-Producing *B. subtilis* Using Continuous Culture," *Biotech. Bioeng.* **29** (1987): 859 – 872.

Kim, S. H. and D. D. Y. Ryu, "Instability Kinetics of *trp* Operon Plasmid ColEl- *trp* in Recombinant *E. coll* MV12[pVH5] and MV12 *trpR*]" *Biotech. Bioeng.* **26** (1984): 497 – 502.

Lee, J. M. and G. An, "Industrial application and genetic engineering of plant cell cultures", *Enzyme Microb. Technol.* **8** (1986): 260 – 265.

Maniatis, T., E. F. Fritsch, and J. Sambrook, *Molecular Cloning: A Laboratory Manual* . Cold Spring Harbor, NY: Cold Spring Harbor Laboratory, 1982

Monbouquette, H. G. and D. F. Ollis, "Even live catalysts die," *Chemtech* (September 1986): 542 – 551.

Parker, C. and D. DiBiasio, "Effect of Growth Rate and Expression Level on Plasmid Stability in *Saccharomyces cerevisiae* ," *Biotech. Bioeng.* **29** (1987): 215 – 221.

Schwartz, L. S., N. B. Jansen, N. W. Y. Ho, and G. T. Tsao, "Plasmid Instability Kinetics of the Yeast S288C pUCKmS [cir +] in Non-Selective and Selective Media," *Biotech. Bioeng.* **32** (1988): 733 – 740.

Weinberg, R. A., "The Molecules of Life," *Scientific American* **253** (October 1985): 48 – 57.

Whiteway, M. S. and A, Ahmed, "Recombinational Instability of Chimeric Plasmid in *Saccharomyces cerevisiae* ," *Molecular & Cellular Biology* (January 1984): 195 – 198.

8

Sterilization

Most industrial fermentations are carried out as pure cultures in which only selected strains are allowed to grow. If foreign microorganisms exist in the medium or any parts of the equipment, the production organisms have to compete with the contaminants for the limited nutrients. The foreign microorganisms can produce harmful products which can limit the growth of the production organisms. Therefore, before starting fermentation, the medium and all fermentation equipment have to be free from any living organisms, in other words, they have to be completely sterilized. Furthermore, the aseptic condition has to be maintained.

8.1 STERILIZATION METHODS

Sterilization of fermentation media or equipment can be accomplished by destroying all living organisms by means of heat (moist or dry), chemical agents, radiation (ultraviolet or X-rays), and mechanical means (some or ultrasonic vibrations). Another approach is to remove the living organisms by means of filtration or high-speed centrifugation.

Heat is the most widely used means of sterilization, which can be employed for both liquid medium and heatable solid objects. It can be applied as dry or moist heat (steam). The moist heat is more effective than the dry heat, because the intrinsic heat resistance of vegetative bacterial cells is greatly increased in a completely dry state. As a result the death rate is much lower for the dry cells than for moist ones. The heat conduction in dry air is also less rapid than in steam. Therefore, dry heat is used only for the sterilization of glassware or heatable solid materials. By pressurizing a vessel, the steam temperature can be increased significantly above the boiling point of water. Laboratory autoclaves are commonly operated at a steam pressure of about 30 psia, which corresponds to 121°C. Even bacterial spores are rapidly killed at 121 °C.

Chemical agents can be used to kill microorganisms as the result of their oxidizing or alkylating abilities. However, they cannot be used for the sterilization of medium because the residual chemical can inhibit thefermentation organisms. Chemical agents are frequently employed for *disinfection* that commonly implies the treatment to remove or reduce the risk from pathogenic organisms. Some of the major antimicrobial chemical agents are (Pelczar and Reid, 1972): phenol and phenolic compounds (phenol, cresol, orthophenylphenol), alcohol (ethyl, methyl), halogens (iodine, hypochlorites, chloramines), detergents, dyes, quaternary ammonium compounds, acids, alkalies, and gaseous chemosterilizers (ethylene oxide, β-propiolactone, formaldehyde).

Many cellular materials absorb ultraviolet light, leading to DNA damage and consequently to cell death. Wavelengths around 265 nm have the highest bactericidal efficiency. However, ultraviolet rays have very little ability to penetrate matter. Therefore, their use is limited to the reduction of microbial population in a room where sterility needs to be maintained, such as hospital operating rooms or clean chambers in a laboratory. X-rays are lethal to microorganisms and have penetration ability. However, they are impractical as sterilization tools due to their expense and safety concerns.

Sonic or ultrasonic waves of sufficient intensity can disrupt and kill cells. This technique is usually employed in the disruption of cells for the purpose of extracting intracellular constituents rather than as a sterilization technique.

Filtration is most effectively employed for the removal of microorganisms from air or other gases. In the case of liquid solutions, it is used with thermolabile medium or products, that is, those easily destroyed by heat, such as human and animal serums and enzymes.

Among the techniques discussed, moist heat is the most economical and efficient for the general sterilization requirements of fermentation. Therefore, the following four sections describe cell death kinetics and sterilization operations utilizing moist heat.

8.2 THERMAL DEATH KINETICS

Thermal death of microorganisms at a particular temperature can be described by first-order kinetics:

$$\frac{dn}{dt} = -k_d n \tag{8.1}$$

where k_d is specific death rate, the value of which depends not only on the type of species but also on the physiological form of cells. For example, the value of k_d for bacterial spores at 121°C is of the order of 1 mm^{-1}, whereas those for vegetative cells vary from 10 to about 10^{10} mm^{-1} depending on the particular organism (Aiba et al., 1973).

Integration of Eq. (8.1) yields

$$\ln \frac{n}{n_0} = -\int_0^t k_d dt \tag{8.2}$$

or

$$n = n_0 \exp\left(-\int_0^t k_d dt\right) \tag{8.3}$$

which shows the exponential decay of the cell population. The temperature dependence of the specific death rate k_d can be assumed to follow the Arrhenius equation:

$$k_d = k_{d_0} \exp\left(-\frac{E_d}{RT}\right) \tag{8.4}$$

where E_d is activation energy, which can be obtained from the slope of the $\ln(k_d)$ versus $1/T$ plot. For example, the activation energy of *E. coli* is 127 kcal/gmole and that of *Bacillus stearothermophilus* Fs 7954 is 68.7 (Aiba et al., 1973).

8.3   DESIGN CRITERION

From Eqs. (8.2) and (8.4), the design criterion for sterilization ∇ can be defined as (Deindoerfer and Humphrey, 1959)

$$\nabla = \ln \frac{n_0}{n} = \int_0^t k_d dt = k_{d_0} \int_0^t \exp\left(-\frac{E_d}{RT}\right) dt \tag{8.5}$$

which is also known as the Del factor, a measure of the size of the job to be accomplished. The Del factor increases as the final number of cells decreases. For example, the Del factor to reduce the number of cells in a fermenter from 10^{10} viable organisms to one is

$$\nabla = \ln \frac{10^{10}}{1} = 23 \tag{8.6}$$

The reduction of the number of cells from 10^{10} to one seems to be impressive. However, even if one organism is left alive, the whole fermenter may be contaminated. Therefore, all viable organisms have to be eliminated. The Del factor to reduce the number of cells to zero is infinity, which means that it is theoretically impossible to ensure the total destruction of the viable cells. Therefore, the final number of cells needs to be expressed as the fraction of one, which is equal to the probability of contamination. For example, $n = 0.001$ means that the chance for a contaminant surviving the sterilization is 1 in 1000. The

Del factor to reduce the number of cells in a fermenter from 10^{10} viable organisms to 0.001 is

$$\Delta = \ln \frac{10^{10}}{0.001} = 30 \tag{8.7}$$

Based on the sterilization criterion calculated, we can design the sterilization unit.

8.4 BATCH STERILIZATION

Sterilization of the medium in a fermenter can be carried out in batch mode by direct steam sparging, by electrical heaters, or by circulating constant pressure condensing steam through heating coil. The sterilization cycles are composed of heating, holding, and cooling. Therefore, the total Del factor required should be equal to the sum of the Del factor for heating, holding and cooling as

$$\nabla_{total} = \nabla_{heat} + \nabla_{hold} + \nabla_{cool} \tag{8.8}$$

The values of ∇_{heat} and ∇_{cool} are determined by the methods used for the heating and cooling. The value of ∇_{hold} is determined by the length of the controlled holding period. The design procedure for the estimation of the holding time is as follows:

1. Calculate the total sterilization criterion, ∇_{total}.
2. Measure the temperature versus time profile during the heating, holding, and cooling cycles of sterilization. If experimental measurements are not practical, theoretical equations for heating and cooling can be employed, which are of linear, exponential, or hyperbolic form depending on the mode of heating and cooling. The suggested equations for different heating and cooling processes are as follows (Deindoerfer and Humphrey, 1959):

    a. For batch heating by direct steam sparging into the medium, the hyperbolic form is used:

$$T = T_0 + \frac{Hm_s t}{c(M + m_s t)} \tag{8.9}$$

    b. For batch heating with a constant rate of heat flow such as electrical heating, the linear form is used:

$$T = T_0 + \frac{qTt}{cM} \tag{8.10}$$

    c. For batch heating with a isothermal heat source such as steam circulation through heating coil, the exponential form is used:

$$T = T_H + (T_0 - T_H) \exp\left(-\frac{UAt}{cM}\right) \tag{8.11}$$

   d. For batch cooling using a continuous nonisothermal heat sink such as passing cooling water through cooling coil, the exponential form is used:

$$T = T_{C_0} + (T_0 - T_{C_0}) \exp\left\{-\left[1 - \exp\left(-\frac{UA}{m_c c}\right)\right]\frac{m_c t}{M}\right\} \tag{8.12}$$

3. Plot the values of k_d as a function of time.

4. Integrate the areas under the k_d-versus-time curve for the heating and the cooling periods to estimate ∇_{heat} and ∇_{cool}, respectively. If using theoretical equations, integrate Eq. (8.5) numerically after substituting in the proper temperature profiles. Then, the holding time can be calculated from

$$t_{\text{hold}} = \frac{\nabla_{\text{total}}}{k_d} = \frac{\nabla_{\text{heat}} + \nabla_{\text{hold}} + \nabla_{\text{cool}}}{k_d} \tag{8.13}$$

Example 8.1

A fermenter containing 40 m^3 of medium (25°C) is going to be sterilized by the direct injection of saturated steam. The typical bacterial count of the medium is about 5×10^{12} m^{-3}, which needs to be reduced to such an extent that the chance for a contaminant surviving the sterilization is 1 in 1,000. The steam (345 kPa, absolute pressure) will be injected with a flow rate of 5,000 kg/hr, which will be stopped when the medium temperature reaches 122°C. During the holding time, the heat loss through the vessel is assumed to be negligible. After a proper holding time, the fermenter will be cooled by passing 100 m^3/hr of 20°C water through the cooling coil in the fermenter until the medium reaches 30°C. The coil has a heat-transfer area of 40 m^2 and for this operation the average overall heat-transfer coefficient (U) for cooling is 2,500 kJ/hr m^2 K. The heat-resistant bacterial spores in the medium can be characterized by an Arrhenius coefficient k_{d_0} of 5.7×10^{39} hr^{-1} and an activation energy (E_d) of 2.834×10^5 kJ/kmol (Deindoerfer and Humphrey, 1959). The heat capacity and density of the medium are 4.187 kJ/kg K and 1,000 kg/m^3, respectively. Estimate the required holding time.

Solution:

The design criterion can be calculated from Eq. (8.5) as

$$\nabla = \ln\frac{n_0}{n} = \ln\left[\frac{(5 \times 10^{12}\text{m}^{-3})(40\text{m}^3)}{1 \times 10^{-3}}\right] = 39.8$$

The direct injection of steam into the medium can be assumed to follow the hyperbolic temperature-time profile of Eq. (8.9), which can be used to calculate the time required to heat the medium from 25°C to 122°C. From the steam table (Felder and Rousseau, 1986), the enthalpy of saturated steam at 345 kPa and water at 25°C is 2,731 and 105 kJ/kg, respectively. Therefore, the enthalpy of the saturated steam at 345 kPa relative to raw medium temperature (25°C) is

$$H = 2.731 - 105 = 2,626 \text{ kJ/kg}$$

From Eq. (8.9),

$$T = T_0 + \frac{(2,626 \text{ kJ/kg})(5,000 \text{ kg/hr})t}{(4.187 \text{ kJ/kg K})[(40 \text{ m}^3)(1,000 \text{ kg/m}^3) + (5,000 \text{ kg/hr})t]}$$

$$= T_0 + \frac{78.4t}{1 + 0.125t}$$

The solution of the preceding equation for t when $T = 395°K$ *and* $T_0 = 298°K$ by using a numerical technique or a trial and error approach yields that the time required to reach 122°C is 1.46 hrs.

Substitution of the previous equation into Eq. (8.5),

$$V_{\text{heat}} = 5.7 \times 10^{39} \int_0^{1.46} \exp\left[\frac{-2.834 \times 10^5}{8.318}\left(298 + \frac{78.4t}{1 + 0.125\,t}\right)^{-1}\right] dt$$

Numerical integration of the preceding equation by using Advanced Continuous Simulation Language (ACSL) or other method yields

$$V_{\text{heat}} = 14.8$$

During the cooling process, the change of temperature can be approximated by Eq. (8.12) as

$$T = 293 + 102 \exp(-0.531t)$$

Solving for t when the final temperature is 303°K yields $t = 4.38$ hrs. Substitution of the previous equation into Eq. (8.5) gives

$$V_{\text{cool}} = 5.7 \times 10^{39} \int_0^{4.38} \exp\left[\frac{-2.834 \times 10^5}{8.318\,[293 + 102\exp(-0.531\,t)]}\right] dt = 17.6$$

Therefore, the Del factor for the holding time is

$$V_{\text{hold}} = V_{\text{total}} - V_{\text{heat}} - V_{\text{cool}} = 39.8 - 14.8 - 17.6 = 7.4$$

At 122°C, the thermal death constant (k_d) is 197.6 hr^{-1} from Eq. (8.4). Therefore, the holding time is

$$t_{\text{hold}} = \frac{\nabla_{\text{hold}}}{k_d} = \frac{7.4}{197.6} = 0.037 \text{ hr} = 2.25 \text{ min}$$

For this example, most of the sterilization was accomplished during the heating and cooling periods.

8.5 CONTINUOUS STERILIZATION

Sterilization can be carried out in a continuous mode rather than in batches. Continuous sterilization offers several advantages:

1. It simplifies production planning, thus allowing maximum plant utilization and minimum delays.
2. It provides reproducible conditions.
3. It can be operated at a high temperature (140°C instead of 121°C in batch sterilization); therefore, the sterilization time can be shortened (holding time of 1 to 2 minutes).
4. It requires less steam by recovering heat from the sterilized medium. As a result, it also requires less cooling water.
5. It is easier to automate the process; thus, it is less labor intensive.

A continuous sterilizer consists of three main sections: heating, holding, and cooling.

Heating Section: Methods of heating can be categorized into two types: direct steam injection and indirect heating in shell-and-tube or plate-and-frame heat exchanger. Direct heating is more effective than indirect heating because there is no barrier between the medium and the heat source. The steam injector heats the medium to the peak sterilization temperature quickly. Therefore, sterilization during the heating period is negligible.

For indirect heating, the plate-and-frame heat exchanger is generally more effective than the shell-and-tube type for heat transfer due to its larger heat-transfer area. However, the former is limited to lower pressures (normally less than 20 atm) due to its weak structural strength compared with the latter. The plate-and-frame type is also favorable for the sterilization of a high viscous system.

The temperature change with respect to residence time (τ_{lold}) as the medium passes through an isothermal heat source can be approximated as (Deindoerfer and Humphrey, 1959b),

$$T_{C_2} = T_H - (T_H - T_{C_1})\exp\left(-\frac{UA\bar{\tau}_{\text{heat}}}{cW}\right) \qquad (8.14)$$

For heating using a countercurrent heat source of equal flow rate and heat capacity,

$$T_{C_2} = T_{C_1} - \frac{\Delta TUA\bar{\tau}_{\text{heat}}}{cW} \tag{8.15}$$

Holding Section: The heated medium passes through a holding section, which is usually composed of a long tube. The holding section is maintained in adiabatic conditions. If the heat loss in the section is negligible, the temperature can be assumed to be constant. The average residence time in the holding section is

$$\bar{\tau}_{\text{hold}} = \frac{L}{\bar{u}} \tag{8.16}$$

from which the Del factor can be estimated as

$$\nabla_{\text{hold}} = \ln \frac{n_0}{n} = k_d \bar{\tau}_{\text{hold}} = k_{d_0} \exp\left(-\frac{E_d}{RT}\right)\bar{\tau}_{\text{hold}} \tag{8.17}$$

where n_0 is the number of cells at the beginning of the holding section.

If the medium in a holding section behaves as ideal plug flow, the residence time of the medium in the section will be exactly the same for all the medium. Therefore, the degree of the sterilization will be uniform. However, the slippage due to the viscous nature of fluid, the friction of the pipe wall, and turbulent eddies of the flowing fluid causes the deviation of the ideal plug flow. The resulting velocity profile has its maximum value at the centerline of a pipe, whereas it has its minimum value in the vicinity of the pipe wall. For laminar flow of Newtonian fluids through a smooth round pipe, the ratio of the average fluid velocity to the maximum velocity $\overline{line}\ u/u_m$ is 0.5. The ratio changes rapidly from 0.5 to about 0.75 when laminar flow changes to turbulent, and then increases gradually to 0.87 when the Reynolds number is about 10^6 (McCabe et al., 1985). As a result, if we use the mean velocity in calculating the required residence time for sterilization, some portion of medium will be understerilized, which may cause a serious contamination problem.

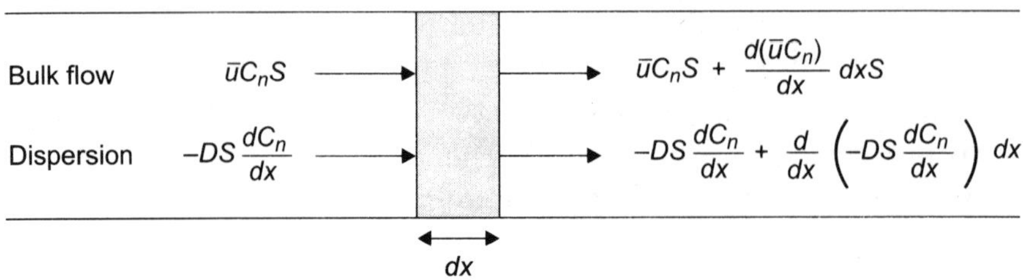

Fig. 8.1 Material balances around an elementary section in a holding tube.

The deviation from ideal plug flow due to the axial mixing can be described by the dispersion model (Levenspiel, 1972). Let's look at the differential element with a thickness dx in a holding tube as shown in Figure 8.1. The basic material balance for the microorganisms suspended in the medium is

$$\text{In} - \text{Out} - \text{Killed by Sterilization} = \text{Accumulation} \qquad (8.18)$$

At steady state, the accumulation term is equal to zero. The input and output of the microorganisms into or out of the element have both a bulk flow and an axial diffusion condition. The number of microorganisms entering minus those leaving by bulk flow is

$$\bar{u}C_nS - \left[\bar{u}C_nS + \frac{d(uC_n)}{dx}Sdx\right] \qquad (8.19)$$

Analogous to the molecular diffusion, the x-directional flux of microorganisms suspended in a medium due to the axial mixing can be represented as

$$J_n = -D\frac{dC_n}{dx} \qquad (8.20)$$

where D is the axial dispersion coefficient, characterized by the degree of backmixing during flow. The mechanism of axial dispersion may be molecular or turbulent. If D is zero, the velocity distribution approaches that of the ideal plug flow. At the other extreme, if D is infinitely large, the fluid in the pipe is well mixed like a fully mixed vessel. For the turbulent flow, the dispersion coefficient is correlated as a function of Reynolds number as shown in Figure 8.2.

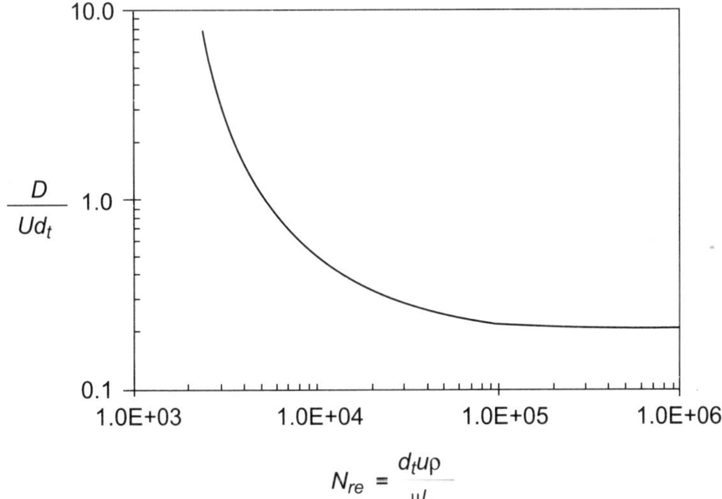

Fig. 8.2 Correlation for D/ud, as a function of Reynolds number. (Levenspiel, 1958).

The number of microorganisms entering and leaving by axial dispersion is

$$-DS\frac{dC_n}{dx} - \left[-DS\frac{dC_n}{dx} + \frac{d}{dx}\left(-DS\frac{dC_n}{dx}\right)dx\right] \tag{8.21}$$

The number of cells killed by sterilization is $k_d C_n S dx$. Therefore, by substituting Eqs. (8.19) and (8.21) into Eq. (8.18) and simplifying,

$$\frac{d}{dx}\left(D\frac{dC_n}{dx}\right) - \frac{d(\bar{u}C_n)}{dx} - k_d C = 0 \tag{8.22}$$

For the constant D and $\bar{u}$, Eq. (8.22) can be modified into a dimensionless form,

$$\frac{d^2 C_n'}{dx'^2} - N_{Pe}\frac{dC_n'}{dx'} - N_{Pe}\frac{k_d L}{\bar{u}}C_n' = 0 \tag{8.23}$$

where

$$C_n' = \frac{C_n}{C_{n_0}} \qquad x' = \frac{x}{L} \qquad N_{Pe} = \frac{\bar{u}L}{D}$$

N_{Pe} is known as Péclet number. When $N_{Pe} = \infty$, it is ideal plug flow. The boundary conditions for the solution of Eq. (8.23) are

$$\frac{dC_n'}{dx'} + N_{Pe}(1 - C_n') = 0 \qquad \text{at } x' = 0 \tag{8.24}$$

$$\frac{dC_n'}{dx'} = 0 \qquad \text{at } x' = l$$

The solution of Eq. (8.23) (Wehner and Wilhelm, 1956) is

$$\left. C_n' \right|_{x'=1} = \frac{4\varphi\exp\left(\dfrac{N_{Pe}}{2}\right)}{(1+\varphi)^2\exp\left(\dfrac{\varphi N_{Pe}}{2}\right) - (1-\varphi)^2\exp\left(-\dfrac{\varphi N_{Pe}}{2}\right)} \tag{8.25}$$

where

$$\varphi = \sqrt{1 + \frac{4k_d L/\bar{u}}{N_{Pe}}}$$

Example 8.2

A continuous sterilizer with a steam injector and a flash cooler will be employed to sterilize medium continuously with the flow rate of $2 \text{ m}^3/\text{hr}$. The time for heating and cooling is negligible with this type of sterilizer. The typical bacterial count of the medium is about

5×10^{12} m^{-3}, which needs to be reduced to such an extent that only one organism can survive during two months of continuous operation. The heat-resistant bacterial spores in the medium can be characterized by an Arrhemus coefficient (k_{d0}) of 5.7×10^{39} hr^{-1} and an activation energy (E_d) of 2.834×10^5 kJ/kmol (Deindoerfer and Humphrey, 1959). The sterilizer will be constructed with the pipe with an inner diameter of 0.102 m. Steam at 600 kPa (gage pressure) is available to bring the sterilizer to an operating temperature of 125°C. The physical properties of this medium at 125°C are $c = 4.187$ kJ/kg K, $\rho = 1000$ kg/m^3, and $\mu = 4$ k/m hr.

a. What length should the pipe be in the sterilizer if you assume ideal plug flow?
b. What length should the pipe be in the sterilizer if the effect of axial dispersion is considered?

Solution:

a. The design criterion can be calculated from Eq. (8.5) as

$$\nabla = \ln \frac{n_0}{n} = \ln \ (5 \times 10^{12} \text{m}^{-3})(2\text{m}^3/\text{hr})(24\text{hr}/\text{day})(60\text{days}) = 37.2$$

Since the temperature at the holding section is constant, Eq. (8.5) is simplified to

$$\nabla = k_d \tau_{\text{hold}}$$

From the given data, k_d can be calculated by using Eq. (8.4) to yield

$$k_d = 378.6 \text{ hr}^{-1}$$

Therefore,

$$\tau_{\text{hold}} = \frac{\nabla}{k_d} = \frac{37.2}{387.6} = 0.0983 \text{ hr}$$

The velocity of medium is

$$\bar{u} = \frac{2 \text{ m}^3/\text{hr}}{\dfrac{\pi}{4} 0.102^2 \text{ m}^2} = 245 \text{ m/hr}$$

The length of the sterilizer is

$$L = u\tau_{\text{hold}} = 24.1\text{m}$$

b. The Reynolds number for the medium flow is

$$N_{\text{Re}} = \frac{0.102\,(245)(1000)}{4} = 6.24 \times 10^3$$

From Figure 8.2 $\qquad \dfrac{D}{\bar{u}dt} \approx 0.8 \qquad$ for $N_{\text{Re}} = 6.24 \times 10^3$

Therefore,

$$D \approx 0.8\,\bar{u}\,dt = 20 \text{ m}^2/\text{h}$$

Now, substituting all the values given and calculated in this problem to Eq. (8.25) will result in an equation with only one unknown, L, which can be solved by using any non-linear equation solver:

$$L = 26.8$$

Therefore, the holding section should be 26.8 m, which is 2.7 m longer than the result from the assumption of ideal plug flow.

Cooling Section: For the cooling section, a quench cooler with adequate heat removal capacity is effective. Another technique is to inject the hot medium through an expansion valve into a vacuum chamber, which is known as flash cooling. Both of these take a very short time; therefore, the sterilization during the cooling period can be assumed to be negligible.

A shell-and-tube or a plate-and-frame heat exchanger can also be employed for cooling. The temperature versus residence time relationship for cooling using an isothermal heat sink is

$$T_{H_2} = T_C - (T_C - T_{H_1})\exp\left(-\frac{UA\bar{\tau}_{\text{cool}}}{cW}\right) \tag{8.27}$$

For cooling using a countercurrent heat sink of equal flow rate and heat capacity,

$$T_{H_2} = T_{H_1} - \frac{\Delta T UA\bar{\tau}_{\text{cool}}}{cW} \tag{8.28}$$

8.6   AIR STERILIZATION

For aerobic fermentations, air needs to be supplied continuously. Typical aeration rates for aerobic fermentation are 0.5 – 1.0 vvm (air volume per liquid volume per minute). This requires an enormous amount of air. Therefore, not only the medium but also the air must be free of microbial contaminants. All of the sterilization techniques discussed for medium can also be employed for air. However, sterilization of air by means of heat is economically impractical and is also ineffective due to the low heat-transfer efficiency of air compared with those of liquids. The most effective technique for air sterilization is filtration using fibrous or membrane filters.

The cotton plug, routinely used as a closure for tubes or flasks of sterile solution, is a good example of removal of microorganisms from air by a fibrous filter. A simple air filter can be made by packing cotton into a column. However, with cotton filters the pressure drop is high and wetting can be a breeding ground for the contamination.

Therefore, glass fibers are favorable as filter medium because they give a lower pressure drop and are less liable to wetting or combustion. Modern fibrous filter systems are cylinders made from bonded borosilicate microfibers sheathed in reinforcing polypropylene mesh in which the layers increase in fineness and density from the center outward. This type of design can deliver over $3 \text{ m}^3/\text{s}$ of sterile air at 0.1 bar of pressure drop (Quesnel, 1987).

With fibrous filters, airborne particles are collected by the mechanisms of impact on, interception, and diffusion.

Impaction: When an air stream containing particles flows around a cylindrical collector, the particle will follow the streamlines until they diverge around the collector. The particles because of their mass will have sufficient momentum to continue to move toward the cylinder and break through the streamlines, as shown in Figure 8.3. The collection efficiency by this inertial impaction mechanism is the function of the Stokes and the Reynolds number as:

$$\eta_{\text{imp}} = f\left(N_{\text{St}},\, N_{\text{Re}}\right) = \left(\frac{C_f \rho_p d_p^2\, v_0}{18 \mu D_c},\; \frac{D_c v_0 \rho}{\mu}\right) \tag{8.29}$$

where C_f is known as Cunningham correction factor. The value of C_f can be estimated from the empirical correlation developed by Davies (Strauss, 1975),

$$C_f = 1 + \frac{2\lambda}{d_p}\left[1.257 + 0.400 \exp\left(-1.10 \frac{d_p}{2\lambda}\right)\right] \tag{8.30}$$

where λ is the mean free path of gas molecules based on the Chapman-Enskog equation,

$$\lambda = \left(\frac{\mu}{0.499 \rho}\right)\sqrt{\frac{\pi M_w}{8RT}} \tag{8.31}$$

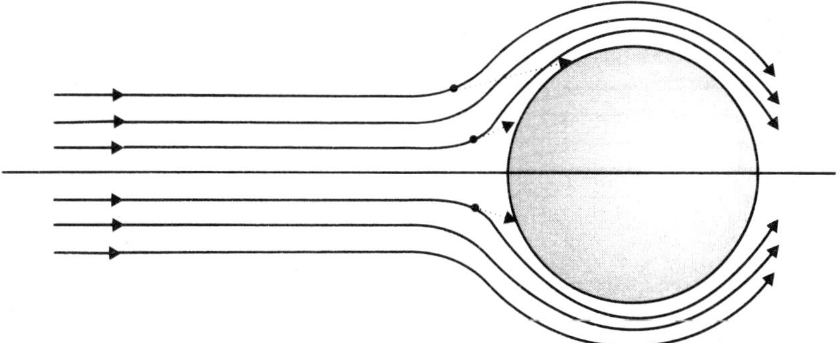

Fig. 8.3 Flow pattern around cylindrical fiber, showing the path of particles collected by inertial impaction.

The efficiency η_{imp} is defined as the fraction of particles approaching the collector which impact. Various correlations are available in the literature. An empirical correlation for the efficiency developed by Thom is (Strauss, 1975):

$$\eta_{\text{imp}} = \frac{N_{\text{St}}^3}{N_{\text{St}}^3 + 0.77 N_{\text{St}}^2 + 0.22} \qquad \text{for } N_{\text{Re}_c} = 10 \qquad (8.32)$$

Another correlation propsed by Friedlander (1967) is

$$\eta_{\text{imp}} = 0.075 \, N_{\text{St}}^{1.2} \qquad (8.33)$$

The efficiency increases with increasing particle diameter or air flow velocity.

Interception: The inertial impaction model assumed particles had mass, and hence inertia, but no size. An interception mechanism is considered where the particle has size, but no mass, and so they can follow the streamlines of the air around the collector. If a streamline which they are following passes close enough to the surface of the fiber, the particles will contact the fiber and be removed (Figure 8.4). The interception efficiency depends on the ratio of the particle diameter to the cylindrical collector diameter ($\kappa = d_p/D_c$):

$$\eta_{\text{int}} = \frac{1}{2.002 - \ln N_{\text{Re}_c}} \left[(1+\kappa) \ln(1+\kappa) - \frac{\kappa(2+\kappa)}{2(1+\kappa)} \right] \qquad (8.34)$$

which was developed by using Langmuir's viscous flow equation (Strauss, 1975). The ratio κ is known as interception parameter. The collection efficiency by interception increases with the increase of the particle size.

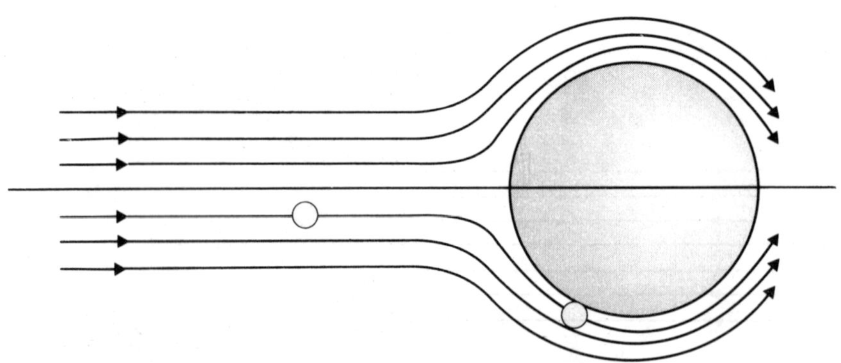

Fig. 8.4 Flow pattern around cylindrical fiber, showing the interception collection mechanism.

Diffusion: Particles smaller than about 1 micron in diameter exhibit a Brownian motion which is sufficiently intense to produce diffusion. If a streamline containing these particles is sufficiently close to the collector, the particles may hit the collector and be removed. Contrary to the previous two mechanisms, the collection efficiency by diffusion increases with decreasing particle size or air velocity. The typical size of particles collected by this mechanism is less than about 0.5 μm. The efficiency of collection by diffusion can be estimated by an equation analogous to Langmuir's equation, Eq. (8.34), as (Strauss, 1975):

$$\eta_{\text{dif}} = \frac{1}{2.002 - \ln N_{\text{Re}_c}} \left[(1 + Z)\ln(1 + Z) - \frac{Z(2 + Z)}{2(1 + Z)} \right] \qquad (8.35)$$

where Z is the diffusion parameter defined as

$$Z = \left[2.24(2.002 - \ln N_{\text{Re}_c}) \frac{D_{Br}}{vD_c} \right]^{\frac{1}{3}} \qquad (8.36)$$

Friedlander (1967) suggested the following correlation

$$\eta_{\text{dif}} = 1.3 N_{Pe}^{-(2/3)} + 0.7\kappa^2 \qquad (8.37)$$

where N_{Pe} is Péclet number, an important dimensionless parameter in the theory of convective diffusion. It is defined as

$$N_{Pe} = \frac{v_c D_c}{D_{Br}} = N_{Re} N_{Sc} \qquad (8.38)$$

where N_{Sc} is the Schmidt number which is defined as

$$N_{Sc} = \frac{\mu}{\rho D_{Br}} \qquad (8.39)$$

The diffusivity due to Brownian motion for submicron size particles can be estimated from

$$D_{Br} = \frac{C_f kT}{3\pi\mu d_p} \qquad (8.40)$$

where k is Boltzmann's constant (1.38054×10^{-9} J/K).

Combined Mechanisms: The total collection efficiency of a fibrous filter is obtained from the combined effect of the preceding three mechanisms. One straightforward way to combine the collection efficiencies of the different mechanisms is to add them together, but this implies that a particle can be collected more than once, which does not make sense. A better approach is to use the following correlation,

$$\eta_c = 1 - (1 - \eta_{\text{imp}})\, 1 - \eta_{\text{int}})\, (1 - \eta_{\text{dif}}) \qquad (8.41)$$

which allows only the particles not collected by one mechanism to be collected by the others. Substitution of Eqs. (8.32), (8.34), and (8.35) into Eq. (8.41) will result in the correlation for the collection efficiency by the combined mechanisms. Pasceri and Friedlander (1960) correlated the combined collection efficiency as

$$\eta_c = \frac{6}{N_{sc}^{2/3} N_{Re_c}^{0.5}} + 3\kappa^2 N_{Re_c}^{0.5} \tag{8.42}$$

As mentioned earlier, with an increase of the superficial air velocity (v_0), η_{imp} and η_{int} increase whereas η_{dif} decreases. Therefore, the combined collection efficiency normally decreases to reach a minimum point and then increases with increasing superficial air velocity.

Effect of Multiple Layers and Packing: All correlations for the collection efficiency discussed so far are based on the ideal case of a single cylindrical collector. Now, let's examine a filter unit consisting of randomly oriented multiple layers. Consider an area (A) of filter at a right angle to the gas flow and with a depth dh. If the packing density α is defined as the volume of fiber per unit volume of filter bed, the velocity within the filter void space is equal to

$$v = \frac{v_0}{(1-\alpha)} \tag{8.43}$$

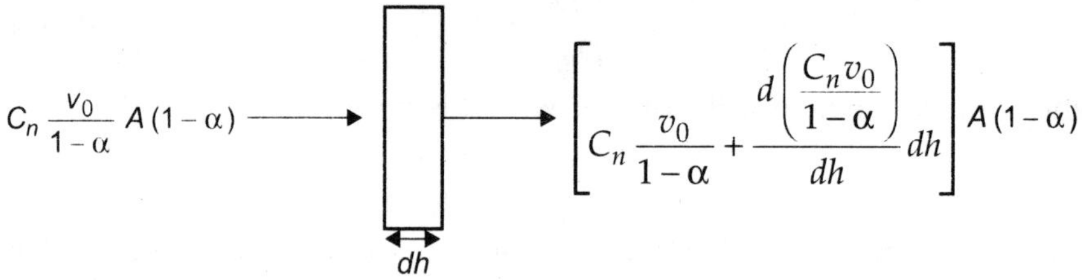

$$C_n \frac{v_0}{1-\alpha} A(1-\alpha) \longrightarrow \qquad \longrightarrow \left[C_n \frac{v_0}{1-\alpha} + \frac{d\left(\dfrac{C_n v_0}{1-\alpha}\right)}{dh} dh \right] A(1-\alpha)$$

$$dh$$

Fig. 8.5 Shell balances around a differential element of a filter.

A mass balance on the particles for the control volume (Figure 8.5) results in

$$\text{Input} - \text{Output} = \text{Collected by the filter} \tag{8.44}$$

$$C_n \frac{v_0}{1-\alpha} A(1-\alpha) - \left[C_n \frac{v_0}{1-\alpha} + \frac{d\left(\dfrac{C_n v_0}{1-\alpha}\right)}{dh} dh \right] A(1-\alpha)$$

$$= \frac{v_0}{1-a} C_n (A dh) \eta_c D_c L \tag{8.45}$$

where L is the length of cylindrical fiber per unit volume of filter bed, which is related to the packing density α and the average collector diameter D_c as

$$\alpha = \frac{\pi D_c^2 L}{4} \tag{8.46}$$

Simplifying Eq. (8.45) and substituting Eq. (8.46) for L gives

$$-\frac{dC_n}{C_n} = \frac{4\alpha \eta_c}{\pi D_c (1 - \alpha)} dh \tag{4.47}$$

which, on integration, results

$$\ln \frac{C_n}{C_{n_0}} = -\frac{4B}{\pi D_c} \left(\frac{\alpha}{1 - \alpha} \right) \eta_c \tag{8.48}$$

where B is the filter depth. Therefore, the collection efficiency for the filter bed can be estimated as

$$\eta_f = 1 - \frac{C_n}{C_{n_0}} = 1 - \exp\left[-\frac{4B}{\pi D_c} \left(\frac{\alpha}{1 - \alpha} \right) \eta_c \right] \tag{8.49}$$

When fibers are packed together in a filter bed, the velocity will be increased and the flow pattern will be changed, which increases the collection efficiency from impaction and interception. Chen (1955) has determined fiber interference effects experimentally and suggests

$$\eta_\alpha = \eta_f (1 + 4.5\alpha) \tag{8.50}$$

which is applicable for $\alpha < 0.1$ and $\eta_f < 1/(1 + 4.5\,\alpha)$

In summary, in order to estimate the collection efficiency of a filter bed, you have to calculate: η_c by using either Eq. (8.41) or Eq. (8.42), η_f using Eq. (8.49), and η_a using Eq. (8.50). However, it should be noted that the predictions of the collection efficiency from various correlations vary widely due to empiricism or the oversimplification in developing the models represent a complex collection mechanism. Furthermore, since the effect of the gas velocity on the collection efficiency is large, the collection efficiency can decrease significantly by increasing or decreasing the gas velocity. To insure sterility in a fermenter system, a conservative approach needs to be taken which considers the minimal efficiency conditions due to possible velocity fluctuations and prediction error from various correlations. In assessing the filtration job to be accomplished, Humphrey (1960) recommended that the design should permit only a one-m-a-thousand chance of a single contaminant penetrating the filter during the entire course of the fermentation.

Example 8.3

A filter bed of glass fibers ($Dc = 15$ μm, the bed depth $B = 10$ cm, and packing density $\alpha = 0.03$) is being used to sterilize air (20°C, 1 atm) with an undisturbed upstream velocity, v_0, of 10 cm/s. The air stream contains 5,000 bacteria per cubic meter ($d_p = 1$ μm and $\rho_p = 1$ g/cm).

  a. Estimate the single fiber collection efficiency by mertial impaction, by interception, and by diffusion.

  b. Estimate the single fiber collection efficiency based on combined mechanisms by using Eq. (8.41) and Eq. (8. 42) and compare the results.

  c. Estimate the collection efficiency (η_α) of the filter bed.

  d. Show how the superficial velocity v_0 affects the various single fiber collection efficiencies.

Solution:

  a. The velocity within the filter void space is from Eq. (8.43)

$$v = \frac{v_0}{(1-\alpha)} = \frac{10}{1-0.03} = 10.3 \text{ cm/s}$$

To estimate η_{imp} by using Eq. (8.32) or Eq. (8.33), we need to calculate the Reynolds and the Stokes number from the given condition and the physical properties of air at 20°C and 1 atm ($\rho = 1.20 \times 10^{-3}$ g/cm^3, $\mu = 1.8 \times 10^{-4}$ g/cm s)

$$N_{Re_c} = \frac{D_c v \rho}{\mu} = \frac{(1.5 \times 10^{-3})(10.3)(1.20 \times 10^{-3})}{1.80 \times 10^{-4}} = 0.103$$

where all units are in the cgs system. The mean free path λ and the Cunningham correction factor can be estimated from Eq. (8.31) and Eq. (8.30) as:

$$\lambda = \left[\frac{1.80 \times 10^{-4}}{0.499(1.20 \times 10^{-3})}\right]\sqrt{\frac{\pi(29)}{8(8.314 \times 10^7)(293)}} = 6.50 \times 10^{-6} \text{ cm}$$

$$C_f = 1 + \frac{2\lambda}{d_p}\left[1.257 + 0.400\exp\left(-1.10\frac{d_p}{2\lambda}\right)\right] = 1.16$$

The Stokes number is

$$N_{St} = \frac{C_f \rho_p d_p^2 v}{18\mu D_c} = \frac{1.16(1)(1 \times 10^{-4})^2(10.3)}{18(1.80 \times 10^{-4})(1.5 \times 10^{-3})} = 0.0247$$

Therefore, from Eq. (8.33), the single fiber collection efficiency by inertial impaction is

$$\eta_{imp} = 0.075 N_{St}^{1.2} = 0.075(0.0247)^{1.2} = 8.82 \times 10^{-4}$$

The single fiber collection efficiency by interception is from Eq. (8.34) for $\kappa = d_p/D_c = 0.0667$ and $N_{Re_c} = 0.103$,

$$h_{int} = \frac{1}{2.002 - \ln N_{Re_c}} \left[(1 + \kappa)\ln(1 + \kappa) - \frac{\kappa(2 + \kappa)}{2(1 + \kappa)} \right] = 0.96 \times 10^{-4}$$

The diffusivity due to the Brownian motion can be estimated from Eq. (8.40),

$$D_{Br} = \frac{C_f kT}{3\pi\mu d_p} = \frac{1.16(1.38 \times 10^{-6})293}{3\pi(1.80 \times 10^{-4})(1 \times 10^{-4})}$$

$$= 2.\frac{\delta y}{\delta x} \frac{\partial^2 \Omega}{\partial u^2} 77 \times 10^{-5}\, cm^2/s$$

Therefore, the Péclet number is

$$N_{Pe} = \frac{D_c v}{D_{Br}} = \frac{1.5 \times 10^{-3}(10.3)}{2.77 \times 10^{-5}} = 5.57 \times 10^{4}$$

The single fiber efficiency by diffusion can be estimated from Eq. (8.37),

$$\eta_{dif} = 1.3 N_{Pe}^{-(2/3)} + 0.7\kappa^2 = 4.00 \times 10^{-3}$$

Therefore, the single fiber collection efficiency by inertial impaction, interception, and diffusion is 8.82×10^{-4}, 9.96×10^{-4}, and 4.00×10^{-3}, respectively.

b. The combined collection efficiency can be estimated from Eq. (8.41) as:

$$\eta_c = 1 - (1 - 8.82 \times 10^{-4})(1 - 9.96 \times 10^{-4})(1 - 4.00 \times 10^{-3}) = 0.0059$$

Instead, if we use Eq. (8.42),

$$N_{Sc} = \frac{\mu}{\rho D_{Br}} = \frac{1.80 \times 10^{-4}}{(1.20 \times 10^{-3})(2.77 \times 10^{-7})} = 5.41 \times 10^{5}$$

Therefore,

$$\eta_c = \frac{6}{(5.41 \times 10^{5})^{2/3}(0.103)^{0.5}} + 3(0.0667)^2(0.103)^{0.5} = 0.0071$$

In this example, diffusion is predominant over impaction and interception. If we use Eq. (8.32) and Eq. (8.35) for the estimation of η_{imp} and η_{dif}, respectively, $\eta_{inp} = 6.82 \times 10^{-3}$ and $\eta_{dif} = 6.97 \times 10^{-4}$, which are significantly lower than those predicted from Eq. (8.33) and Eq. (8.37). Strauss (1975) compared the predicted values from various correlations and found that the diffusion collection efficiency η_{dif} from Eq. (8.35) tends to predict the lower value than the experimental values.

c. If we choose to use $\eta_c = 0.0059$ which was predicted from Eqs. (8. 33), (8.34), (8.37), and (8.41), the collection efficiency for the filter bed can be estimated from Eq. (8.49) as

$$\eta_f = 1 - \exp\left[-\frac{4(10)}{\pi(1.5 \times 10^{-3})}\left(\frac{0.03}{1-0.03}\right)0.0059 \right] = 0.79$$

and the collection efficiency including the interference effects is from Eq. (8.50),

$$\eta_\alpha = 0.79 \left[1 + 4.5 \, (0.03)\right]$$

d. Similarly, we can calculate various collection efficiencies at various values of v_0. As shown in Table 8.1, with an increase of v_0, η_{imp} and η_{int} increased whereas η_{dif} decreased. Therefore, the combined collection efficiency η_c decreased initially, reached a minimum value, and then increased with increasing v_0. The minimum collection efficiency was reached when v_0 was about 5 cm/s.

Table 8.1 Various Collection Efficiencies for 1 μm Particle

v_o cm/s	$\eta_{irnp} \times 10^3$ Eq. (8.33)	$\eta_{int} \times 10^3$ Eq. (8.34)	$\eta_{dif} \times 10^3$ Eq. (8.37)	$\eta_c \times 10^3$ Eq. (8.41)	$\eta_c \times 10^3$ Eq. (8.42)
0.1	0.004	0.479	22.3	22.8	28.6
1	0.056	0.647	7.25	7.941	0.3
5	0.0384	0.857	4.53	5.76	7.01
10	0.883	0.996	4.00	5.87	7.10
15	1.44	1.10	3.79	6.32	7.54
20	2.03	1.19	3.67	6.88	8.05
50	6.09	1.60	3.42	11.1	10.8

8.7 NOMENCLATURE

A surface area across which heat transfer occurs during sterilization, m^2

B filter bed depth, m

C_f Cunningham correction factor, dimensionless

C_n cell number density, number of cells/m^3

c specific heat of medium, J/kg K

D_c collector diameter, m

d_p particle diameter, m

d_t pipe diameter, m

D axial dispersion coefficient, m^2/s

D_{Br} diffusivity due to the Brownian motion, m^2/s

E_d activation energy for thermal cell destruction in Arrhenius equation, J/kmol

H enthalpy of steam relative to raw medium temperature, J/kg

J_n flux of microorganisms due to axial dispersion, m^{-2}s^{-1}

k Boltzmann's constant, 1.3805×10^{-23} J/K or 1.3805×1^{-16} erg/K

k_d specific death rate, s^{-1} or kg/m^{3}s

L length of holding section, m

M initial mass of medium in batch sterilizer, kg

M_w molecular weight of gas molecules, kg/kmol

m_s steam mass flow rate, kg/s

m_c coolant mass flow rate, kg/s

N_{pe} Péclet number (/line $\bar{u}\,L/D$ or $v_0 D_c/D_{Br}$), dimensionless

N_{Re} Reynolds number ($d_t u \rho/\mu_L$), dimensionless

N_{Rec} collector Reynolds number ($D_c v_0 \rho/\mu$), dimensionless

N_{Sc} Schmidt number ($\mu/\rho D_{Br}$), dimensionless

N_{St} Stokes number ($C_f \rho_p d_p^2 v_0/18\,\mu Dc$), dimensionless

n number of cells in a system

q rate of heat transfer, J/s

R gas constant, 8.314×10^3 J/kmol K or 8.314×10^7 erg/mol K

S cross-sectional area of a pipe, m^{-2}

T absolute temperature, °K

T_0 initial absolute temperature of medium, °K

T_C absolute temperature of heat sink, °K

T_{C_0} initial absolute temperature of heat sink, °K

T_H absolute temperature of heat source, °K

t time, s

t_d doubling time, s

U overall heat-transfer coefficient, J/s m^2 K.

u velocity, m/s

$\bar{u}$ velocity, m/s

u_m maximum velocity, m/s

v fluid velocity within the filter void space, m/s

v_0 undisturbed upstream fluid velocity, m/s

W mass of medium in a sterilizer, kg

x x-directional distance, m

Z the diffusion parameter defined in Eq. (8.36), dimensionless

α packing density defined as the volume of fiber per unit volume of filter bed, dimensionless

η collection efficiency, dimensionless

κ the ratio of the particle and the collector diameter (d_p/D_c), dimensionless

λ the mean free path of gas molecules, m

μ fluid viscosity, kg/m s

μ_L liquid viscosity, kg/m s

ρ density, kg/m^3

ρ_p density of particles, kg/m^3

τ residence time, s

$\bar{\tau}$ average residence time, s

∇ design criterion for sterilization, dimensionless

8.8   PROBLEMS

8.1 A fermenter containing 10 m^3 of medium (25°C) is going to be sterilized by passing saturated steam (500 kPa, gage pressure) through the coil in the fermenter. The typical bacterial count of the medium is about 3×10^{12} m^{-3}, which needs to be reduced to such an extent that the chance for a contaminant surviving the sterilization is 1 in 100. The fermenter will be heated until the medium reaches 115°C. During the holding time, the heat loss through the vessel is assumed to be negligible. After the proper holding time, the fermenter will be cooled by passing 20 m^3/hr of 25°C water through the coil in the fermenter until the medium reaches 40°C. The coils have a heat-transfer area of 40 m^2 and for this operation the average overall heat-transfer coefficient (U) for heating and cooling are 5,500 and 2,500 kJ/hr m^{2}K, respectively. The heat resistant bacterial spores in the medium can be characterized by an Arrhenius coefficient (k_{d_0}) of 5.7×10^{39} hr^{-1} and an activation energy (E_d) of 2.834×10^5 kJ/kmol (Demdoerfer and Humphrey, 1959). The heat capacity and density of the medium are 4.187 kJ/kgK and 1,000 kg/m^3, respectively. Estimate the required holding time.

8.2 A continuous sterilizer with a steam injector and a flash cooler will be employed to sterilize medium continuously. The time for heating and cooling is negligible with this type of sterilizer. The typical bacterial count of the medium is about 5×10^{12} m^{-3}, which needs to be reduced to such an extent that only one organism can survive during the three months of continuous operation. The heat resistant bacterial spores in the medium can be characterized by an Arrhenius coefficient (k_{d_0}) of 5.7×10^{39} hr^{-1} and an activation energy (E_d) of 2.834×10^5 kJ/kmol (Deindoerfer and Humphrey, 1959). The holding section of the sterilizer will be constructed with 20 m-long pipe with an inner diameter of 0.078 m. Steam at 600 kPa (gage pressure) is available to bring the sterilizer to an operating temperature of 125°C. The physical properties of this medium at 125°C are c = 4.187 kJ/kgK and ρ =1,000 kg/m^3, and μ = 4 kg/m hr.

   a. How much medium can be sterilized per hour if you assume ideal plug flow?

   b. How much medium can be sterilized per hour if the effect of axial dispersion is considered?

8.3 You need to design a filter for a 10,000-gallon fermenter that will be aerated at a rate of 535 ft^3/min (at 20°C and 1 atm). The bacterial count in the air is 80 per ft^3. Average size of the bacteria is 1 μm with density of 1.08 g/cm^3. You are going to use glass fibers (D_c = 15 μm) with packing density α = 0.03. The cross-sectional area of the filter will be designed to give a superficial air velocity v_0 of 5 ft/s.

   a. What depth of the filter would you recommend to prevent contamination?

   b. How is the answer in (a) changed if v_0 is decreased to 1 ft/s? Explain the results.

8.4 Based on the combined mechanisms of impaction, interception, and diffusion, a minimum efficiency should result for spheres depositing on cylindrical collectors.

   a. Develop an expression for the particle size corresponding to this minimum efficiency. As an approximation, ignore the effect of particle size on the Cunningham correction for slip.

   b. What is the particle size corresponding to the minimum efficiency for the filter bed of glass fibers described in Example 8.3?

8.9 REFERENCES

Aiba, S., A. E. Humphrey and N. F. Millis, *Biochemical Engineering* (2nd ed.), pp. 242 – 246. Tokyo, Japan: University of Tokyo Press, 1973.

Chen, C. Y., "Filtration of Aerosols by Fibrous Media," *Chem. Rev.* **55** (1955): 595 – 623.

Deindoerfer, F. H. and A. E. Humphrey, "Analytical Method for Calculating Heat Sterilization Time," *Appl. Micro.* **7** (1959a): 256 – 264.

Deindoerfer, F. H. and A. E. Humphrey, "Principles in the Design of Continuous Sterilizers," *Appl. Micro.* **7** (1959b): 264 – 270.

Felder, R. M. and R. W. Rousseau, *Elementary Principles of Chemical Processes* (2nd ed.) pp. 630 – 635. New York, NY: John Wiley & Sons, 1986.

Friedlander, S. K., "Aerosol Filtration by Fibrous Filters," in *Biochemical and Biological Engineering Science,* vol 1., ed. N. Blakebrough. London, England: Academic Press, Inc., 1967, pp. 49 – 67.

Humphrey, A. E., "Air Sterilization," *Adv. Appi Micro.* **2** (1960): 301 – 311.

Levenspiel, O., "Longitudinal Mixing of Fluids Flowing in Circular Pipes,"*Ind. Eng. Chem.* **50** (1958): 343 – 346.

Levenspiel, O., *Chemical Reaction Engineering* (2nd ed.), p. 272. New York, NY: John Wiley & Sons, 1972.

McCabe, W. L., J. C. Smith, and P. Harriott, *Unit Operations of Chemical Engineering* (4th ed.), pp. 76 – 90. New York, NY: McGraw-Hill Book Co., 1985.

Pasceri, R. E. and S. K. Friedlander, "The Efficiency of Fibrous Aerosol Filters," *Can. J. Chem. Eng.,* **38** (1960): 212 – 213.

Pelczar, M. J. and R. D. Reid, *Microbiology,* pp. 441 – 461. New York, NY: McGraw-Hill Book Co., 1972.

Quesnel, L. B., "Sterilization and Sterility," in *Basic Biotechnology,* eds. J. Bu'lock and B. Kristiansen. London, England: Academic Press, 1987, pp. 197 – 215.

Strauss, W., *Industrial Gas Cleaning* (2nd ed.), pp. 182, 278 – 297. Oxford, England: Pergamon Press Ltd., 1975.

Wehner, J. F. and R, H. Wilhelm, "Boundary Conditions of Flow Reactor," *Chem. Eng. Sci.* **6** (1956): 89 – 93.

9

Agitation and Aeration

9.1 INTRODUCTION

One of the most important factors to consider in designing a fermenter is the provision for adequate mixing of its contents. The main objectives of mixing in fermentation are to disperse the air bubbles, to suspend the microorganisms (or animal and plant tissues), and to enhance heat and mass transfer in the medium.

Since most nutrients are highly soluble in water, very little mixing is required during fermentation just to mix the medium as microorganisms consume nutrients. However, dissolved oxygen in the medium is an exception because its solubility in a fermentation medium is very low, while its demand for the growth of aerobic microorganisms is high.

For example, when the oxygen is provided from air, the typical maximum concentration of oxygen in aqueous solution is on the order of 6 to 8 mg/L. Oxygen requirement of cells is, although it can vary widely depending on microorganisms, on the order of 1 g/L h. Even though a fermentation medium is fully saturated with oxygen, the dissolved oxygen will be consumed in less than one minute by organisms if not provided continuously. Adequate oxygen supply to cells is often critical in aerobic fermentation. Even temporary depletion of oxygen can damage cells irreversibly. Therefore, gaseous oxygen must be supplied continuously to meet the requirements for high oxygen needs of microorganisms, and the oxygen transfer can be a major limiting step for cell growth and metabolism.

Mixing provided by a laboratory shaker apparatus is adequate to cultivate microorganisms in flasks or test tubes. Rotary or reciprocating action of a shaker is effective to provide gentle mixing and surface aeration. For bench-, pilot-, and production-scale fermenters, the mixing is usually provided by mechanical agitation with or without aeration. The most widely used arrangement is the radial-flow impeller with six flat blades mounted on a disk (Figure 9.1), which is called flat-blade disk turbine or Rushton turbine.

Radial-flow impellers (paddles and turbines) produce flow radially from the turbine blades toward the side of the vessel, where the flow splits into two directions: one part goes upward along the side, back to the center along the liquid surface, and down to the impeller region along the agitating shaft; and the other goes downward along the side and bottom, then back to the impeller region. On the other hand, the axial flow impellers (propellers and pitched blade paddles) generate flow downward to the tank bottom, then up the side and back down the center to the impeller region. Therefore, the flat-blade disk turbine has the advantage of limiting the short-circuiting of gas along the drive shaft by forcing the gas, introduced from below, along the path into the discharge jet.

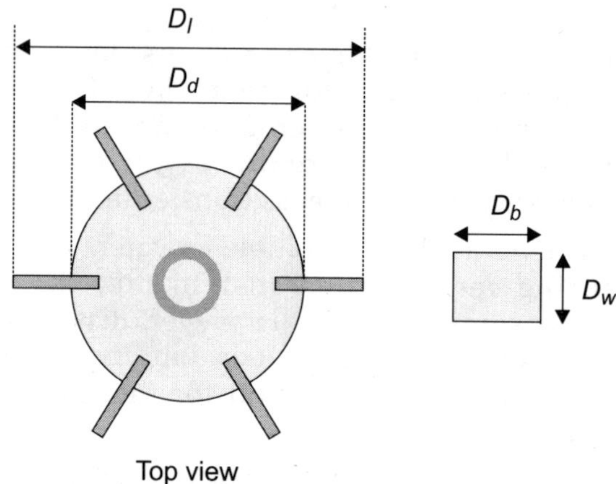

Fig. 9.1 Typical dimensions of flat-blade disk turbine
$d_l : D_b : D_w : D_d = 1 : 0.25 : 0.2 : 0.67$.

Mass-Transfer Path: The path of gaseous substrate from a gas bubble to an organelle in a microorganism can be divided into several steps (Figure 9.2) as follows:

1. Transfer from bulk gas in a bubble to a relatively unmixed gas layer

2. Diffusion through the relatively unmixed gas layer

3. Diffusion through the relatively unmixed liquid layer surrounding the bubble

4. Transfer from the relatively unmixed liquid layer to the bulk liquid

5. Transfer from the bulk liquid to the relatively unmixed liquid layer surrounding a microorganism

6. Diffusion through the relatively unmixed liquid layer

7. Diffusion from the surface of a microorganism to an organelle in which oxygen is consumed

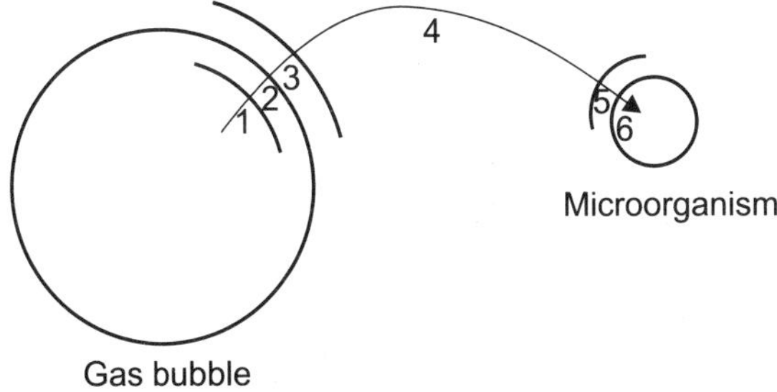

Fig. 9.2 Schematic diagram of the path of a gaseous substrate to an organelle in a cell.

Steps 3 and 5, the diffusion through the relatively unmixed liquid layers of the bubble and the microorganism, are the slowest among those outlined previously and, as a result, control the overall mass-transfer rate. Agitation and aeration enhance the rate of mass transfer in these steps and increase the mterfacial area of both gas and liquid.

In this chapter, we study various correlations for gas-liquid mass transfer, interfacial area, bubble size, gas hold-up, agitation power consumption, and volumetric mass-transfer coefficient, which are vital tools for the design and operation of fermenter systems. Criteria for the scale-up and shear sensitive mixing are also presented. First of all, let's review basic mass-transfer concepts important in understanding gas-liquid mass transfer in a fermentation system.

9.2 BASIC MASS-TRANSFER CONCEPTS

9.2. 1 Molecular Diffusion in Liquids

When the concentration of a component varies from one point to another, the component has a tendency to flow in the direction that will reduce the local differences in concentration.

Molar flux of a component A relative to the average molal velocity of all constituent J_A is proportional to the concentration gradient dC_A/dz as

$$J_A = -D_{AB}\frac{dC_A}{dz} \tag{9.1}$$

which is Fick's first law written for the z-direction. The D_{AB} in Eq. (9.1) is the diffusivity of component A through B, which is a measure of its diffusive mobility.

Molar flux relative to stationary coordinate N_A is equal to

$$N_A = \frac{C_A}{C}(N_A + N_B) - D_{AB}\frac{dC_A}{dz} \tag{9.2}$$

where C is total concentration of components A and B and N_B is the molar flux of B relative to stationary coordinate. The first term of the right hand side of Eq. (9.2) is the flux due to bulk flow, and the second term is due to the diffusion. For dilute solution of A,

$$N_A \approx J_A \qquad (9.3)$$

Diffusivity: The kinetic theory of liquids is much less advanced than that of gases. Therefore, the correlation for diffusivities in liquids is not as reliable as that for gases. Among several correlations reported, the Wilke-Chang correlation (Wilke and Chang, 1955) is the most widely used for dilute solutions of nonelectrolytes,

$$D^o_{AB} = \frac{1.117 \times 10^{-16} (\xi M_B)^{0.5} T}{\mu V_{bA}^{0.6}} \qquad (9.4)$$

When the solvent is water, Skelland (1974) recommends the use of the correlation developed by Othmer and Thakar (1953).

$$D^o_{AB} = \frac{1.112 \times 10^{-13}}{\mu^{1.1} V_{bA}^{0.6}} \qquad (9.5)$$

The preceding two correlations are not dimensionally consistent; therefore, the equations are for use with the units of each term as SI unit as follows:

D^o_{AB} diffusivity of A in B, in a very dilute solution, m^2/s

M_B molecular weight of component B, kg/kmol

T temperature, °K

μ solution viscosity, $kg/m^3 s$

V_{bA} solute molecular volume at normal boiling point, $m^3/kmol$ 0.0256 $m^3/kmol$ for oxygen [See Perry and Chilton (p.3 – 233, 1973) for extensive table]

ξ association factor for the solvent: 2.26 for water, 1.9 for methanol, 1.5 for ethanol, 1.0 for unassociated solvents, such as benzene and ethyl ether.

Example 9.1

Estimate the diffusivity for oxygen in water at 25°C. Compare the predictions from the Wilke-Chang and Othmer-Thakar correlations with the experimental value of 2.5×10^{-9} m^2/s (Perry and Chilton, p. 3 – 225, 1973). Convert the experimental value to that corresponding to a temperature of 40°C.

Solution:

Oxygen is designated as component A, and water, component B. The molecular volume of oxygen V_{BA} is 0.0256 $m^3/kmol$. The association factor for water ξ is 2.26. The viscosity of water at 25°C is 8.904×10^{-4} kg/m s (CRC Handbook of Chemistry and Physics , p. F-38, 1983). In Eq. (9.4)

$$D^o_{AB} = \frac{1.173 \times 10^{-16}\,[2.26(18)]^{0.5}\,209}{(8.904 \times 10^{-4})^{1.1}(0.0256)^{0.6}} = 2.25 \times 10^{-9}\,m^2/s$$

InEq. (9.5)

$$D^o_{AB} = \frac{1.112 \times 10^{-13}}{(8.904 \times 10^{-4})^{1.1}(0.0256)^{0.6}} = 2.27 \times 10^{-9}\,m^2/s$$

If we define the error between these predictions and the experimental value as

$$\% \text{ error} = \frac{(D^o_{AB})_{\text{predicted}} - (D^o_{AB})_{\text{experimental}}}{(D^o_{AB})_{\text{experimental}}} \times 100$$

The resulting errors are –9.6 percent and -9.2 percent for Eqs. (9.4) and (9.5), respectively. Since the estimated possible error for the experimental value is ± 20 percent (Perry and Chilton, p. 3 – 225, 1973), the estimated values from both equations are satisfactory.

Eq. (9.4) suggests that the quantity $D^o_{AB}\,\mu/T$ is constant for a given liquid system. Though this is an approximation, we may use it here to estimate the diffusivity at 40°C. Since the viscosity of water at 40°C is 6.529×10^{-4} kg/m s from the handbook,

$$D^o_{AB} \text{ at } 40°C = 2.5 \times 10^{-9}\left(\frac{8.904 \times 10^{-4}}{6.529 \times 10^{-4}}\right)\left(\frac{313}{298}\right) = 3.58 \times 10^{-9}\,m^2/s$$

If we use Eq. (9.5), $D^o_{4B}\,\mu^{11}$ is constant,

$$D^o_{AB} \text{ at } 40°C = 2.5 \times 10^{-9}\left(\frac{8.904 \times 10^{-4}}{6.529 \times 10^{-4}}\right)^{1.1} = 3.52 \times 10^{-9}\,m^2/s$$

9.2.2 Mass-Transfer Coefficient

The mass flux, the rate of mass transfer q_G per unit area, is proportional to a concentration difference. If a solute transfers from the gas to the liquid phase, its mass flux from the gas phase to the interface N_G is

$$N_G = \frac{q_G}{A} = k_G(C_G - C_{G_1}) \tag{9.6}$$

where C_G and C_{Gi}, is the gas-side concentration at the bulk and the interface, respectively, as shown in Figure 9.3. K_G is the individual mass-transfer coefficient for the gas phase and A is the mterfacial area.

Similarly, the liquid-side phase mass flux N_L is

$$N_L = \frac{q_L}{A} = k_L(C_{L_1} - C_L) \tag{9.7}$$

where k_L is the individual mass-transfer coefficient for the liquid phase.

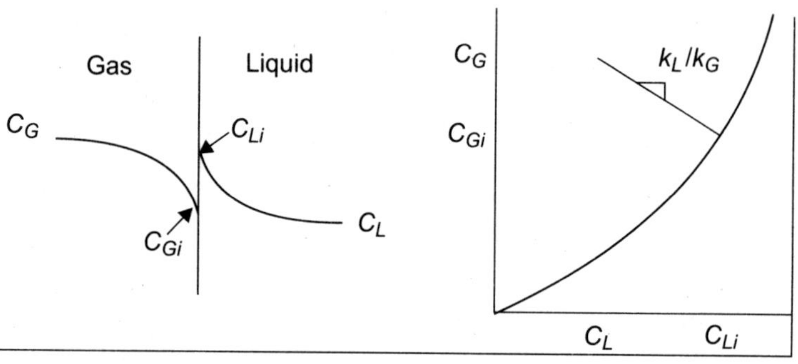

Fig, 9.3   Concentration profile near a gas-liquid interface and an equilibrium curve.

Since the amount of solute transferred from the gas phase to the interface must equal that from the interface to the liquid phase,

$$N_G = N_L \tag{9.8}$$

Substitution of Eq. (9.6) and Eq. (9.7) into Eq. (9.8) gives

$$\frac{C_G - C_{G_i}}{C_L - C_{L_i}} = \frac{k_L}{k_G} \tag{9.9}$$

which is equal to the slope of the curve connecting the (C_L, C_G) and (C_{L_1}, C_{G_1}), as shown in Figure 9.3.

It is hard to determine the mass-transfer coefficient according to Eq. (9.6) or Eq. (9.7) because we cannot measure the interfacial concentrations, C_{L_1}, or C_{G_1}. Therefore, it is convenient to define the overall mass-transfer coefficient as follows:

$$N_G = N_L = K_G (C_H - C_G^*) = K_L (C_L^* - C_L) \tag{9.10}$$

where C_G^* is the gas-side concentration which would be in equilibrium with the existing liquid phase concentration. Similarly, C_L^* is the liquid-side concentration which would be in equilibrium with the existing gas-phase concentration. These can be easily read from the equilibrium curve as shown in Figure 9.4. The newly defined K_G and K_L are overall mass-transfer coefficients for the gas and liquid sides, respectively.

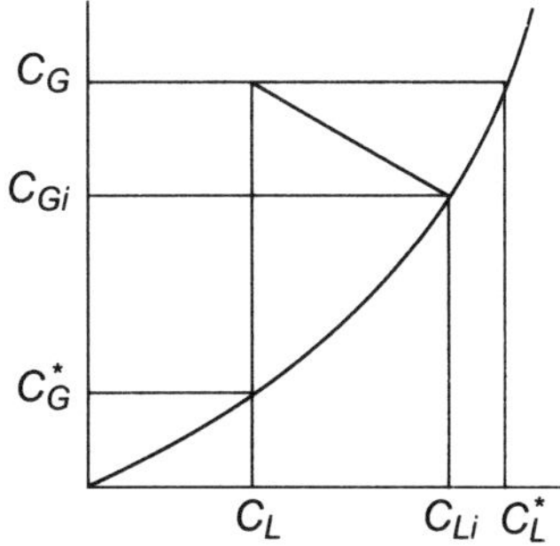

Fig. 9.4   The equilibrium curve explaining the meaning of C_G^* and C_L^*

Example 9.2

Derive the relationship between the overall mass-transfer coefficient for liquid phase K_L and the individual mass-transfer coefficients, k_L and k_G. How can this relationship be simplified for sparingly soluble gases?

Solution:

According to Eqs. (9.7) and (9.10),

$$k_L(C_{Li} - C_L) = K_L(C_L^* - C_L) \tag{9.11}$$

Therefore, by rearranging Eq. (9.11)

$$\frac{1}{k_L} = \frac{1}{k_L}\frac{C_L^* - C_L}{C_{L_i} - C_L}$$

$$= \frac{1}{k_L}\frac{(C_{L_i} - C_L) + (C_L^* - C_{L_i})}{C_{L_i} - C_L} \tag{9.12}$$

$$= \frac{1}{k_L} + \frac{1}{k_L}\frac{(C_L^* - C_{L_i})}{C_{L_i} - C_L}$$

Since

$$k_L(C_{L_i} - C_L) = k_G(C_G - C_{G_i}) \tag{9.13}$$

By substituting Eq. (9.13) to Eq. (9.12), we obtain

$$\frac{1}{K_L} = \frac{1}{k_L} + \frac{1}{k_G}\frac{C_L^* - C_{L_i}}{C_G - C_{G_i}} = \frac{1}{k_L} + \frac{1}{k_G M} \tag{9.14}$$

which is the relationship between K_L, k_L, and k_G. M is the slope of the line connecting (C_{L_1}, C_{G_1}) and (C_L^*, C_G) as shown in Figure 9.4.

For sparingly soluble gases, the slope of the equilibrium curve is very steep; therefore, M is much greater than 1 and from Eq. (9.14)

$$K_L \approx k_L \qquad (9.15)$$

Similarly, for the gas-phase mass-transfer coefficient,

$$K_G \approx k_G \qquad (9.16)$$

9.2.3 Mechanism of Mass Transfer

Several different mechanisms have been proposed to provide a basis for a theory of interphase mass transfer. The three best known are the two-film theory, the penetration theory, and the surface renewal theory.

The *two-film theory* supposes that the entire resistance to transfer is contained in two fictitious films on either side of the interface, in which transfer occurs by molecular diffusion. This model leads to the conclusion that the mass-transfer coefficient k_L is proportional to the diffusivity D_{AB} and inversely proportional to the film thickness z_f as

$$k_L = \frac{D_{AB}}{z_f} \qquad (9.17)$$

Penetration theory (Higbie, 1935)assumes that turbulent eddies travel from the bulk of the phase to the interface where they remain for a constant exposure time t_e. The solute is assumed to penetrate into a given eddy during its stay at the interface by a process of unsteady-state molecular diffusion. This model predicts that the mass-transfer coefficient is directly proportional to the square root of molecular diffusivity

$$k_L = 2\left(\frac{D_{AB}}{\pi t_e}\right)^{1/2} \qquad (9.18)$$

Surface renewal theory (Danckwerts, 1951) proposes that there is an infinite range of ages for elements of the surface and the surface age distribution function $\phi(t)$ can be expressed as

$$\phi(t) = se^{st} \qquad (9.19)$$

where s is the fractional rate of surface renewal. This theory predicts that again the mass-transfer coefficient is proportional to the square root of the molecular diffusivity

$$k_L = (sD_{AB})^{1/2} \qquad (9.20)$$

All these theories require knowledge of one unknown parameter, the effective film thickness z_f the exposure time 4, or the fractional rate of surface renewal s. Little is known about these properties, so as theories, all three are incomplete. However, these theories help us to visualize the mechanism of mass transfer at the interface and also to know the exponential dependency of molecular diffusivity on the mass-transfer coefficient.

9.3   CORRELATION FOR MASS-TRANSFER COEFFICIENT

Mass-transfer coefficient is a function of physical properties and vessel geometry. Because of the complexity of hydrodynamics in multiphase mixing, it is difficult, if not impossible, to derive a useful correlation based on a purely theoretical basis. It is common to obtain an empirical correlation for the mass-transfer coefficient by fitting experimental data. The correlations are usually expressed by dimensionless groups since they are dimensionally consistent and also useful for scale-up processes. The dimensionless group important for correlations can be derived by using Buckingham Pi theory as shown in the following example.

Example 9.3

The mass transfer coefficient k_L of oxygen transfer in fermenters is a function of Sauter mean diameter D_{32}, diffusivity D_{AB}, and density ρ_c viscosity μ_c of continuous phase (liquid phase). Sauter-mean diameter D_{32} can be calculated from measured drop-size distribution from the following relationship,

$$D_{32} = \frac{\displaystyle\sum_{i=1}^{n} n_i D_i^3}{\displaystyle\sum_{i=1}^{n} n_i D_i^2} \tag{9.21}$$

Determine appropriate dimensionless parameters that can relate the mass transfer coefficient by applying the Buckingham-Pi theorem.

Solution:

The first step of Buckingham-Pi theorem is to count the total number of parameters. In this case, there are five parameters: k_L, D_{32}, D_{AB}, ρ_c, and μ_c, all of which can be expressed with three principle units: mass M, length L, and time T. Therefore,

    Number of parameter: $n = 5$
    Number of principle dimension: $r = 3$

In developing the dimensionless groups, every dimensionless group will contain $r = 3$ of repeating parameters and the total number of dimensioness group will be $n - m$ as

    Number of repeating parameter: $m = r = 3$
    Number of dimensionless group: $n - m = 5 - 3 = 2$

Therefore, we need to choose three repeating parameters. Since we want to develop a correlation for k_L, we want it to be only one dimensionless group and therefore, cannot be a repeating parameter. You can choose any three out of D_{32}, D_{AB}, ρ, and μ, although you may end up different set of dimensionless groups depending on how you select them. If we choose D_{32}, D_{AB}, and ρ as repeating parameters, two dimensionless groups can be constructed as

$$\Pi_1 = D_{32}^a D_{AB}^b \rho_c^c k_L \tag{9.22}$$

$$\Pi_2 = D_{32}^a D_{AB}^{b'} \rho_L^{c'} \mu_c \tag{9.23}$$

Now, the exponents of the above equations can be determined so that both groups can be dimensionless by substituting each parameter with their dimensions as

$$\Pi_i = L^a (L^2/T)^b (M/L^3)^c (L/T) \tag{9.24}$$

Collecting all exponents for M, T, and L unit,

$$M: \qquad c = 0$$
$$T: \qquad -6-1 = 0 \qquad\qquad b = -1$$
$$L: \qquad a + 2b - 3c = 0 \qquad\qquad a = 1$$

Therefore, the first dimensionless group is

$$\Pi_1 = D_{32}^0 D_{AB}^{-1} \rho_c^0 k_L = \frac{k_L D_{32}}{D_{AB}} \tag{9.25}$$

which is known as Sherwood number N_{Sh}. Similarly,

$$\Pi_2 = D_{32}^1 D_{AB}^{-1} \rho_c^{-0} \mu_L = \frac{\mu_c}{D_{AB}\rho_c} \tag{9.26}$$

which is known as Schmidt number, N_{Sc}.

Earlier studies in mass transfer between the gas-liquid phase reported the volumetric mass-transfer coefficient $k_L a$. Since $k_L a$ is the combination of two experimental parameters, mass-transfer coefficient and mterfacial area, it is difficult to identify which parameter is responsible for the change of $k_L a$ when we change the operating condition of a fermenter. Calderbank and Moo-Young (1961) separated $k_L a$ by measuring interfacial area and correlated mass-transfer coefficients in gas-liquid dispersions in mixing vessels, and sieve and sintered plate column, as follows:

1. For small bubbles less than 2.5 mm in diameter,

$$k_L = 0.31 \, N_{Sc}^{-2/3} \left(\frac{\Delta\rho\mu_c g}{\rho_c^2} \right)^{1/3} \tag{9.27}$$

or
$$N_{Sh} = 0.31 N_{Sc}^{1/3} N_{Gr}^{1/3} \tag{9.28}$$

where N_{Gr} is known as Grashof number and defined as

$$\text{Grashof number, } N_{Gr} = \frac{D_{32}^3 \rho_c g \Delta\rho}{\mu_c^2} \tag{9.29}$$

The more general forms which can be applied for both small rigid sphere bubble and suspended solid particle are

$$k_L = \frac{2D_{AB}}{D_{32}} + 0.31 N_{Sc}^{-2/3} \left(\frac{\Delta\rho\mu_c g}{\rho_c^2} \right)^{1/3} \tag{9.30}$$

or

$$N_{Sh} = 2.0 + 0.31\, N_{Sc}^{1/3}\, N_{Gr}^{1/3} \tag{9.31}$$

Eqs. (9.30) and (9.31) were confirmed by Calderbank and Jones (1961), for mass transfer to and from dispersions of low-density solid particles in agitated liquids which were designed to simulate mass transfer to microorganisms in fermenters.

2. For bubbles larger than 2.5 mm in diameter,

$$k_L = 0.42\, N_{Sc}^{-1/2} \left(\frac{\Delta\rho\mu_c g}{\rho_c^2} \right)^{1/3} \tag{9.32}$$

or

$$N_{Sh} = 0.42\, N_{Sc}^{1/2}\, N_{Gr}^{1/3} \tag{9.33}$$

Based on the three theories reviewed in the previous section, the exponential dependency of molecular diffusivity on the mass-transfer coefficient is expected to be some value between 0.5 and 1. It is interesting to note that the exponential dependency of molecular diffusivity in the preceding correlations is 2/3 or 1/2, which is within the range predicted by the theories.

Example 9.4

Estimate the mass-transfer coefficient for the oxygen dissolution in water 25°C in a mixing vessel equipped with flat-blade disk turbine and sparger by using Calderbank and Moo-Young's correlations.

Solution:

The diffusivity of the oxygen in water 25°C is 2.5×10^{-9} m^2/s (Example .1). The viscosity and density of water at 25°C is 8.904×10^{-4} kg/m s (*CRC Handbook of Chemistry and Physics*, p. F-38, 1983) and 997.08 kg/m^3 (Perry and Chilton, p. 3 – 71, 1973), respectively. The density of air can be calculated from the ideal gas law,

$$\rho_{air} = \frac{PM}{RT} = \frac{1.01325 \times 10^5 (29)}{8.314 \times 10^3 (298)} = 1.186 \text{ kg/m}^3$$

Therefore the Schmidt number,

$$N_{Sc} = \frac{\mu}{\rho D_{AB}} = \frac{8.904 \times 10^{-4}}{997.08(2.5 \times 10^9)} = 357.2$$

Substituting in Eq. (9.27) for small bubbles,

$$k_L = 0.31(357.2)^{-2/3} \left[\frac{(997.08 - 1.186)(8.904 \times 10^{-4})(9.81)}{(997.08)^2} \right]^{1/3}$$

$$= 1.27 \times 10^{-4} \text{ m/s}$$

Substituting in Eq. (9.32) for large bubbles,

$$k_L = 0.42(357.2)^{-1/2} \left[\frac{(997.08 - 1.186)(8.904 \times 10^{-4})(9.81)}{(997.08)^2} \right]^{1/3}$$

$$= 4.58 \times 10^{-4} \text{ m/s}$$

Therefore, for the air-water system, Eqs. (9.27) and (9.32) predict that the mass-transfer coefficients for small and large bubbles are 1.27×10^{-4} and 4.58×10^{-4} m/s, respectively, which are independent of power consumption and gas-flow rate.

Eq. (9.32) predicts that the the mass-transfer coefficient for the oxygen dissolution in water 25°C in a mixing vessel is 4.58×10^{-4} m/s, regardless of the power consumption and gas-flow rate as illustrated in the previous example problem. Lopes De Figueiredo and Calderbank (1978) reported later that the value of k_L varies from 7.3×10^{-4} to 3.4×10^{-3} m/s, depending on the power dissipation by impeller per unit volume ($P_{m/v}$) as

$$k_L \propto \left(\frac{P_m}{v} \right)^{0.33} \tag{9.34}$$

This dependence of k_L on $P_{m/v}$ was also reported by Prasher and Wills (1973) based on the absorption of CO_2 into water in an agitated vessel, as follows:

$$k_L = 0.592 \, D_{AB}^{1/2} \left(\frac{P_m}{v \mu_c} \right)^{0.25} \tag{9.35}$$

which is dimensionally consistent. However, this equation is limited in its use because the correlation is based on only one gas-liquid system.

Akita and Yoshida (1974) evaluated the liquid-phase mass-transfer coefficient based on the oxygen absorption into several liquids of different physical properties using bubble columns without mechanical agitation. Their correlation for k_L is

$$\frac{k_L D_{32}}{D_{AB}} = 0.5 \left(\frac{v_c}{D_{AB}} \right)^{1/2} \left(\frac{g D_{32}^{3}}{v_c^{2}} \right)^{1/4} \left(\frac{g D_{32}^{2} \rho_c}{\sigma} \right)^{3/8} \tag{9.36}$$

where v_c and σ are the kinematic viscosity of the continuous phase and interfacial tension, respectively. Eq. (9.36) is applicable for column diameters of up to 0.6 m, superficial gas velocities up to 25 m/s, and gas hold-ups to 30 percent.

9.4   MEASUREMENT OF INTERFACIAL AREA

To calculate the gas absorption rate q_L for Eq. (9.7), we need to know the gas-liquid interfacial area, which can be measured employing several techniques such as photography, light transmission, and laser optics.

The interfacial area per unit volume can be calculated from the Sauter-mean diameter D_{32} and the volume fraction of gas-phase H, as follows:

$$a = \frac{6H}{D_{32}} \tag{9.37}$$

The Sauter-mean diameter, a surface-volume mean, can be calculated by measuring drop sizes directly from photographs of a dispersion according to Eq. (9.21).

Photographic measurement of drop sizes is the most straightforward method among many techniques because it does not require calibration. However, taking a clear picture may be difficult, and reading the picture is tedious and time consuming. Pictures can be taken through the base or the sidewall of a mixing vessel. To eliminate the distortion due to the curved surface of a vessel wall, the vessel can be immersed in a rectangular tank, or a water pocket can be installed on the wall (Skelland and Lee, 1981). One limitation of this approach is that the measurement of drop size is limited to the regions near the wall, which may not represent the overall dispersion in a fermenter. Another method is to take pictures by immersing the extension tube with a objective lense in the tank as described by Hong and Lee (1983).

Drop size distribution can be indirectly measured by using the light-transmission technique. When a beam of light is passed through a gas-liquid dispersion, light is scattered by the gas bubbles. It was found that a plot of the extinction ratio (reciprocal of light transmittance, $1/T$) against interfacial area per unit volume of dispersion a, gave a straight line, as follows (Vermeulen et al., 1955; Calderbank, 1958):

$$\frac{1}{T} = m_1 + m_2 a \tag{9.38}$$

In theory, m_1 is unity and m_2 is a constant independent of drop-size distribution as long as all the bubbles are approximately spherical.

The light transmission technique is most frequently used for the determination of average bubble size in gas-liquid dispersion. It has the advantages of quick measurement and on-line operation. The probes are usually made of mirror-treated glass rods (Vermeulen et al., 1955), internally blackened tubes with mirrors (Calderbank, 1958), or fiber optic light guide (Hong and Lee, 1983).

9.5 CORRELATIONS FOR A AND D_{32}

9.5.1 Gas Sparging with No Mechanical Agitation

Leaving the vicinity of a sparger, the bubbles may break up or coalesce with others until an equilibrium size distribution is reached. A stable size is achieved when turbulent fluctuations and surface tension forces are in balance (Calderbank, 1959).

Akita and Yoshida (1974) determined the bubble-size distribution in bubble columns using a photographic technique. The gas was sparged through perforated plates and single orifices, while various liquids were used. The following correlation was proposed for the Sauter-mean diameter:

$$\frac{D_{32}}{D_C} = 26\left(\frac{g\, D_C^2 \rho_c}{\sigma}\right)^{-0.5}\left(\frac{g D_C^3}{v_c^2}\right)^{-0.12}\left(\frac{V_s}{\sqrt{g D_C}}\right)^{-0.12} \tag{9.39}$$

and for the interfacial area

$$a D_c = \frac{1}{3}\left(\frac{g\, D_C^2 \rho_L}{\sigma}\right)^{0.5}\left(\frac{g D_C^3}{v_L^2}\right)^{0.1} H^{1.13} \tag{9.40}$$

where D_C is bubble column diameter and V_S is superfical gas velocity, which is gas flow rate Q divided by the tank cross-sectional area. Eqs. (9.39) and (9.40) are based on data in columns of up to 0.3 m in diameter and up to superficial gas velocities of about 0.07 m/s.

9.5.2 Gas Sparging with Mechanical Agitation

Calderbank (1958) correlated the mterfacial areas for the gas-liquid dispersion agitated by a flat-blade disk turbine as follows:

1. For $V_S < 0.02$ m/s,

$$a_0 = 1.44 \left[\frac{(P_m/v)^{0.4} \rho_c^{0.2}}{\sigma^{0.6}} \right] \left(\frac{V_S}{V_t} \right)^{1/2} \tag{9.41}$$

for $N_{\mathrm{Re}_i}^{0.7} \left(\frac{ND_I}{V_S} \right)^{0.3} < 20{,}000$

where N_{Re} is the impeller Reynolds number defined as

$$N_{\mathrm{Re}_i} = \frac{D_I^2 N \rho_c}{\mu_c} \tag{9.42}$$

The interfacial area for $N_{\mathrm{Re}_i}^{0.7} (ND_I / V_S)^{0.3} > 20{,}000$ can be calculated from the interfacial area a_0 obtained from Eq. (9.41) by using the following relationship.

$$\log_{10} \left(\frac{2.3a}{a_0} \right) = (1.95 \times 10^{-5}) N_{\mathrm{Re}_i}^{0.7} \left(\frac{ND_I}{V_S} \right)^{0.3} \tag{9.43}$$

1. For $V_S > 0.02$ m/s, Miller (1974) modified Eq. (9.41) by replacing the aerated power input by mechanical agitation P_m with the effective power input P_e, and the terminal velocity V_t with the sum of the superficial gas velocity and the terminal velocity $V_S + V_t$. The effective power input P_e combines both gas sparging and mechanical agitation energy contributions. The modified equation is

$$a_0 = 1.44 \left[\frac{(P_e/v)^{0.4} \rho_c^{0.2}}{\sigma^{0.6}} \right] \left(\frac{V_S}{V_t + V_S} \right)^{1/2} \tag{9.44}$$

Calderbank (1958) also correlated the Sauter-mean diameter for the gas-liquid dispersion agitated by a flat-blade disk turbine impeller as follows:

1. For dispersion of air in pure water,

$$D_{32} = 4.15 \left[\frac{\sigma^{0.6}}{(P_m/v)^{0.4} \rho_c^{0.2}} \right] H^{0.5} + 9.0 \times 10^{-4} \tag{9.45}$$

2. In electrolyte solutions (NaCl, Na_2SO_4, and Na_3PO_4),

$$D_{32} = 2.25 \left[\frac{\sigma^{0.6}}{(P_m/v)^{0.4} \rho_c^{0.2}} \right] H^{0.65} \left(\frac{\mu_d}{\mu_c} \right)^{0.25} \tag{9.46}$$

3. In alcoholic solution (aliphatic alcohols),

$$D_{32} = 1.90 \left[\frac{\sigma^{0.6}}{(P_m/v)^{0.4} \rho_c^{0.2}} \right] H^{0.65} \left(\frac{\mu_d}{\mu_c} \right)^{0.25} \tag{9.47}$$

where all constants are dimensionless except 9.0×10^{-4}m in Eq. (9.45). Again the preceding three equations can be modified for high gas flow rate ($V_S > 0.02$ m/s) by replacing P_m by P_e, as suggested by Miller (1974).

9.6 GAS HOLD-UP

Gas hold-up is one of the most important parameters characterizing the hydrodynamics in a fermenter. Gas hold-up depends mainly on the superficial gas velocity and the power consumption, and often is very sensitive to the physical properties of the liquid. Gas hold-up can be determined easily by measuring level of the aerated liquid during operation Z_F and that of clear liquid Z_L. Thus, the average fractional gas hold-up H is given as

$$H = \frac{Z_F - Z_L}{Z_F} \tag{9.48}$$

Gas Sparging with No Mechanical Agitation: In a two-phase system where the continuous phase remains in place, the hold-up is related to superficial gas velocity V_S and bubble rise velocity V_t (Sridhar and Potter, 1980):

$$H = \frac{V_s}{V_s + V_t} \tag{9.49}$$

Akita and Yoshida (1973) correlated the gas hold-up for the absorption of oxygen in various aqueous solutions in bubble columns, as follows:

$$\frac{H}{(1-H)^4} = 0.20 \left(\frac{gD_C^2\rho_c}{\sigma} \right)^{1/8} \left(\frac{gD_C^3}{v_c^2} \right)^{1/12} \left(\frac{V_s}{\sqrt{gD_C}} \right) \tag{9.50}$$

Gas Sparging with Mechanical Agitation: Calderbank (1958) correlated gas hold-up for the gas-liquid dispersion agitated by a flat-blade disk turbine impeller as

$$H = \left(\frac{V_s H}{V_t} \right)^{1/2} + (2.16 \times 10^{-4}) \left[\frac{(P_m/v)^{0.4}\rho_c^{0.2}}{\sigma^{0.6}} \right] \left(\frac{V_s}{V_t} \right)^{1/2} \tag{9.51}$$

where 2.16×10^{-4} has a unit (m) and $V_t = 0.265$ m/s when the bubble size is in the range of 2–5 mm diameter. The preceding equation can be obtained by combining Eqs. (9.41) and Eq. (9.45) by means of Eq. (9.37).

For high superficial gas velocities ($V_S > 0.02$ m/s), replace P_m and V_t of Eq. (9.51) with effective power input P_e and $V_t + V_S$, respectively (Miller, 1974).

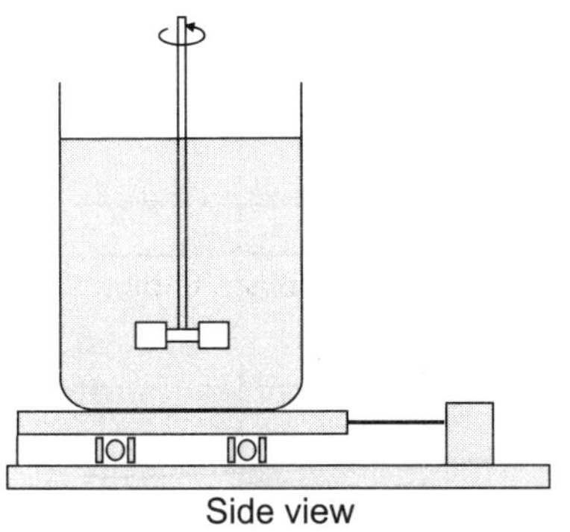
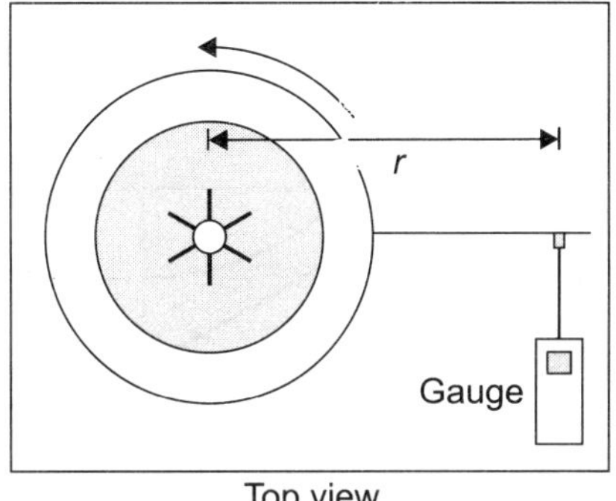

Fig. 9.5 A torque table to measure the power consumption for mechanical agitation.

9.7 POWER CONSUMPTION

The power consumption for mechanical agitation can be measured using a torque table as shown in Figure 9.5. The torque table is constructed by placing a thrust bearing between a base and a circular plate, and the force required to prevent rotation of the turntable during agitation F is measured. The power consumption P can be calculated by the following formula

$$P = 2\pi r N F \qquad (9.52)$$

where N is the agitation speed, and r is the distance from the axis to the point of the force measurement.

Power consumption by agitation is a function of physical properties, operating condition, and vessel and impeller geometry. Dimensional analysis provides the following relationship:

$$\frac{P}{\rho N^3 D_I^5} = f\left(\frac{\rho N D_1^2}{\mu}, \frac{N^2 D_I}{g}, \frac{D_T}{D_I}, \frac{H}{D_I}, \frac{D_W}{D_I}, \dots\right) \qquad (9.53)$$

The dimensionless group in the left-hand side of Eq. (9.53) is known as power number N_P, which is the ratio of drag force on impeller to inertial force. The first term of the right-hand side of Eq. (9.53) is the impeller Reynolds number N_{Re_i} which is the ratio of inertial force to viscous force, and the second term is the Froude number N_{Fr} which takes into account gravity forces. The gravity force affects the power consumption due to the formation of the vortex in an agitating vessel. The vortex formation can be prevented by installing baffles.

For fully baffled geometrically similar systems, the effect of the Froude number on the power consumption is negligible and all the length ratios in Eq. (9.53) are constant. Therefore, Eq. (9.53) is simplified to

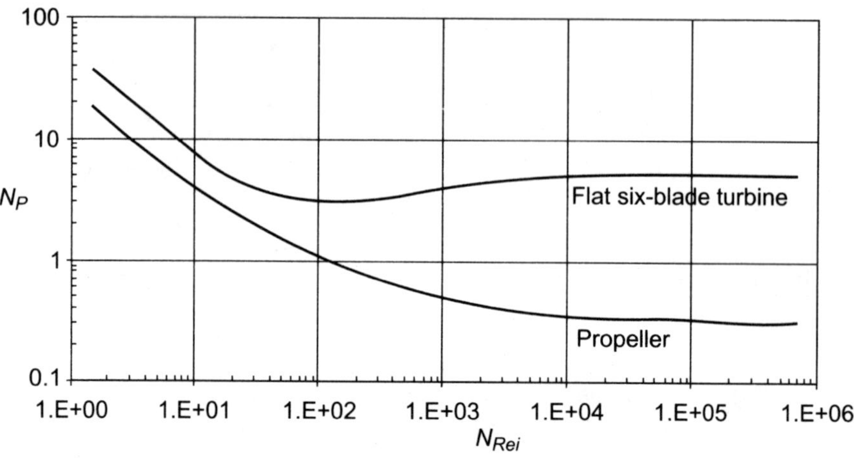

Fig. 9.6 Power number-Reynolds number correlation in an agitator with four baffles (each 0. $1D_T$). (Rushton, et al., 1950)

$$N_p = \alpha \, (N_{Re})^\beta \tag{9.54}$$

Figure 9.6 shows Power number-Reynolds number correlation in an agitator with four baffles (Rushton et al., 1950) for three different types of impellers. The power number decreases with an increase of the Reynolds number and reaches a constant value when the Reynolds number is larger than 10,000. At this point, the power number is independent of the Reynolds number. For the normal operating condition of gas-liquid contact, the Reynolds number is usually larger than 10,000. For example, for a 3-inch impeller with an agitation speed of 150 rpm, the impeller Reynolds number is 16,225 when the liquid is water. Therefore, Eq. (9.54) is simplified to

$$N_p = \text{constant} \quad \text{for } N_{Re} > 10,000 \tag{9.55}$$

The power required by an impeller in a gas sparged system P_m is usually less than the power required by the impeller operating at the same speed in a gas-free liquids P_{mo}. The P_m for the flat-blade disk turbine can be calculated from P_{mo} (Nagata, 1975), as follows:

$$\log_{10} = \frac{P_m}{P_{mo}} = -192\left(\frac{D_I}{D_T}\right)^{4.38}\left(\frac{D_I^{\,2}N}{v}\right)^{0.115}\left(\frac{D_I N^2}{9}\right)^{1.96\left(\frac{D_I}{D_T}\right)}\left(\frac{Q}{ND_I^{\,3}}\right) \tag{9.56}$$

Example 9.5

A cylindrical tank (1.22 m diameter) is filled with water to an operating level equal to the tank diameter. The tank is equipped with four equally spaced baffles whose width is one tenth of the tank diameter. The tank is agitated with a 0.36 m diameter flat six-blade disk turbine. The impeller rotational speed is 2.8 rps. The air enters through an open-ended tube situated below the impeller and its volumetric flow rate is 0.00416 m^3/s at 1.08 atm and 25°C.

Calculate the following properties and compare the calculated values with those experimental data reported by Chandrasekharan and Calderbank (1981): $P_m = 697$ W; $H = 0.02$; $k_L a = 0.0217$ s^{-1}

   a. Power requirement
   b. Gas hold-up
   c. Sauter-mean diameter
   d. Interfacial area
   e. Volumetric mass-transfer coefficient

Solution:

a. Power requirement: The viscosity and density of water at 25°C is 8.904×10^{-4} kg/m^3s (CRC Handbook of Chemistry and Physics , p. F-38, 1983) and 997.08 kg/m^3 (Perry and Chilton, p. 3-71, 1973), respectively. Therefore, the Reynolds number is

$$N_{Re} = \frac{\rho N D_I^2}{\mu} = \frac{(997.08)(2.8)(0.36)^2}{8.904 \times 10^{-4}} = 406{,}357$$

which is much larger than 10,000, above which the power number is constant at 6. Thus,

$$P_{mo} = 6\rho \, N^3 D_I^5 = 6(997.08)(2.8)^3(0.36)^5 = 794 \text{ W}$$

The power required in the gas-sparged system is from Eq. (9.56)

$$\log_{10} = -\frac{P_m}{P_{mo}} 192\left(\frac{0.36}{1.22}\right)^{4.38}\left(\frac{0.36^2(2.8)}{8.93 \times 10^{-7}}\right)^{0.115}$$
$$\left(\frac{(0.36)2.8^2}{9.81}\right)^{1.96\left(\frac{0.36}{1.22}\right)}\left(\frac{0.00416}{(2.8)0.36^3}\right)$$

Therefore, $P_m = 687$ W

b. Gas hold-up: The interfacial tension for the air-water interface is 0.07197 kg/s^2 (CRC Handbook of Chemistry and Physics , p. F-33, 1983). The volume of the the dispersion is

$$v = \frac{\pi}{4} 1.22^2(1.\,22) = 1.43 m^3$$

The superficial gas velocity is

$$v = \frac{4Q}{\pi DT^2} = \frac{4(0.00416)}{\pi 1.22^2} = 0.00356 \text{ m/s}$$

Substituting these values into Eq. (9.51) gives

$$H = \left(\frac{0.00356H}{0.265}\right)^{1/2} + 2.16 \times 10^{-4}\left[\frac{(687/1.43)^{0.4}997.08^{0.2}}{0.07197^{0.6}}\right]\left(\frac{0.00356}{0.265}\right)^{1/2}$$

The solution of the preceding equation for H gives

$$H = 0.023$$

c. Sauter-mean diameter: In Eq. (9.45)

$$D_{32} = 4.15\left[\frac{0.07197^{0.6}}{(687/1.43)^{0.4}997.08^{0.2}}\right]0.023^{0.5} + 9.0 \times 10^{-4}$$

$$= 0.00366m = 3.9 \text{ mm}$$

d. Interfacial area a: In Eq. (9.37)

$$a = \frac{6H}{D_{32}} = \frac{6\,(0.023)}{0.00366} = 37.6 \text{ m}^{-1}$$

e. Volumetric mass-transfer coefficient: Since the average size of bubbles is 4 mm, we should use Eq. (9.32) . Then, from Example. 4

$$k_L = 4.58 \times 10^{-4}\text{m/s}$$

Therefore,

$$k_L a = 4.58 \times 10^{-4}\,(37.7) = 0.017\text{s}^{-1}$$

The preceding estimated values compare well with those experimental values. The percent errors as defined in Example .1 are –1.4 percent for the power consumption, 15 percent for the gas hold-up, and –21.7 percent for the volumetric mass-transfer coefficient.

9.8 DETERMINATION OF OXYGEN-ABSORPTION RATE

To estimate the design parameters for oxygen uptake in a fermenter, you can use the correlations presented in the previous sections, which can be applicable to a wide range of gas-liquid systems in addition to the air-water system. However, the calculation procedure is lengthy and the predicted value from those correlations can vary widely. Sometimes, you may be unable to find suitable correlations which will be applicable to your type and size of ferment ers. In such cases, you can measure the oxygen-transfer rate yourself or use correlations based on those experiments.

The oxygen absorption rate per unit volume q_a/v can be estimated by

$$\frac{q_a}{v} = K_L a(C_L^* - C_L) = k_L a(C_L^* - C_L) \qquad (9.57)$$

Since the oxygen is sparingly soluble gas, the overall mass-transfer coefficient K_L is equal to the individual mass-transfer coefficient K_L. Our objective in fermenter design is to maximize the oxygen transfer rate with the minimum power consumption necessary to agitate the fluid, and also minimum air flow rate. To maximize the oxygen absorption rate, we have to maximize K_L, a, $C_L^* - C_L$. However, the concentration difference is quite limited for us to control because the value of C_L^* is limited by its very low maximum solubility. Therefore, the main parameters of interest in design are the mass-transfer coefficient and the mterfacial area.

Table 9.1 lists the solubility of oxygen at 1 atm in water at various temperatures. The value is the maximum concentration of oxygen in water when it is in equilibrium with pure oxygen. This solubility decreases with the addition of acid or salt as shown in Table 9.2.

Normally, we use air to supply the oxygen demand of fermenters. The maximum concentration of oxygen in water which is in equilibrium with air C^*L at atmospheric pressure is about one fifth of the solubility listed, according to the Henry's law,

Table 9.1 Solubility of Oxygen in Water at 1 atm.[a]

Temperature °C	Solubility	
	mmol O_2/L	mg O_2/L
0	2.18	69.8
10	1.70	54.5
15	1.54	49.3
20	1.38	44.2
25	1.26	40.3
30	1.16	37.1
35	1.09	34.9
40	1.03	33.0

[a] Data from International Critical Tables, Vol. III. New York: McGraw-Hill Book Co., 1928, p. 271.

$$C_L^* = \frac{p_{O_2}}{H_{O_2}(T)} \qquad (9.58)$$

where p_{O_2} is the partial pressure of oxygen and $H_{O_2}(T)$ *is* Henry's law constant of oxygen at a temperature, T. The value of Henry's law constant can be obtained from the solubilies listed in Table 9.2. For example, at 25 °C, C_L^* is 1.26 mmol/L and p_{O_2} is 1 atm because it is pure oxygen. By substituting these values into Eq. (9.58), we obtain H_{O_2} (25°C) is 0.793 atm L/mmol. Therefore, the equilibrium concentration of oxygen for the air-water contact at 25°C will be

$$C_L^* = \frac{0.209 \text{ atm}}{0.793 \text{ atm L/mmol}} \; 0.264 \text{ mmoL} = O_2/L - 8.43 \text{ mg/L}$$

Ideally, oxygen-transfer rates should be measured in a fermenter which contains the nutrient broth and microorganisms during the actual fermentation process. However, it is difficult to carry out such a task due to the complicated nature of the medium and the ever changing rheology during cell growth. A common strategy is to use a synthetic system which approximates fermentation conditions.

Table 9.2 Solubility of Oxygen in Solution of Salt or Acid at 25°C.[a]

Conc.	Solubility, mmol O_2/L		
mol/L	HCl	H_2SO_4	NaCl
0.0	1.26	1.26	1.26
0.5	1.21	1.21	1.07
1.0	1.16	1.12	0.89
2.0	1.12	1.02	0.71

[a] Data from F.Todt, Electrochemishe Sauer-stoffmess-ungen, Berlin: W. de Guy & Co., 1958

9.8.1. Sodium Sulfite Oxidation Method

The sodium sulfite oxidation method (Cooper et al., 1944) is based on the oxidation of sodium sulfite to sodium sulfate in the presence of catalyst (Cu^{++} or Co^{++}) as

$$Na_2SO_3 + \frac{1}{2}O_2 \xrightarrow{\;Cu^{++} \; or \; Co^{++}\;} Na_2SO_4 \qquad (9.59)$$

This reaction has following characteristics to be qualified for the measurement of the oxygen-transfer rate:

1. The rate of this reaction is independent of the concentration of sodium sulfite within the range of 0.04 to 1 N.

2. The rate of reaction is much faster than the oxygen transfer rate; therefore, the rate of oxidation is controlled by the rate of mass transfer alone.

To measure the oxygen-transfer rate in a fermenter, fill the fermenter with a 1 N sodium sulfite solution containing at least 0.003 M Cu^{++} ion. Turn on the air and start a timer when the first bubbles of air emerge from the sparger. Allow the oxidation to continue for

4 to 20 minutes, after which, stop the air stream, agitator, and timer at the same instant, and take a sample. Mix each sample with an excess of freshly pipetted standard iodine reagent. Titrate with standard sodium thiosulfate solution ($Na_2S_2O_3$) to a starch indicator end point. Once the oxygen uptake is measured, the $k_L a$ may be calculated by using Eq. (9.57) where C_L is zero and C_L^* is the oxygen equilibrium concentration.

The sodium sulfite oxidation technique has its limitation in the fact that the solution cannot approximate the physical and chemical properties of a fermentation broth. An additional problem is that this technique requires high ionic concentrations (1 to 2 mol/L), the presence of which can affect the interfacial area and, in a lesser degree, the mass-transfer coefficient (Van't Riet, 1979). However, this technique is helpful in comparing the performance of fermenters and studying the effect of scale-up and operating conditions.

Example 9.6

To measure $k_L a$, a fermenter was filled with 10 L of 0.5 M sodium sulfite solution containing 0.003 M Cu^{++} ion and the air sparger was turned on. After exactly 10 minutes, the air flow was stopped and a 10 mL sample was taken and titrated. The concentration of the sodium sulfite in the sample was found to be 0.21 mol/L. The experiment was carried out at 25°C and 1 atm. Calculate the oxygen uptake and $k_L a$.

Solution:

The amount of sodium sulfite reacted for 10 minutes is

$$0.5 - 0.21 = 0.29 \text{ mol/L}$$

According to the stoichiometric relation, Eq. (9.59) the amount of oxygen required to react 0.29 mol/L is

$$0.29^{12} = 0.145 \text{mol/L}$$

Therefore, the oxygen uptake is

$$(0.145 \text{ mole } O_2/L)\left(\frac{32\text{g } O_2/\text{mol}}{600\,\text{s}}\right) = 7.73 \; ´ \; 10^{-3} \text{ g/Ls}$$

The solubility of oxygen in equilibrium with air can be estimated by Eq. (9.58) as

$$C^*{}_L = \frac{p_{O_2}}{H_{O_2}(T)} = \frac{(1\text{atm})(0.209\,\text{mol } O_2/\text{mol air})}{(793\,\text{atm L/mol})(1\,\text{mol}/32\text{g})} = 8.43 \times 10^{-4} \text{g/L}$$

Therefore, the value of $k_L a$ is, according to the Eq. (9.57),

$$k_L a = \frac{q_a/v}{C_L^* - C_L} = \frac{7.73 \times 10^{-3}\,\text{g/Ls}}{(8.43 \times 10^{-3}\,\text{g/L} - 0)}\ 0.917\ \text{s}^{-1}$$

9.8.2   Dynamic Gassing-out Technique

This technique (Van't Riet, 1979) monitors the change of the oxygen concentration while an oxygen-rich liquid is deoxygenated by passing nitrogen through it. Polarographic electrode is usually used to measure the concentration. The mass balance in a vessel gives

$$\frac{dC_L(t)}{dt} = k_L a\,[C_L^* - C_L(t)] \tag{9.60}$$

Integration of the preceding equation between t_1 and t_2 results in

$$k_L a = \frac{\ln\!\left[\dfrac{C_L^* - C_L(t_1)}{C_L^* - C_L(t_2)}\right]}{t_2 - t_1} \tag{9.61}$$

from which $k_L a$ can be calculated based on the measured values of $C_L(t_1)$ and $C_L(t_2)$.

9.8.3   Direct Measurement

In this technique, we directly measure the oxygen content of the gas stream entering and leaving the fermenter by using gaseous oxygen analyzer. The oxygen uptake can be calculated as

$$q_a = (Q_{in} C_{o_2,in} - Q_{out}\, C_{o_2,out}) \tag{9.62}$$

where Q is the gas flow rate.

Once the oxygen uptake is measured, the $k_L a$ can be calculated by using Eq. (9.57), where C_L, is the oxygen concentration of the liquid in a fermenter and C_L^* is the concentration of the oxygen which would be in equilibrium with the gas stream. The oxygen concentration of the liquid in a fermenter can be measured by an on-line oxygen sensor. If the size of the fermenter is rather small (less than 50 L), the variation of the $C_L^* - C_L$ in the fermenter is fairly small. However, if the size of a fermenter is very large, the variation can be significant. In this case, the log-mean value of $C_L^* - C_L$ of the inlet and outlet of the gas stream can be used as

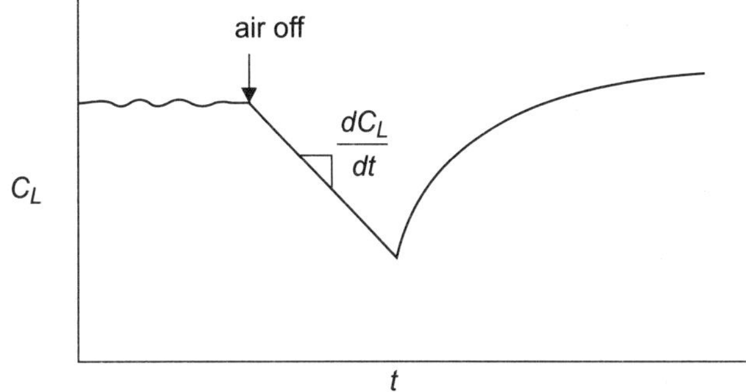

Fig. 9.7 Dynamic technique for the determination of $k_L a$.

$$(C_L^* - C_L)_{LM} = \frac{(C_L^* - C_L)_{in} - (C_L^* - C_L)_{out}}{\ln[(C_L^* - C_L)_{in}/(C_L^* - C_L)_{out}]} \tag{9.63}$$

9.8.4 Dynamic Technique

By using the dynamic technique (Taguchi and Humphrey, 1966), we can estimate the $k_L a$ value for the oxygen transfer during an actual fermentation process with real culture medium and microorganisms. This technique is based on the oxygen material balance in an aerated batch fermenter while microorganisms are actively growing as

$$\frac{dC_L}{dt} = k_L a(C_L^* - C_L) - r_{O_2} C_X \tag{9.64}$$

where r_{O_2} is cell respiration rate [g O_2/g cell h].

While the dissolved oxygen level of the fermenter is steady, if you suddenly turn off the air supply, the oxygen concentration will be decreased (Figure 9.7) with the following rate

$$\frac{dC_L}{dt} = r_{O_2} C_X \tag{9.65}$$

since $k_L a$ in Eq. (9.64) is equal to zero. Therefore, by measuring the slope of the C_L vs. t curve, we can estimate $r_{O_2} C_X$. If you turn on the airflow again, the dissolved oxygen concentration will be increased according to Eq. (9.64), which can be rearranged to result in a linear relationship as

$$C_L = C_L^* - \frac{1}{k_L a}\left(\frac{dC_L}{dt} + r_{O_2} C_X\right) \tag{9.66}$$

The plot of C_L versus $dC_L/dt + r_{O_2}$ will result in a straight line which has the slope of $-1/(k_L a)$ and the y-intercept of C_L^*.

9.9 CORRELATION FOR $K_L a$

9.9.1 Bubble Column

Akita and Yoshida (1973) correlated the volumetric mass-transfer coefficient $k_L a$ for the absorption of oxygen in various aqueous solutions in bubble columns, as follows:

$$k_L a = 0.6 \, D_{AB}^{0.5} \, V_c^{-0.12} \left(\frac{\sigma}{\rho_c} \right)^{-0.62} D_T^{0.17} \, g^{0.93} \, H^{1.1} \tag{9.67}$$

which applies to columns with less effective spargers.

In bubble columns, for $0 < V_S < 0.15$ m/s and $100 < P_g/v < 1100$ W/m^3, Botton et al. (1980) correlated the $k_L a$ as

$$\frac{k_L a}{0.08} = \left(\frac{P_g/v}{800} \right)^{0.75} \tag{9.68}$$

where P_g is the gas power input, which can be calculated from

$$\frac{P_g}{v} = 800 \left(\frac{V_s}{0.1} \right)^{0.75} \tag{9.69}$$

9.9.2 Mechanically Agitated Vessel

For aerated mixing vessels in an aqueous solution, the mass-transfer coefficient is proportional to the power consumption (Lopes De Figueiredo and Calderbank, 1978) as

$$k_L \propto \left(\frac{P_m}{v} \right)^{0.33} \tag{9.70}$$

The interfacial area for the aerated mixing vessel is a function of agitation conditions. Therefore, according to Eq. (9.41)

$$a \propto \left(\frac{P_m}{v} \right)^{0.4} V_s^{0.5} \tag{9.71}$$

Therefore, by combining the above two equations, $k_L a$ will be

$$k_L a \propto \left(\frac{P_m}{v} \right)^{0.77} V_s^{0.5} \tag{9.72}$$

Numerous studies for the correlations of $k_L a$ have been reported and their results have the general form as

$$k_L a = b_1 \left(\frac{P_m}{v} \right)^{b_2} V_s^{b_3} \tag{9.73}$$

where b_1, b_1, and b_1 vary considerably depending on the geometry of the system, the range of variables covered, and the experimental method used. The values of b_2 and b_3 are generally between 0 to 1 and 0.43 to 0.95, respectively, as tabulated by Sideman et al. (1966).

Van't Riet (1979) reviewed the data obtained by various investigators and correlated them as follows:

1. For "coalescing" air-water dispersion,

$$k_L a = 0.026 \left(\frac{P_m}{v} \right)^{0.4} V_s^{0.54} \tag{9.74}$$

1. For "noncoalescmg" air-electrolyte solution dispersions,

$$k_L a = 0.002 \left(\frac{P_m}{v} \right)^{0.7} V_s^{0.2} \tag{9.75}$$

both of which are applicable for the volume up to 2.6 m^3; for a wide variety of agitator types, sizes, and D_I/D_T ratios; and $500 < P_{m/v} < 10,000$ W/m^3. These correlations are accurate within approximately 20 percent to 40 percent.

Example 9.7

Estimate the volumetric mass-transfer coefficient $k_L a$ for the gas-liquid contactor described in Example .4 by using the correlation for $k_L a$ in this section.

Solution:

From Example .4, the reactor volume v is 1.43 m^3, the superficial gas velocity V_S is 0.00356 m/s, and power consumption P_m is 687 W. By substituting these values into Eq. (9.74),

$$k_L a = 0.026 \left(\frac{687}{1.43} \right)^{0.4} 0.00356^{0.5} = 0.018 \text{ s}^{-1}$$

9.10   SCALE-UP

9.10.1   Similitude

For the optimum design of a production-scale fermentation system (prototype), we must translate the data on a small scale (model) to the large scale. The fundamental requirement for scale-up is that the model and prototype should be similar to each other.

Two kinds of conditions must be satisfied to insure similarity between model and prototype. They are:

1. Geometric similarity of the physical boundaries: The model and the prototype must be the same shape, and all linear dimensions of the model must be related to the corresponding dimensions of the prototype by a constant scale factor.
2. Dynamic similarity of the flow fields: The ratio of flow velocities of corresponding fluid particles is the same in model and prototype as well as the ratio of all forces acting on corresponding fluid particles. When dynamic similarity of two flow fields with geometrically similar boundaries is achieved, the flow fields exhibit geometrically similar flow patterns.

The first requirement is obvious and easy to accomplish, but the second is difficult to understand and also to accomplish and needs explanation. For example, if forces that may act on a fluid element in a fermenter during agitation are the viscosity force F_V, drag force on impeller F_D, and gravity force F_G, each can be expressed with characteristic quantities associated with the agitating system. According to Newton's equation of viscosity, viscosity force is

$$F_V = \mu \left(\frac{du}{dy} \right) A \qquad (9.76)$$

where du/dy is velocity gradient and A is the area on which the viscosity force acts. For the agitating system, the fluid dynamics involved are too complex to calculate a wide range of velocity gradients present. However, it can be assumed that the average velocity gradient is proportional to agitation speed N and the area A is to D_I^2, which results.

$$F_V = \propto \mu \, ND_I^2 \qquad (9.77)$$

The drag force F_D can be characterized in an agitating system as

$$F_D \propto \frac{P_{mo}}{D_I N} \qquad (9.78)$$

Since gravity force F_G is equal to mass m times gravity constant g,

$$F_G \propto \rho D_1^3 g \qquad (9.79)$$

The summation of all forces is equal to the inertial force F_I as,

$$\Sigma F = F_V + F_D + F_G = F_I \propto \rho D_I^4 N^2 \qquad (9.80)$$

Then dynamic similarity between a model (m) and a prototype (p) is achieved if

$$\frac{(F_V)_m}{(F_V)_p} = \frac{(F_D)_m}{(F_D)_p} = \frac{(F_G)_m}{(F_G)_p} = \frac{(F_I)_m}{(F_I)_p} \qquad (9.81)$$

or in dimensionless forms:

$$\left(\frac{F_1}{F_V}\right)_p = \left(\frac{F_1}{F_V}\right)_m \quad \left(\frac{F_I}{F_D}\right)_p = \left(\frac{F_I}{F_D}\right)_m \quad \left(\frac{F_I}{F_G}\right)_p = \left(\frac{F_I}{F_G}\right)_m \qquad (9.82)$$

The ratio of inertial force to viscosity force is

$$\frac{F_I}{F_V} = \frac{\rho D_I^{\,4} N^2}{\mu \, N D_I^{\,2}} = \frac{\rho D_I^{\,2} N}{\mu} = N_{\mathrm{Rei}} \qquad (9.83)$$

which is the Reynolds number. Similarly,

$$\frac{F_I}{F_V} = \frac{\rho D_I^{\,4} N^2}{P_{mo}/D_I N} = \frac{\rho N^3 D_I^{\,5}}{P_{mo}} = \frac{1}{N_P} \qquad (9.84)$$

$$\frac{F_I}{F_G} = \frac{\rho D_I^{\,4} N^2}{\rho D_I^{\,3} g} = \frac{D_I N^2}{g} = N_{Fr} \qquad (9.85)$$

Dynamic similarity is achieved when the values of the nondimensional parameters are the same at geometrically similar locations.

$$(N_{\mathrm{Rei}})p = (N_{\mathrm{Rei}})_m$$
$$(N_P)_P = (N_P)_m \qquad (9.86)$$
$$(N_{Fr})_p = (N_{Fr})_m$$

Therefore, using dimensionless parameters for the correlation of data has advantages not only for the consistency of units, but also for the scale-up purposes.

However, it is difficult, if not impossible, to satisfy the dynamic similarity when more than one dimensionless group is involved in a system, which creates the needs of scale-up criteria. The following example addresses this problem.

Example 9.8

The power consumption by an agitator in an unbaffled vessel can be expressed as

$$\frac{P}{\rho N^3 D_I^{\,5}} = f\left(\frac{\rho N D_I^{\,2}}{\mu}, \frac{N^2 D_I}{g}\right)$$

Can you determine the power consumption and impeller speed of a 1,000-gallon fermenter based on the findings of the optimum condition from a geometrically similar one-gallon vessel? If you cannot, can you scale up by using a different fluid system?

Solution:

Since $V_p/V_m = 1{,}000$, the scale ratio is,

$$\frac{(D_I)_p}{(D_I)_m} = 1{,}000^{1/3} = 10 \tag{9.87}$$

To achieve dynamic similarity, the three dimensionless numbers for the prototype and the model must be equal, as follows:

$$\left(\frac{P_{mo}}{\rho N^3 D_I{}^5}\right)_p = \left(\frac{P_{mo}}{\rho N^3 D_I{}^5}\right)_m \tag{9.88}$$

$$\left(\frac{\rho N D_I{}^2}{\mu}\right)_p = \left(\frac{P N D_1^2}{\mu}\right)_m \tag{9.89}$$

$$\left(\frac{N^2 D_I}{g}\right)_p = \left(\frac{N^2 D_I}{g}\right)_m \tag{9.90}$$

If you use the same fluid for the model and the prototype, $\rho_p = \rho_m$ and $\mu_p = \mu_m$. Canceling out the same physical properties and substituting Eq. (9.87) to Eq. (9.88) yields

$$(P_{mo})_p = 10^5 (P_{mo})_m \left(\frac{N_p}{N_m}\right)^3 \tag{9.91}$$

The equality of the Reynolds number requires

$$N_p = 0.01\, N_m \tag{9.92}$$

On the other hand, the equality of the Froude number requires

$$N_P = \frac{1}{\sqrt{10}}\, N_m \tag{9.93}$$

which is conflicting with the previous requirement for the equality of the Reynolds number. Therefore, it is impossible to satisfy the requirement of the dynamic similarity unless you use different fluid systems.

If $\rho_p \neq \rho_m$ and $\mu_p \neq \mu_m$, to satisfy Eqs. (9.89) and (9. 90), the following relationship must hold.

$$\left(\frac{\mu}{\rho}\right)_m = \frac{1}{31.6}\left(\frac{\mu}{\rho}\right)_p \tag{9.94}$$

Therefore, if the kinematic viscosity of the prototype is similar to that of water, the kinematic viscosity of the fluid, which needs to be employed for the model, should be 1/31.6 of the kinematic viscosity of water. It is impossible to find the fluid whose kinematic viscosity is that small. As a conclusion, if all three dimensionless groups are important, it is impossible to satisfy the dynamic similarity.

The previous example problem illustrates the difficulties involved in the scale-up of the findings of small-scale results. Therefore, we need to reduce the number of dimensionless parameters involved to as few as possible, and we also need to determine which is the most important parameter, so that we may set this parameter constant. However, even though only one dimensionless parameter may be involved, we may need to define the scale-up criteria.

As an example, for a fully baffled vessel when $N_{Re} > 10{,}000$, the power number is constant according to Eq. (9.55). For a geometrically similar vessel, the dynamic similarity will be satisfied by

$$\left(\frac{P_{mo}}{\rho N^3 D_I{}^5}\right)_p = \left(\frac{P_{mo}}{\rho N^3 D_I{}^5}\right)_m \tag{9.95}$$

If the fluid employed for the prototype and the model remains the same, the power consumption in the prototype is

$$(P_{mo})_p = (P_{mo})_m \left(\frac{N_p}{N_m}\right)^3 \left[\frac{(D_I)_m}{(D_I)_p}\right]^5 \tag{9.96}$$

where $(D_I)_p/(D_I)_m$ is equal to the scale ratio. With a known scale ratio and known operating conditions of a model, we are still unable to predict the operating conditions of a prototype because there are two unknown variables, P_{mo} and N. Therefore, we need to have a certain criteria which can be used as a basis.

9.10.2 Criteria of Scale-Up

Most often, power consumption per unit volume P_{mo}/v is employed as a criterion for scale-up. In this case, to satisfy the equality of power numbers of a model and a prototype,

$$\left(\frac{P_{mo}}{D_I{}^3}\right)_p = \left(\frac{P_{mo}}{D_I{}^3}\right)_m \left(\frac{N_p}{N_m}\right)^3 \left[\frac{(D_I)_p}{(D_I)_m}\right]^2 \tag{9.97}$$

Note that P_{mo}/D_I^3) represents the power per volume because the liquid volume is proportional to D_I^3 for the geometrically similar vessels. For the constant P_{mo}/D_I^3,

$$\left(\frac{N_p}{N_m}\right)^3 = \left[\frac{(D_I)_m}{(D_I)_p}\right]^2 \tag{9.98}$$

As a result, if we consider scale-up from a 20-gallon to a 2,500-gallon agitated vessel, the scale ratio is equal to 5, and the impeller speed of the prototype will be

$$N_p = \left[\frac{(D_I)_m}{(D_I)_p}\right]^{2/3} N_m = 0.34\, N_m \qquad (9.99)$$

which shows that the impeller speed in a prototype vessel is about one third of that in a model. For constant P_{mo}/v, the Reynolds number and the impeller tip speed cannot be the same. For the scale ratio of 5,

$$(N_{Re_i})_p = 8.5\,(N_{Re_i})_m \qquad (9.100)$$

$$(ND_I)_p = 1.7\,(ND_I)_m \qquad (9.101)$$

Table 9.3 shows the values of properties for a prototype (2,500-gallon) when those for a model (20-gallon) are arbitrarily set as 1.0 (Oldshue, 1966). The parameter values of the prototype depend on the criteria used for the scale-up. The third column shows the parameter values of the prototype, when P_{mo}/v is set constant. The values in the third column seem to be more reasonable than those in the fourth, fifth, and sixth columns, which are calculated based on the constant value *of* Q/v, ND_I, and $ND_I^2\rho/\mu$, respectively. For example when the Reynolds number is set constant for the two scales, the P_{mo}/v reduces to 0.16 percent of the model and actual power consumption P_{mo} also reduces to 20 percent of the model, which is totally unreasonable.

As a conclusion, there is no one scale-up rule that applies to many different kinds of mixing operations. Theoretically we can scale up based on geometrical and dynamic similarities, but it has been shown that it is possible for only a few limited cases. However, some principles for the scale-up are as follows (Oldshue, 1985):

1. It is important to identify which properties are important for the optimum operation of a mixing system. This can be mass transfer, pumping capacity, shear rate, or others. Once the important properties are identified, the system can be scaled up so that those properties can be maintained, which may result in the variation of the less important variables including the geometrical similarity.

2. The major differences between a big tank and a small tank are that the big tank has a longer blend time, a higher maximum impeller shear rate, and a lower average impeller shear rate.

3. For homogeneous chemical reactions, the power per volume can be used as a scale-up criterion. As a rule of thumb, the intensity of agitation can be classified based on the power input per 1,000 gallon as shown in Table 9.4

Table 9.3 Properties of Agitator on Scale-Up[a]

Property	Model 20 gal			Prototype 2,500 gal	
P_{mo}	1.0	125.	3126	25	0.2
P_{mo}/v	1.0	**1.0**	25	0.2	0.0016
N	1.0	0.34	1.0	0.2	0.04
DI	1.0	5.0	5.0	5.0	5.0
Q	1.0	42.5	125	25	5.0
Q/V	1.0	0.34	**1.0**	0.2	0.04
ND_1	1.0	1.7	5.0	**1.0**	0.2
N_{Re}	1.0	8.5	25	5.0	**1.0**

[a] Reprinted with permission from J. Y. Olclshue, "Mixing Scale-Up Techniques" *Biotech. Bioeng.* 8 (1966):3–24. ©1966 by John Wile} & Sons. Inc.. New York. NY.

4. For the scale-up of the gas-liquid contactor, the volumetric mass-transfer coefficient kLa can be used as a scale-up criterion. In general, the volumetric mass-transfer coefficient is approximately correlated to the power per volume. Therefore, constant power per volume can mean a constant kLa.

5. Typical impeller-to-tank diameter ratio D_I/T for fermenters is 0.33 to 0.44. By using a large impeller, adequate mixing can be provided at an agitation speed which does not damage living organisms. Fermenters are not usually operated for an optimum gas-liquid mass transfer because of the shear sensitivity of cells, which is discussed in the next section.

Table 9.4 Criteria of Agitation Intensity

Horsepower per 1000 gal	Agitation Intensity
0.5 – 1	Mild
2 – 3	Vigorous
4 – 10	Intense

9.11 SHEAR-SENSITIVE MIXING

One of the most versatile fermenter systems used industrially is the mechanically agitated fermenter. This type of system is effective in the mixing of fermenter contents, the suspension of cells, the breakup of air bubbles for enhanced oxygenation, and the prevention of forming large cell aggregates. However, the shear generated by the agitator can disrupt the cell membrane and eventually kill some microorganisms (Midler and Fin, 1966), animal cells (Croughan et al., 1987), and plant cells (Hooker et al., 1988). Shear also is responsible for the deactivation of enzymes (Charm and Wong, 1970). As a result, for optimum operation of an agitated fermentation system, we need to understand the hydrodynamics involved in shear sensitive mixing.

For the laminar flow region of Newtonian fluid, shear stress τ is equal to the viscosity μ times the velocity gradient du/dy as

$$\tau = -\mu \frac{du}{dy} \qquad (9.102)$$

which is known as Newton's equation of viscosity. The velocity gradient is also known as shear rate .

For the turbulent flow,

$$\tau = -\eta \frac{du}{dy} \qquad (9.103)$$

where η is eddy viscosity, which is not only dependent upon the physical properties of the fluid, but also the operating conditions. Therefore, to describe the intensity of shear in a turbulent system such as an agitated fermenter, it is easier to estimate shear rate du/dy instead of shear stress.[1] Even though we use the shear rate as a measure of the shear intensity, we should remember that it is the shear stress that ultimately affects the living cells or enzyme. Depending upon the magnitude of viscosity and also whether the flow is laminar or turbulent, there is a wide range of shear stress generated for the same shear rate. However, the shear rate is a good measure for the intensity of the agitation.

In agitated systems, it is difficult, if not impossible, to determine the shear rate, because of the complicated nature of the fluid dynamics generated by impellers. A fluid element in an agitated vessel will go through a wide range of shear rates during agitation: maximum shear when it passes through the impeller region and minimum shear when it passes near the corner of the vessel. Metzner and Otto (1957) developed a general correlation for the average shear rate generated by a flat-blade disk turbine, based on power measurements on non-Newtonian liquids, as:

$$\left(-\frac{du}{dy} \right)_{av} = 13\, N \text{ for } N_{Re_i} < 20 \qquad (9.104)$$

9.12 NOMENCLATURE

A	interfacial area, m^2
a	gas-liquid interfacial area per unit volume of dispersion for low impeller Reynolds numbers, m^{-1}

---

[1] For a non-Newtonian fluid, the viscosity is not constant even for the laminar flow. Therefore, shear rate is easier to estimate than shear stress.

a_0	gas-liquid interfacial area per unit volume of dispersion, m^{-1}
b	constant
C	concentration, $kmol/m^3$
C_D	friction factor and drag coefficient, dimensionless
	D diameter of bubble or solid particle, m
D_{AB}	diffusivity of component A through B, m^2/s
D_{AB}°	diffusivity of component A in a very dilute solution of B, m^2/s
D_{32}	Sauter-mean diameter, m
D_I	impeller diameter, m
D_T	tank diameter, m
D_W	impeller width, m
F	force, N
J_A	molar flux of component A relative to the average molal velocity of all constituents, $kmol/\,m^2s$
K	overall mass-transfer coefficient, m/s
k	individual mass-transfer coefficient, m/s
g	acceleration due to gravity, m/s^2
H	volume fraction of gas phase in dispersion, dimensionless
N	impeller speed, s^{-1}
N_A, N_B	mass flux of A and B relative to stationary coordinate, $kmol/m^2s$
N_{Fr}	Froude number $(D_I N^2/g)$, dimensionless
N_G, N_L	mass flux from gas to liquid phase and from liquid to gas phase, respectively, $kmol/m^2s$
N_{Gr}	Grashof number $(D_{32}^3 \rho_c \Delta\rho g/\mu_c^2)$, dimensionless
N_o	number of sparger orifices or sites
N_P	power number $(P_{mo}/\rho D_I^5 N^3)$, dimensionless
N_{Re_b}	bubble Reynolds number $(D_{BM} V_t \rho_c/\mu_c)$, dimensionless
N_{Re_i}	impeller Reynolds number $(D_I^2 N\rho_c/\mu_c)$, dimensionless
N_{Sc}	Schmidt number $(\mu_c/\rho_c D_{AB})$, dimensionless
N_{Sh}	Sherwood number $(k_L D_{32}/D_{AB})$, dimensionless
P	total pressure, N/m^2
P_e	effective power input by both gas sparging and mechanical agitation, W
P_g	power dissipated by sparged gas, W
P_m	power dissipated by impeller in aerated liquid dispersion, W
P_{mo}	power dissipated by impeller without aeration, W

q	the rate of mass transfer, kmol/s
Q	volumetric gas-flow rate, m^3/s
R	gas constant
r	length of radial arm for dynamometer system, m
s	fractional rate of surface renewal, s^{-1}
t_e	exposure time for the penetration theory, s
u	velocity, m/s
u_0	sparger hole velocity, m/s
V_S	superficial gas velocity, gas-flow rate divided by tank cross sectional area, m/s
V_t	terminal gas-bubble velocity in free rise, m/s
v	volume of liquid, m^3
Z_F, Z_L	level of the aerated liquid during operation, and that of clear liquid, m
Z_f	film thickness in the two-film theory, m
η	0.06, fraction of jet energy transmitted to bulk liquid
Δ_p	difference in density between dispersed (gas) and continuous (liquid) phases, kg/m^3
γ	shear rate, s^{-1}
μ	viscosity, kg/ms
ν	kinematic viscosity, m^2/s
ω	angular velocity, s^{-1}
π	absolute pressure, N/m^2
π_0	pressure at sparger, N/m^2
ρ	density, kg/m^3
$\phi(t)$	surface age distribution function
σ	surface tension, kg/s^2
τ	shear stress, N/m^2

SUBSCRIPT

c	continuous phase or liquid phase
d	dispersed phase or gas phase
G	gas phase
g	gas phase
L	liquid phase

9.13 PROBLEMS

9.1 Derive the relationship between the overall mass transfer coefficient for gas phase K_G and the individual mass-transfer coefficients, k_L and k_G. How can this relationship be simplified for sparingly soluble gases?

9.2 Prove that Eq. (9.27) is the same with Eq. (9.28), and Eq. (9.30) is the same with Eq. (9.31).

9.3 The power consumption by impeller P in geometrically similar fermenters is a function of the diameter D_I and speed N of impeller, density ρ and viscosity μ. of liquid, and acceleration due to gravity g. Determine appropriate dimensionless parameters that can relate the power consumption by applying dimensional analysis using the Buckingham-Pi theorem.

9.4 A cylindrical tank (1.22 m diameter) is filled with water to an operating level equal to the tank diameter. The tank is equipped with four equally spaced baffles, the width of which is one tenth of the tank diameter. The tank is agitated with a 0.36 m diameter, flat-blade disk turbine. The impeller rotational speed is 4.43 rps. The air enters through an open-ended tube situated below the impeller and its volumetric flow rate is 0.0217 m^3/s at 1.08 atm and 25°C. Calculate:

   a. power requirement
   b. gas hold-up
   c. Sauter-mean diameter
   d. mterfacial area
   e. volumetric mass-transfer coefficient

   Compare the preceding calculated results with those experimental values reported by Chandrasekharan and Calderbank (1981): P_m = 2282 W; H= 0.086; $k_L a$ = 0.0823 s^{-1}.

9.5 Estimate the volumetric mass-transfer coefficient k_L a for the gas-liquid contactor described in Problem .4 by using a correlation for k_L a and compare the result with the experimental value.

9.6 The power consumption by an agitator in an unbaffled vessel can be expressed as

$$\frac{P_{mo}}{\rho N^3 D_I{}^5} = f\left(\frac{\rho N D_I{}^2}{\mu}\right)$$

Can you determine the power consumption and impeller speed of a 1,000-gallon fermenter based on findings of the optimum condition from a one-gallon vessel by using the same fluid system? Is your conclusion reasonable? Why or why not?

9.7 The optimum agitation speed for the cultivation of plant cells in a 3-L fermenter equipped with four baffles was found to be 150 rpm.

a. What should be the impeller speed of a geometrically similar 1,000 L fermenter if you scale up based on the same power consumption per unit volume.

b. When the impeller speed suggested by part (a) was employed for the cultivation of a 1,000-L fermenter, the cells do not seem to grow well due to the high shear generated by the impeller even though the impeller speed is lower than 150 rpm of the model system. This may be due to the higher impeller tip speed which is proportional to ND_I. Is this true? Justify your answer with the ratio of the impeller tip speed of the prototype to the model fermenter.

c. If you use the impeller tip speed as the criteria for the scale-up, what will be the impeller speed of the prototype fermenter?

9.8 The typical oxygen demand for yeast cells growing on hydrocarbon is about 3 g per g of dry cell. Design a 10-L stirred fermenter (the diameter and height of fermenter, the type and diameter of impeller) and determine its operating condition (impeller speed and aeration rate) in order to meet the oxygen demand during the peak growth period with the growth rate of 0.5 g dry cell per liter per hour. You can assume that the physical properties of the medium is the same as pure water. You are free to make additional assumptions in order to design the required fermenter.

9.14 REFERENCES

Akita, K. and F. Yoshida, "Gas Holdup and Volumetric Mass Transfer Coefficient in Bubble Column," *I&EC Proc. Des. Dev.* **12** (1973): 76-80.

Akita, K. and F. Yoshida, "Bubble Size, Interfacial Area, and Liquid-Phase Mass Transfer Coefficient in Bubble Columns," *I&EC Proc. Des. Dev.* **13** (1974):84-91.

Botton, R., D. Cosserat, and J. C. Charpentier, "Operating Zone and Scale Up of Mechanically Stirred Gas-Liquid Reactors," *Chem. Eng. Sci.* **35** (1980):82-89.

Bowen, R. L., "Unraveling the Mysteries of Shear-sensitive Mixing Systems," *Chem. Eng.* 9 (1986):55-63.

Calderbank, P. H., "Physical Rate Processes in Industrial Fermentation, Part I: Interfacial Area in Gas-Liquid Contacting with Mechanical Agitation," *Trans. Instn. Chem. Engrs.* **36** (1958):443-459.

Calderbank, P. H., "Physical Rate Processes in Industrial Fermentation, Part II: Mass Transfer Coefficients in Gas-Liquid Contacting with and without Mechanical Agitation," *Trans. Instn. Chem. Engrs.* **37** (1959):173-185.

Calderbank, P. H. and S. J. R. Jones, "Physical Rate Properties in Industrial Fermentation - Part III. Mass Transfer from Fluids to Solid Particles Suspended in Mixing Vessels," *Trans. Instn. Chem. Engrs.* **39** (1961):363-

368.

Calderbank, P. H. and M. B. Moo-Young, "The Continuous Phase Heat and Mass-Transfer Properties of Dispersions," *Chem. Eng. Sci.* **16** (1961):39-54.

Chandrasekharan, K. and P. H. Calderbank, "Further Observations on the Scale-up of Aerated Mixing Vessels," *Chem. Eng. Sci.* **36** (1981):819- 823.

Charm, S. E. and B. L. Wong, "Enzyme Inactivation with Shearing," *Biotechnol. Bioeng.* **12** (1970): 1103-1109.

Cooper, C. M., G. A. Fernstrom, and S. A. Miller, "Performance of Agitated Gas-Liquid Contactors,"*Ind. Eng. Chem.* **36** (1944):504-509.

Croughan, M. S., J.-F. Hamel, and D. I. C. Wang, "Hydrodynamic Effects on Animal Cells Grown in Microcarrier Cultures," *Biotech. Bioeng.* **29** (1987):130-141.

CRC Handbook of Chemistry and Physics. Cleveland, OH: CRC Press, 1983.

Danckwerts, P. V., "Significance of Liquid-Film Coefficients in Gas Absorption," *Ind. Eng. Chem.* **43** (1951): 1460-1467.

Higbie, R. "The rate of absorption of a pure gas into a still liquid during short periods of exposure," *Tram. AIChE* 31 (1935):365-389.

Hong, P. O. and J. M. Lee, "Unsteady- State Liquid-Liquid Dispersions in Agitated Vessels," *I&EC Proc. Des. Dev.* **22** (1983): 130–135.

Hooker, B. S., J. M. Lee, and G. An, "The Response of Plant Tissue Culture to a High Shear Environment," *Enzyme Microb. Technol.* **11** (1989):484-490.

Lopes De Figueiredo, M. M. and P. H. Calderbank, "The Scale-Up of Aerated Mixing Vessels for Specified Oxygen Dissolution Rates," *Chem. Eng. Sci.* **34** (1979): 1333-1338.

Metzner, A. B. and R. E. Otto, "Agitation of Non-Newtonian Fluids," *AIChE J.* 3(1957):3-10.

Midler, Jr., M. and R. K. Finn, "A Model System for Evaluating Shear in the Design of Stirred Fermentors," *Biotech. Bioeng.* **8** (1966):71-84.

Miller, D. N., "Scale-Up of Agitated Vessels Gas-Liquid Mass Transfer," *AIChE. J.* **20** (1974):442-453.

Nagata, S., *Mixing: Principles and Application,* pp. 59-62. New York, NY: John Wiley & Sons, 1975.

Oldshue, J. Y., "Fermentation Mixing Scale-Up Techniques," *Biotech. Bioeng.* **8** (1966):3-24.

Oldshue, J. Y., "Current Trends in Mixer Scale-up Techniques," in *Mixing of Liquids by Mechanical Agitation* , ed. J. J. Ulbrecht and G. K. Patterson. New York: Gordon and Breach Science Publishers, 1985, pp. 309-342.

Othmer, D. F. and M. S. Thakar, "Correlating Diffusion Coefficients in Liquids," *Ind. Eng. Chem.* **45** (1953):589-593.

Perry, R. H. and C. H. Chilton, *Chemical Engineers' Handbook,* (5th ed.). New York, NY: McGraw-Hill Book Co., 1973.

Prasher, B. D. and G. B. Wills, "Mass Transfer in an Agitated Vessel," *I&EC Proc. Des. Dev.* **12** (1973):35 1-354.

Rushton, J. H., E. W. Costich, and H. J. Everett, "Power Characteristics of

Mixing Impellers, Part II," *Chem. Eng. Prog.* **46** (1950):467-476.

Sideman, S., O. Hortacsu, and J. W. Fulton, "Mass Transfer in Gas-Liquid Contacting System," *Ind Eng. Chem.* **58** (1966):32-47.

Skelland, A. H. P., *DiffmionalMass Transfer,* pp. 56-58. New York, NY: JohnWiley& Sons, 1974.

Skelland, A. H. P. and J. M. Lee, "Drop Size and Continuous Phase Mass Transfer in Agitated Vessels," *AIChE J.* **27** (1981):99-111.

Sridhar, T. and O. E. Potter, "Gas Holdup and Bubble Diameters in Pressurized Gas-Liquid Stirred Vessels," */ & E. C. Fundam.* **19** (1980):21-26.

Taguchi, H. and A. E. Humphrey, "Dynamic Measurement of Volumetric Oxygen Transfer Coefficient in Fermentation Systems," *J. Ferment. 7ec/z.*(Jap.) **44** (1966):881-889.

Van't Riet, K., "Review of Measuring Methods and Results in Nonviscous Gas-Liquid Mass Transfer in Stirred Vessels," *I&EC Proc. Des. Dev.* **18**(1979):357-364.

Vermeulen, T., G. M. Williams, and G. E. Langlois, "Interfacial Area in Liquid-Liquid and Gas-Liquid Agitation," *Chem. Eng. Prog.* **51** (1955):85F-94F.

Wilke, C. R. and P. Chang, "Correlation of Diffusion Coefficients in Dilute Solutions," *AIChE J.* **1** (1955):264-270.

10

Downstream Processing

10.1 INTRODUCTION

After successful fermentation or enzyme reactions, desired products must be separated and purified. This final step is commonly known as downstream processing or bioseparation, which can account for up to 60 percent of the total production costs, excluding the cost of the purchased raw materials (Cliffe, 1988).

The fermentation products can be the cells themselves (biomass), components within the fermentation broth (extracellular), or those trapped in cells (intracellular), examples of which are listed in Table 10.1. As shown in Figure 10.1, if the product of our interest is the cell, cells are separated from the fermentation broth and then washed and dried. In the case of extracellular products, after the cells are separated, products in the dilute aqueous medium need to be recovered and purified. The intracellular products can be released by rupturing the cells and then they can be recovered and purified. The downstream processing for enzyme reactions will be similar to the process for extracellular products.

Table 10.1 Examples of Bioprocessing Products and Their Typical Concentrations

Types	Products	Concentration
Cell itself	baker's yeast, single cell protein	30 g/L
Extracellular	alcohols, organic acids, amino acids	100 g/L
Extracellular	enzymes, antibiotics	20 g/L
Intracellular	recombinant DNA proteins	10 g/L

Bioseparation processes make use of many separation techniques commonly used in the chemical process industries. However, bioseparations have distinct characteristics that are not common in the traditional separations of chemical processes. Some of the unique characteristics of bioseparation products can be listed as follows:

1. The products are in dilute concentration in an aqueous medium.

2. The products are usually temperature sensitive.

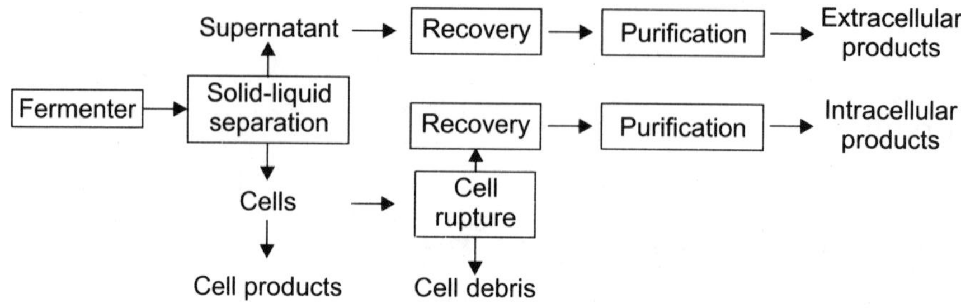

Fig. 10.1 Major process steps in downstream processing.

3. There is a great variety of products to be separated.
4. The products can be intracellular, often as insoluble inclusion bodies.
5. The physical and chemical properties of products are similar to contaminants.
6. Extremely high purity and homogeneity may be needed for human health care.

These characteristics of bioseparations products limit the use of many traditional separation technologies and also require the development of new methods.

10.2 SOLID-LIQUID SEPARATION

The first step in downstream processing is the separation of msolubles from the fermentation broth. The selection of a separation technique depends on the characteristics of solids and the liquid medium.

The solid particles to be separated are mainly cellular mass with the specific gravity of about 1.05 to 1.1, which is not much greater than that of the broth. Shapes of the particles may be spheres, ellipsoids, rods, filaments, or flocculents. Typical sizes for various cells vary widely such as,

bacterial cells: 0.5 – 1 μm
yeast cells: 1 – 7 7 μm
fungal hyphae: 5–15 μm in diameter, 50 – 500 μm in length
suspension animal cells (lymphocytes): 10 – 20 μm suspension
plant cells: 20 – 40 μm

The separation of solid particles from the fermentation broth can be accomplished by filtration or centrifugation.

10.2.1 Filtration

Filtration separates particles by forcing the fluid through a filtering medium on which solids are deposited. Filtration can be divided into several categories depending on the filtering medium used, the range of particle sizes removed, the pressure differences, and the principles of the filtration, such as conventional filtration, microfiltration, ultrafiltration, and reverse osmosis. In this section we limit our discussion on the conventional filtration which involves large particles $(d_p > 10$ μm). This technique is effective for dilute suspension of large and rigid particles. The other filtration techniques are discussed in the purification section at the end of this chapter.

A wide variety of filters are available for the cell recovery. There are generally two major types of filters: pressure and vacuum filters. The detailed descriptions of those filter units can be found in *Chemical Engineers Handbook* (Perry and Chilton, 1973). The two types of filters most used for cell recovery are the filter press and rotary drum filters. A filter press is often employed for the small-scale separation of bacteria and fungi from broths. For large-scale filtration, rotary drum filters are usually used. A common filter medium is the cloth filter made of canvas, wool, synthetic fabrics, metal, or glass fiber.

Assuming the laminar flow across the filter, the rate of filtration (dV_f/dt) can be expressed as a function of pressure drop Δ_p by the modified D'Arcy's equation as (Belter et al., p. 22, 1988),

$$\frac{1}{A}\frac{dV_f}{dt} = \frac{\Delta p}{\mu(L/K)} \tag{10.1}$$

where A is the area of filtering surface, k is D'Arcy's filter cake permeability, and L is the thickness of the filter cake. Eq. (10.1) states that the filtration rate is proportional to the pressure drop and inversely proportional to the filtration resistance L/k, which is the sum of the resistance by the filter medium R_M and that by the cake R_C,

$$\frac{L}{k} = R_M + R_C \tag{10.2}$$

The value of R_M is relatively small compared to R_C, which is proportional to the filtrate volume as

$$\frac{L}{k} \approx R_C = \frac{\alpha \rho_c V_f}{A} \tag{10.3}$$

where α is the specific cake resistance and ρ_c is the mass of cake solids per unit volume of filtrate.

Substitution of Eq. (10.3) into Eq. (10.1) and integration of the resultant equation yields

$$\frac{L}{k} \approx R_C = \frac{\alpha \rho_c V_f}{A} \tag{10.4}$$

which can be used to relate the pressure drop to time when the filtration rate is constant.

Eq. (10.4) shows that the higher the viscosity of a solution, the longer it takes to filter a given amount of solution. The increase of the cake compressibility increases the filtration resistance α, therefore, increases the difficulty of the filtration. Fermentation beers and other biological solutions show non-Newtonian behaviors with high viscosity and form highly compressible filter cakes. This is especially true with mycelial microorganisms. Therefore, biological feeds may require pretreatments such as:

1. Heating to denature proteins

2. Addition of electrolytes to promote coagulation and flocculation

3. Addition of filter aids (diatomaceous earths or perlites) to increase the porosity and to reduce the compressibility of cakes

Although the filtration theory reviewed helps us to understand the relationship between the operating parameters and physical conditions, it is rarely used as a sole basis for design of a filter system because the filtering characteristics must always be determined on the actual slurry in questions (Perry and Chilton, p. **19-57,** 1973). The sizing and scale-up of production-scale filter are usually done by performing *filtration leaf test* procedures (Dorr-Oliver, 1972). From the test data, the filtration capacity can be expressed either in dry pounds of filter cakes per square foot of filter area per hour (lb/ft^2 hr) or in gallons of filtrate per square foot per hour (gal/ft^2 hr), depending on whether the valuable product is the cake or filtrate, respectively. Since the filtration leaf test is performed under ideal conditions, it is common to apply a safety factor to allow the production variation of operating conditions. Safety factor commonly used is 0.65 (Dorr-Oliver, 1972). The following example shows the use of the filtration leaf data for the sizing of production-scale filter unit.

Example 10.1

Filtration leaf test results indicate that the filtration rate of a protein product is 50 dry $lbs/(ft^2$ hr). What size production filter would be required to obtain 100 dry lbs of filter cake per hour?

Solution:

Let's apply safety factor of 0.65, then

$$50 \; \frac{lbs}{ft^2 hr} \times 0.65 = 32.5 \; \frac{lbs}{ft^2 hr}$$

The size of the filter required is,

$$\frac{100\ \text{lbs/hr}}{32.5\ \text{lbs}\,(\text{ft}^2\text{hr})} = 3.08\ \text{ft}^2$$

10.2.2 Centrifugation

Centrifugation is an alternative method when the filtration is ineffective, such as in the case of small particles. Centrifugation requires more expensive equipment than filtration and typically cannot be scaled to the same capacity as filtration equipment.

Two basic types of large-scale centrifuges are the tubular and the disk centrifuge as shown schematically in Figure 10.2. The *tubular centrifuge* consists of a hollow cylindrical rotating element in a stationary casing. The suspension is usually fed through the bottom and clarified liquid is removed from the top leaving the solid deposit on the bowl's wall. The accumulated solids are recovered manually from the bowl. A typical tubular centrifuge has a bowl of 2 to 5 in. in diameter and 9 to 30 in. in height with maximum rotating speed of 15,000 to 50,000 rpm (Ambler, 1979).

The *disk centrifuge* is the type of centrifuge used most often for bioseparations. It has the advantage of continuous operation. It consists of a short, wide bowl 8 to 20 in. in diameter that turns on a vertical axis (Figure 10.2b). The closely spaced cone-shaped discs in the bowl decrease the distance that a suspended particle has to be moved to be captured on the surface and increases the collection efficiencies. In operation, feed liquid enters the bowl at the bottom, flows into the channels and upward past the disks. Solid particles are thrown outward and the clear liquid flows toward the center of the bowl and is discharged through an annular slit. The collected solids can be removed intermittently or continuously.

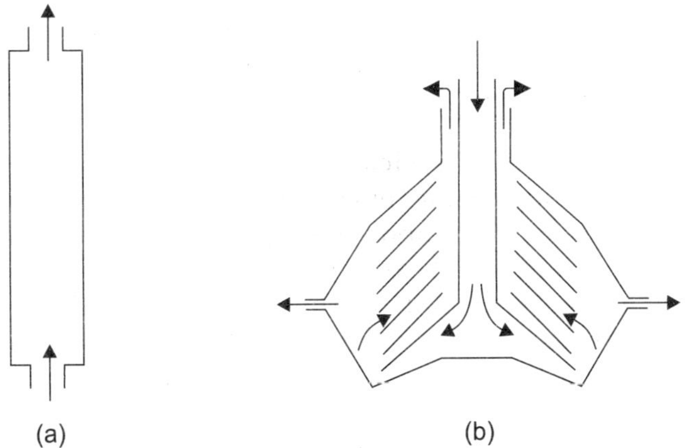

(a)   (b)

Fig. 10.2 Two basic types of centrifuges: (a) tubular and (b) disk.

When a suspension is allowed to stand, the particles will settle slowly under the influence of gravity due to the density difference between the solid and surrounding fluid, a process known as sedimentation. The velocity of a particle increases as it falls and reaches a constant velocity (known as terminal velocity) at which

$$\text{Weight force} - \text{Buoyancy force} = \text{Drag force}$$

The expression for the terminal velocity can be derived from the balance of the forces acting on a particle as,

$$v_t = \frac{d_p^2 (\rho_s - \rho) a}{18\mu} \tag{10.5}$$

which is applicable when the Reynolds number is less than 1 ($d_p v \rho / u < 1$) which is always the case for biological solutes. In the case of a sedimentation process, the acceleration term in Eq. (10.5) is equal to the acceleration due to gravity. Due to the small difference in density between the cells and the broth, simple settling can take a long time unless cells are large or the cells form a large aggregate.

Under the centrifugal force, the acceleration term in Eq. (10.5) becomes

$$a = \omega^2 r \tag{10.6}$$

where ω is the angular velocity and r is the radial distance from the center of a centrifuge to a particle. Therefore, the increase in acceleration by the centrifugal force speeds up the settling process.

10.3 CELL RUPTURE®

Once the cellular materials are separated, those with intracellular proteins need to be ruptured to release their products. Disruption of cellular materials is usually difficult because of the strength of the cell walls and the high osmotic pressure inside. The cell rupture techniques have to be very powerful, but they must be mild enough so that desired components are not damaged. Cells can be ruptured by physical, chemical, or biological methods.

Physical Methods Physical methods include mechanical disruption by milling, homogenization, or ultrasonication. Typical high-speed *bead mills* are composed of a grinding chamber filled with glass or steel beads which are agitated with disks or impellers mounted on a motor-driven shaft. The efficiency of cell disruption in a bead mill depends on the concentration of the cells, the amount and size of beads, and the type and rotation speed of the agitator. The optimum wet solid content for the cell suspension for a bead mill is typically somewhere between 30 percent to 60 percent by volume. The amount of beads in the chamber is 70 percent to 90 percent by volume (Keshavarz et al., 1987). Small beads are generally more efficient, but the smaller the bead, the harder it is to separate them from ground solids. Cell disruption by bead mills is inexpensive and can be operated on a large scale.

A *high-pressure homogenizer* is a positive displacement pump with an adjustable orifice valve. It is one of the most widely used methods for large-scale cell disruption. The pump pressurizes the cell suspension (about 50 percent wet cell concentration) to approximately 400 to 550 bar and then rapidly releases it through a special discharge valve, creating very high shear rates. Refrigerated cooling to 4 or 5°C is necessary to compensate for the heat generated during the adiabatic compression and the homogenization steps (Cliffe, 1988).

An *ultrasonicator* generates sound waves above 16 kHz, which causes pressure fluctuations to form oscillating bubbles that implode violently generating shock waves. Cell disruption by an ultrasonicator is effective with most cell suspensions and is widely used in the laboratory. However, it is impractical to be used on a large scale due to its high operating cost.

Chemical Methods   Chemical methods of cell rupture include the treatment of cells with detergents (surfactants), alkalis, organic solvents, or by osmotic shock. The use of chemical methods requires that the product be insensitive to the harsh environment created by the chemicals. After cell disruption, the chemicals must be easily separable or they must be compatible with the products.

Surfactants disrupt the cell wall by solubilizing the lipids in the wall. Sodium dodecylsulfate (SDS), sodium sulfonate, Triton X-100, and sodium taurocholate are examples of the surfactants often employed in the laboratory. *Alkali treatment* disrupts the cell walls in a number of ways including the saponification of lipids. Alkali treatment is inexpensive and effective, but it is so harsh that it may denature the protein products. *Organic solvents* such as toluene can also rupture the cell wall by penetrating the cell wall lipids, swelling the wall. When red blood cells or a number of other animal cells are dumped into pure water, the cells can swell and burst due to the *osmotic flow* of water into the cells.

Biological Methods   Enzymatic digestion of the cell wall is a good example of biological cell disruption. It is an effective method that is also very selective and gentle, but its high cost makes it impractical to be used for large-scale operations.

10.4   RECOVERY

After solid and liquid are separated (and cells are disrupted in the case of mtracellular products), we obtain a dilute aqueous solution, from which products have to be recovered (or concentrated) and purified. Recovery and purification cannot be divided clearly because some techniques are employed by both. However, among the many separation processes used for the recovery step, extraction and adsorption can be exclusively categorized as recovery and are explained in this section.

10.4.1 Extraction

Extraction is the process of separating the constituents (solutes) of a liquid solution (feed) by contact with another insoluble liquid (solvent). During the liquid-liquid contact, the solutes will be distributed differently between the two liquid phases. By choosing a suitable solvent, you can selectively extract the desired products out of the feed solution into the solvent phase. After the extraction is completed, the solvent-rich phase is called the *extract* and the residual liquid from which solute has been removed is called the *raffinate*.

The effectiveness of a solvent can be measured by the distribution coefficient K,

$$K = \frac{y^*}{x} \qquad (10.7)$$

where y^* is the mass fraction of the solute in the extract phase at equilibrium and x is that in the raffinate phase. A system with a large K value is desirable, since it requires less solvent and produces a more concentrated extract phase. The K value can be increased by selecting the optimum pH. For example, the K value for penicillin F between the liquid phases composed of water and amyl acetate is 32 at pH 4.0, however, it drops to 0.06 at pH 6.0 (Belter et al., p. 102, 1988). The addition of countenons such as acetate and butyrate can also increase the K value dramatically.

Another measure for the effectiveness of solvent is the selectivity β,

$$\beta = \frac{y_a^*/x_a}{y_b^*/x_b} \qquad (10.8)$$

which is the ratio of the distribution coefficient of solute a and that of solute b. For all useful extraction operations, the selectivity must be larger than 1. Other requirements for a good solvent include the mutual insolubility of the two liquid systems, easy recoverability, a large density difference between the two phases, nontoxicity, and low cost.

The most widely used extractor is the mixer-settler that is a cylindrical vessel with one or several agitators. The vessel is usually equipped with four equally spaced baffles to prevent the vortex formation. A radial-flow type agitator such as a flat-bladed turbine is better for the extraction process than an axial-flow type such as a propeller because the relatively low-density difference between the two liquid systems does not require a strong downward flow toward the bottom of the vessel.

The mixer and settler can be combined in one vessel or separated. In the combined case, the agitator is turned off after extraction so that the two phases can separate. For continuous operation, the mixer and settler are separated. A settler is nothing but a large tank.

Single-stage Extraction

Extraction can be carried out as a single-stage operation either in batch or continuous mode. Figure 10.3 shows the flow diagram for a single-stage mixer-settler. For single-stage extraction design, we need to estimate the concentration of solute in the extract and in the raffinate with a given input condition. The overall material balance for the mixer-settler yields

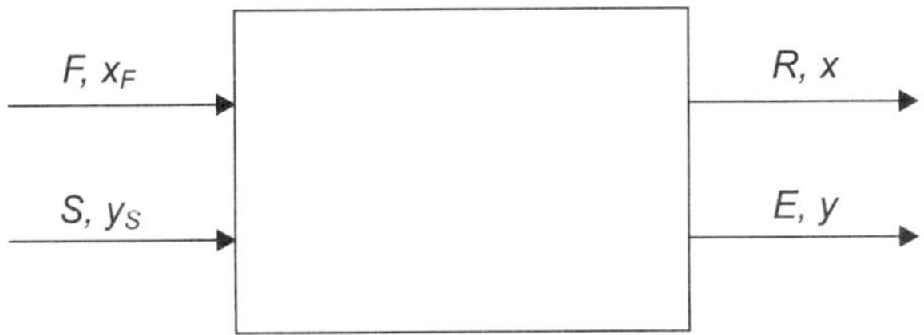

Fig 10.3 Flow diagram for the single-stage mixer settler.

$$F + S = R - E \tag{10.9}$$

and the material balance for a solute gives

$$Fx_F + Sy_S = Rx + Ey \tag{10.10}$$

If we assume the mixer-settler has reached an equilibrium,

$$y = Kx \tag{10.11}$$

If inlet conditions (F, S, x_F, and y_S) are known, we have four unknown variables (R, E, x, and y). However, since we have only three equations, we need additional information to be able to solve for the unknown variables, which are the equilibrium data of the ternary system: solute, solvent, and diluent, which are usually described graphically in triangular coordinates (Treybal, 1980).

In bioseparations, the solute concentration in the feed is usually low, therefore, the changes of the extract and the raffinate streams are negligible. We can assume that $F = R$ and $S = E$. In that case, we have only two unknown variables, x and y, so we can solve it to obtain,

$$y = \frac{Rx_F + Ey_S}{R/K + E} \tag{10.12}$$

If pure solvent is used, $J_S = 0$ and Eq. (10.12) can be simplified to

$$y = \frac{Kx_F}{1 + K(E/R)} \tag{10.13}$$

which shows that the solute concentration in the extract depends on the distribution coefficient K and the ratio of the extract and the rafmate E/R.

Another parameter that is important in extraction is the recoverability of solute. Even though Y is high, if the recoverability of solute is low, the extractor cannot be regarded as an efficient unit. The recoverability γ can be defined as the fraction of the solute recovered by an extractor as

$$\gamma = \frac{Ey}{Rx_F} \tag{10.14}$$

Substitution of Eq. (10.13) into Eq. (10.14) gives

$$\gamma = \frac{K(E/R)}{1 + K(E/R)} \tag{10.15}$$

With a given system of constant K, a decrease of E/R increases y, but decreases γ. Therefore, an optimum operation condition has to be determined based on the various factors affecting the economy of the separation processes, such as the value of products, equipment costs, and operating costs. It is interesting to note that Y depends on the ratio E/R, but not on the values of E and R. Can we increase E and R indefinitely to maintain the same y as long as E/R is constant for a continuous extractor? The answer is "no." We should remember that Eq. (10.13) is based on the assumption that the extractor is in equilibrium. Therefore, the increase of E and R will shorten the residence time; as a result, the extractor cannot be operated in equilibrium and y will decrease.

Multistage Extraction

The optimum recoverability of solute by a single-stage extractor is determined based on the K value and the E/R ratio. To increase the recoverability further, several extractors can be connected crosscurrently or countercurrently.

Multistage Crosscurrent Extraction In multistage crosscurrent extraction, the raffinate is successively contacted with fresh solvent which can be done continuously or in batch (Figure 10.4). With the given flow rates of the feed and the solvent streams, the concentrations of the extracts and the raffmates can be estimated in the same manner as a single-stage extractor.

Multistage Countercurrent Extraction The two phases can be contacted countercurrently as shown in Figure 10.5. The countercurrent contact is more efficient than the crosscurrent contact due to the good distributions of the concentration driving force for the mass transfer between stages. The disadvantage of this scheme is that it cannot be operated in a batch mode. The overall material balance for the countercurrent extractor with N_p stages gives

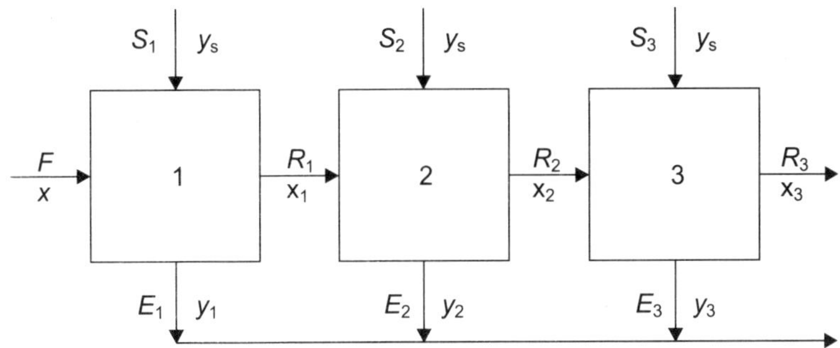

Fig. **10.4** Flow diagram for the crosscurrent multistage mixer-settler.

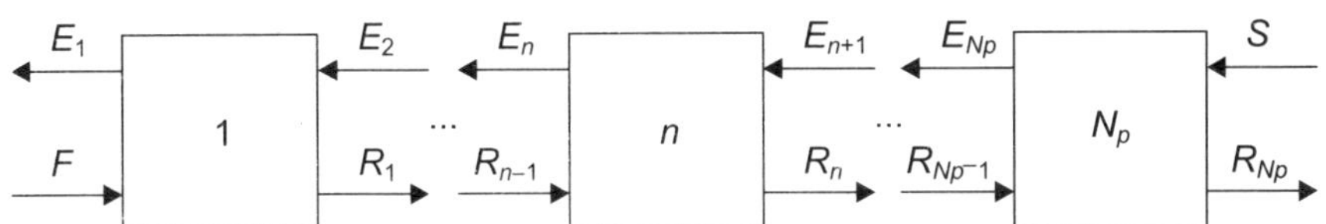

Fig 10.5 Flow diagram for the countercurrent multistage mixer-settler.

$$F + S = R_{N_p} + E_1 \tag{10.16}$$

The material balance for the solute yields

$$Fx_F + Sy_S = R_{N_p}x_{N_p} + E_1y_1 \tag{10.17}$$

If the flow rates of the raffmate and extract streams are constant, $F = R_{N_p} = R$ and $S = E_1 = E$. Therefore, Eq. (10.17) can be rearranged to

$$y_1 = \frac{R}{E}\left(x_F - x_{N_p}\right) + y_S \tag{10.18}$$

which can be used to estimate y_1 with the known feed conditions ($F = R$, $S = E$, x_F, y_S) and the separation required (x_{N_p}). The concentrations of the intermediate streams can be estimated from the solute balance for the first n stages as

$$y_{n+1} = y_1 - \frac{R}{E}(x_F - x_n) \tag{10.19}$$

which is operating line for this extractor. For a linear equilibrium relationship $y_n = Kx_n$,

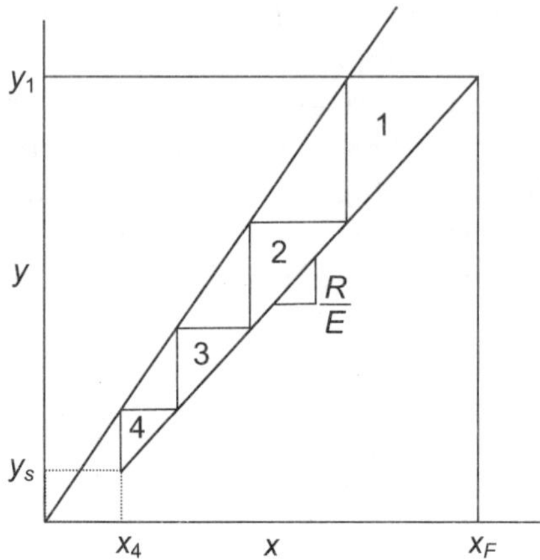

Fig. 10.6   The graphical estimation of the number of ideal stages for multistage countercurrent extraction.

$$y_{n+1} = y_1 - \frac{R}{E}\left(x_F - \frac{y_n}{K}\right) \tag{10.20}$$

which can be used to calculate y_{n+1} with a known value of y_n starting from y_1. The total number of ideal stages required to accomplish a certain job (N_p) can be estimated by continuing this calculation for y_{n+1} until $y_{n+1} \leq y_s$. Another convenient equation that can be used to calculate N_p is (Treybal, 1980),

$$N_p = \frac{\ln\left[\dfrac{x_F - y_S/K}{x_{N_p} - y_S/K}\left(1 - \dfrac{R}{KE}\right) + \dfrac{R}{KE}\right]}{\ln\left(\dfrac{KE}{R}\right)} \tag{10.21}$$

The number of ideal stages and intermediate concentrations of the extract and the raffmate phases can be estimated graphically as shown in Figure 10. 6. The equilibrium line ($y^* = Kx$) and the operating line, Eq. (10.19), are plotted. The operating line passes a point (x_F, y_1) and has the slope of R/E. Then, steps are drawn as shown in Figure 10.6 until the final step passes a point (x_{N_p}, y_S).

Example 10.2

Penicillin F is to be extracted from the clarified fermentation beer by using pure amyl acetate as solvent at pH 4.0. The distribution coefficient K of the system was found to be 32. The initial concentration of penicillin in the feed is 400 mg/L. The flow rates of the feed and the solvent streams are 500 L/hr and 30 L/hr, respectively.

   a. How many ideal stages (countercurrent contact) are required to recover 97 percent of penicillin in the feed?

   b. If you use three countercurrent stages, what will be the percent recovery?

   c. If you use three crosscurrent stages with equal solvent flow rate (10 L/hr each), what will be the percent recovery?

Solution:

   a. The mass fraction of the solute in the feed can be calculated by assuming the density of the feed is the same as water,

$$x_F = \frac{400}{10^6} = 4.0 \times 10^{-4}$$

Since pure solvent was used, $y_S = 0$. For 97 percent recovery of the solute,

$$x_{N_p} = x_F (1 - 0.97) = 1.2 \times 10^{-5}$$

and

$$\frac{R}{KE} = \frac{500}{32(30)} = 0.52$$

From Eq. (10.21)

$$N_p = \frac{\ln\left[\dfrac{4.0 \times 10^{-4}}{1.2 \times 10^{-5}} (1 - 0.52) + 0.52 \right]}{\ln(1/0.52)} = 4.3$$

The total number of ideal stages is 4.3.

   b. Since $N_p = 3$, from Eq. (10.21),

$$3 = \frac{\ln\left[\dfrac{4.0 \times 10^{-4}}{x_3} (1 - 0.52) + 0.52 \right]}{\ln(1/0.52)}$$

Solving the preceding equation for x_3 yields

$$x_3 = 2.9 \times 10^{-5}$$

Therefore, the percent recovery of the solute is

$$\gamma = \frac{x_F - x_3}{x_F} \times 100 = 93\%$$

c. The concentrations of the extract and the raffinate leaving the first extractor can be calculated from Eq. (10.13) and the equilibrium relationship as

$$y = \frac{Kx_F}{1 + K(E/R)} = \frac{32(40 \times 10^{-4})}{1 + 32(10/500)} = 7.8 \times 10^{-3}$$

$$x_1 = \frac{y_1}{K} = 2.4 \times 10^{-4}$$

Those streams leaving the second and third stages can be calculated from Eq. (10.13) by replacing x_F with x_1 and x_2, respectively.

$$y_2 = 4.8 \times 10^{-3}, \quad x_2 = 1.5 \times 10^{-4}$$

$$y_3 = 2.9 \times 10^{-3}, \quad x_2 = 9.1 \times 10^{-4}$$

(T^5 Therefore, the percent recovery of the solute is

$$\gamma = \frac{x_F - x_3}{x_F} \times 100 = 77\%$$

which is lower than that of the countercurrent contact.

10.4.2 Adsorption

A specific substance in solution can be adsorbed selectively by certain solids as a result of either the physical or the chemical interactions between the molecules of the solid and substance adsorbed. Since the adsorption is very selective while the solute loading on the solid surface is limited, adsorption is an effective method for separation of very dilutely dissolved substances.

Adsorption can be classified into three categories: conventional adsorption, ion exchange, and affinity adsorption.

Conventional Adsorption   Conventional adsorption is a reversible process as the result of intermolecular forces of attraction (van der Waals forces) between the molecules of the solid and the substance adsorbed. Among many adsorbents available in industry, activated carbons are most often used in bioseparations. Activated carbons are made by mixing organic matter (such as fruit pits and sawdust) with inorganic substances (such as calcium chloride), and the carbonizing and activating with hot air or steam. They are often used to eliminate the trace quantities of impurities from potable water or processing liquids. However, they are also being used for the isolation of valuable products from the fermentation broth by adsorption and then recovery by elution.

Ion-exchange   An ion-exchange resin is composed of three basic components: a polymeric network (such as styrene-divmyl-benzene, acrylate, methacrylate, plyamine, cellulose, or dextran), ionic functional groups which are permanently attached to this network (which may be amons or cations), and counterions. For example, strong-acid cation-exchange resins contain fixed charges like $-SO_3^-$, which are prepared by sulfonating the benzene rings in the polymer. Fixed ionic sites in the resin are balanced by a like number of charged ions of the opposite charge (countenons) to maintain electrical neutrality.

When a solution containing positively charged ions contacts the cation-exchange resins, the positively charged ions will replace the counter ions and be separated from the solution which is the principle of the separation by ion-exchange. To illustrate the principle of ion-exchange, consider the reaction of the hydrogen form of a cation-exchange resin (H^+R^-) with sodium hydroxide

$$H^+R^- + Na^+OH^- \rightarrow Na^+R^- + HOH \tag{10.22}$$

In this reaction, the positively charged ions in the solution (Na^+) replaced the counterion (H^+). In the case of a hydroxide-form anion-exchange resin (R^+OH^-) with hydrochloric acid,

$$R^+OH^- + H^+Cl^- \rightarrow R^+Cl^- + HOH \tag{10.23}$$

where the negatively charged ions in the solution (Cl^-) replaced the counter ion (OH^-).

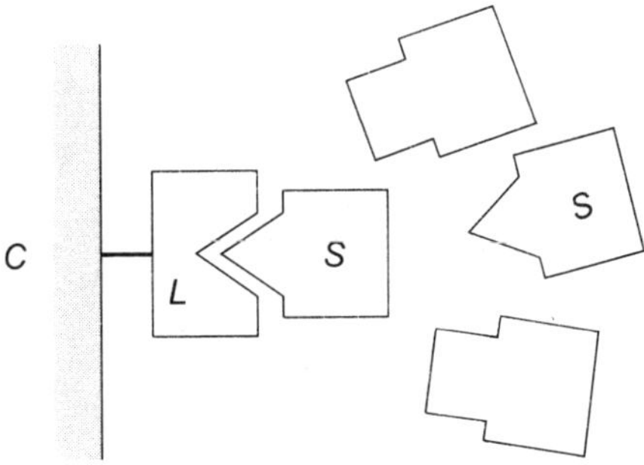

Fig. 10.7   Principle of affinity adsorption. The specific ligand (L), such as enzyme inhibitor and antigen, is immobilized on a water-insoluble carrier (C). A solute (S) can react selectively with the affinity ligand.

Affinity Adsorption   Affinity adsorption is based on the chemical interaction between a solute and a ligand which is attached to the surface of the carrier particle by covalent or ionic bonds. The principle is illustrated in Figure 10.7.

Examples of ligand/solute pairs are (Schmidt-Kastner and Gölker, 1987):

enzyme inhibitors — enzymes
antigens or haptens — antibodies
glycoprotems or polysacchandes — lectins
complementary base sequences – nucleic acids
receptors — hormons
carrier proteins — vitamins

Affinity adsorption offers high selectivity in many bioseparations. However, the high cost of the resin is a major disadvantage and limits its industrial use.

Adsorption Isotherm

For separation by adsorption, adsorption capacity is often the most important parameter because it determines how much adsorbent is required to accomplish a certain task. For the adsorption of a variety of antibiotics, steroids, and hormons, the adsorption isotherm relating the amount of solute bound to solid and that in solvent can be described by the empirical Freundlich equation.

$$Y^* = bX^c \tag{10.24}$$

where Y^* is the equilibrium value of the mass of solute adsorbed per mass of adsorbent and X is the mass fraction of solute in the diluent phase in solute-free basis.[1] The constants b and c are determined experimentally by plotting log Y^* versus log X

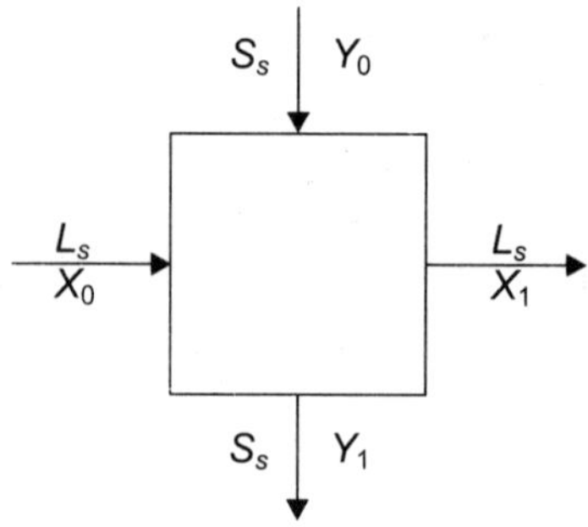

Fig. 10.8　Flow diagram for the single-stage contact filtration unit.

Another correlation often employed to correlate adsorption data for proteins is the Langmuir isotherm,

---

[1] Since the concentration in the adsorbent phase Y is expressed as a solute-free basis, the concentration in the diluent phase X is also expressed on a solute-free basis for uniformity. However, for dilute solutions, the difference between X (mass fraction on a solute-free basis) and x (mass fraction in solution including solute) is negligible.

$$Y^* = \frac{Y_{max}X}{K_L + X}$$ (10.25)

where Y_{max} is the maximum amount of solute adsorbed per mass of adsorbent, and K_L is a constant.

Adsorption Operation

Adsorption can be carried out by stagewise or continuous-contacting methods. The stagewise operation of adsorption is called contact filtration because the liquid and the solid are contacted in a mixer and then the solid is separated from the solution by filtration.

Single-stage Adsorption   Contact filtration can be carried out as a single-stage operation (Figure 10.8) either in a batch or a continuous mode. If we assume that the amount of liquid retained with the solid is negligible, the material balance for a solute gives

$$S_S\,(Y_1 - Y_0) = L_S(X_0 - X_1)$$ (10.26)

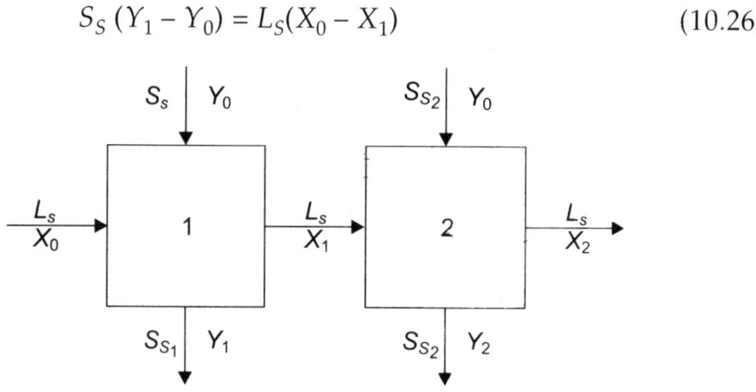

Fig. 10.9   Two-stage crosscurrent adsorption

or $$\frac{S_S}{L_S} = \frac{X_0 - X_1}{Y_1 - Y_0}$$ (10.27)

where S_S and L_S are the mass of adsorbent and of diluent, respectively. If we also assume that the contact filtration unit is operated at equilibrium and that the equilibrium relationship can be approximated by the Freundlich or Langmuir equation, the adsorbent-solution ratio (S_S/L_S) can be estimated from Eq. (10.27) after substituting in the equilibrium relationship for Y_1.

Multistage Crosscurrent Adsorption   The amount of adsorbent required for the separation of a given amount of solute can be decreased by employing multistage crosscurrent contact, which is usually operated in batch mode, although continuous operation is also possible. The required adsorbent is further decreased by increasing the number of stages. However, it is seldom economical to use more than two stages due to the increased operating costs of the additional stages.

If the equilibrium isotherm can be expressed by the Freundlich equation and fresh adsorbent is used in each stage ($Y_0 = 0$), the total amount of adsorbent used for a two-stage crosscurrent adsorption unit (Figure 10.9) is

$$\frac{S_{S_1} + S_{S_2}}{L_S} = \frac{X_0 - X_1}{bX_1^c} + \frac{X_1 - X_2}{bX_2^c} \tag{10.28}$$

For the minimum total adsorbent, f

$$\frac{d}{dX_1}\left(\frac{S_{S_1} + S_{S_2}}{L_S}\right) = 0 \tag{10.29}$$

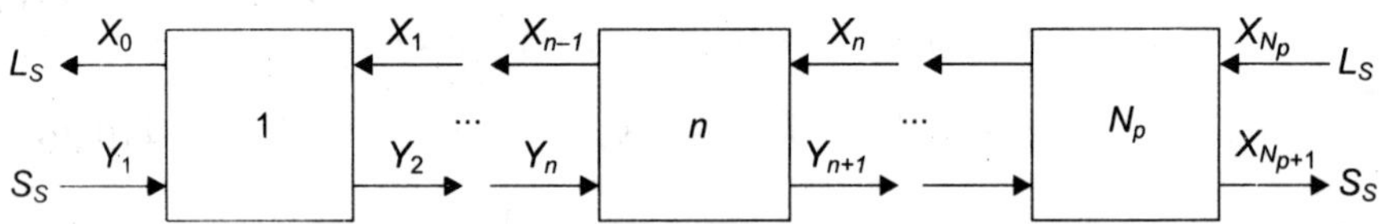

Fig. 10.10 Multistage countercurrent adsorption

Therefore, from Eqs. (10.28)and (10.29), the relationship for the intermediate concentrations for the minimum total adsorbent can be obtained as

$$\left(\frac{X_1}{X_2}\right)^c - c\left(\frac{X_0}{X_1}\right) = 1 - c \tag{10.30}$$

Multistage Countercurrent Adsorption The economy of the absorbent can be even further improved by multistage countercurrent adsorption (Figure 10.10). Like extraction, the disadvantage of this method is that it cannot be operated in a batch mode. Since the equilibrium line is not usually linear for adsorption, it is more convenient to use a graphical method to determine the number of ideal stages required or the intermediate concentrations rather than using an analytical method. The solute balance about the n stages is

$$L_S X_0 + S_S Y_{n+1} = L_S X_n + S_S Y_1 \tag{10.31}$$

which is the operating line.

The number of ideal stages can be estimated by drawing steps between the equilibrium curve and the operating line as shown in Figure 10.11. The intermediate concentrations can be also read from the figure.

Example 10.3

We are planning to isolate an antibiotic from a fermentation broth (10 L) by using activated carbon. The concentration of the antibiotic is 1.1×10^{-6} g per g water. Ninety-five percent of the antibiotic in solution needs to be recovered. Absorption studies at the operating condition gave the following result.

$X \times 10^6$ (g solute/g water)	0.1	0.3	0.6	0.9	1.2
$Y^* \times 10^3$ (g solute/g carbon)	1.3	1.7	2.3	2.4	2.6

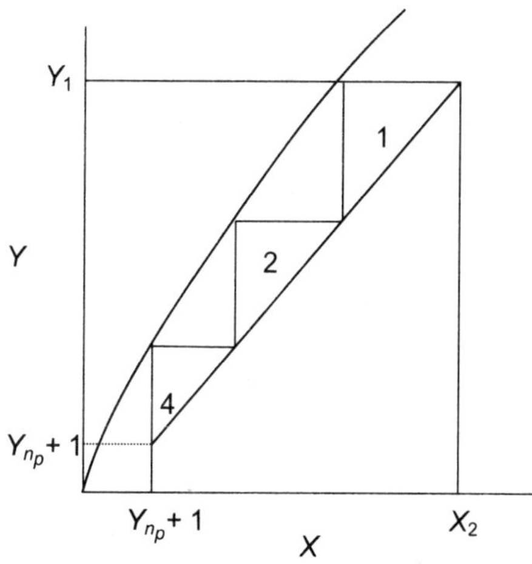

Fig. 10.11 The graphical estimation of the number of ideal stages for countercurrent multistage adsorption.

a. Which isotherm (Freundlich, Langmuir, or linear) fits the data best? Determine the experimental constants for the isotherm.

b. If we use single-stage contact filtration, how much absorbent is needed?

c. If we use a two-stage crosscurrent unit, what is the minimum total amount of absorbent? How much absorbent is introduced to each stage?

d. If we use a two-stage countercurrent unit, how much total absorbent is needed?

Solution:

a. In order to test which isotherm fits the data best, linear regression analysis was carried out for the adsorption data. The results are as follows:The mass fraction of the solute in the feed can be calculated by assuming the density of the feed is the same as water,

Isotherm	Plot	Slope	Intercept	Corr. Coeff.
Freundlich	ln X vs. ln Y^*	0.29	−0.87	0.99
Langmuir	$1/X$ vs. $1/Y^*$	4.1×10^{-5}	381	0.97
Linear	X vs. Y^*	1160	0.0013	0.95

The results show that any isotherm can fit the data very well. If we choose the the Freundlich equation, which fits the adsorption data the best, the slope and the intercept of the log$X$ versus logY^* plot yields the equation,

$$Y^* = 0.13X^{0.29}$$

b. The concentration of the inlet stream was given as $X_0 = 1.1 \times 10^{-6}$. Since 95 percent of the solute in the input liquid stream is recovered, the concentration of the outlet stream will be

$$X_2 = 0.05X_0 = 5.5 \times 10^{-8}$$

From Eq. (10.27) and the Freundlich equation, Eq. (10.24), the amount of the absorbent required is

$$S_S = L_S \frac{X_0 - X_2}{bX_2^c} = 9.6 \text{ g}$$

c. The minimum usage of absorbent is realized when Eq. (10.30) is satisfied,

$$\left(\frac{X}{5.5 \times 10^{-8}}\right)^{0.29} - \frac{0.29(1.1 \times 10^{-6})}{X_1} = 1 - 0.29$$

The solution of the preceding equation yields $X_1 = 3.3 \times 10^{-7}$. Therefore, the amount of adsorbent introduced into the first and the second stages

$$S_{S_1} = L_S \frac{X_0 - X_1}{bX_1^c} = 4.2 \text{ g}$$

$$S_{S_2} = L_S \frac{X_1 - X_2}{bX_2^c} = 2.5 \text{ g}$$

The total amount of adsorbent is 6.7g which is significantly less than what was calculated for single-stage contact filtration. It should be noted that the minimum amount of adsorbent usage is realized when a larger amount of the adsorbent is fed to the first stage followed by the smaller amount to the second stage.

d. For two-stage countercurrent adsorption, a solute balance of both stages yields

$$S_S = L_S \frac{X_0 - X_2}{bX_1^c}$$

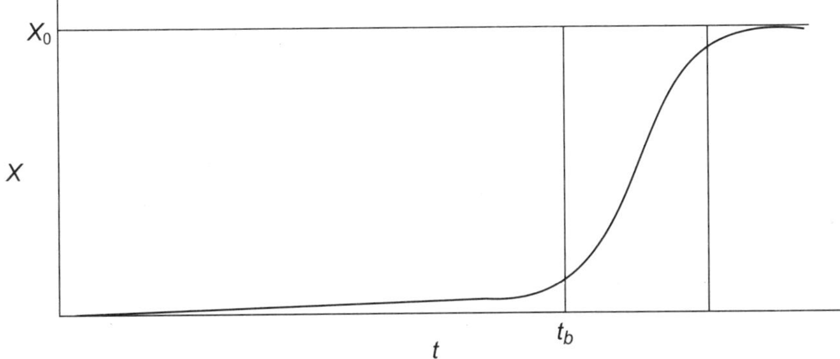

Fig. 10.12 Breakthrough curve for fixed-bed adsorption.

A solute balance for the second stage only yields

$$S_S = L_S \frac{X_1 - X_2}{bX_2^c}$$

Equating the two equations results in

$$\frac{X_0 - X_2}{bX_1^c} = \frac{X_1 - X_2}{bX_2^c}$$

Inserting the given values and solving will yield $X_1 = 5.8 \times 10^{-7}$. The amount of adsorbent required is

$$S_s = L_S \frac{X_1 - X_2}{bX_1^c} = 4.9 \text{ g}$$

Therefore, further reduction of adsorbent usage can be realized with the countercurrent system.

Fixed-Bed Adsorption The adsorbent can be packed in a cylindrical column (or tube in small scale) and the diluent passed through for the selective adsorption of the solute of interest. It is one of the most common methods for adsorption in both laboratory or industrial scale separations.

One simple way to analyze the performance of a fixed-bed adsorber is to prepare a breakthrough curve (Figure 10.12) by measuring the solute concentration of the effluent as a function of time. As the solution enters the column, most of the solute will be adsorbed in the uppermost layer of solid. The adsorption front will move downward as the adsorption progresses. The solute concentration of the effluent will be virtually free of solute until the adsorption front reaches the bottom of the bed, and then the concentration will start to rise sharply. At this point (t_b in Figure 10.12), known as the break point, the whole adsorbent is saturated with solute except the adsorption layer at the very bottom. If the solution continues to flow, the solute concentration will continue to increase until it is the same as the inlet concentration (X_0). Normally, the adsorption is stopped at the break point and the adsorbed material is eluted by washing the bed with solvent at conditions suitable for desorption. The amount of solute lost with the diluent can be estimated from the graphical integration of the breakthrough curve.

10.5 PURIFICATION

After a product is recovered or isolated, it may need to be purified further. The purification can be accomplished by numerous methods such as precipitation, chromatography, electrophoresis, and ultrafiltration.

10.5.1 Precipitation

Precipitation is widely used for the recovery of proteins or antibiotics. It can be induced by the addition of salts, organic solvents, or heat.

The addition of salt precipitates proteins because the protein solubility is reduced markedly by the increase of salt concentration in solution. Precipitation is effective and relatively inexpensive. It causes little denaturation. Ammonium sulphate is the most commonly employed salt. The disadvantage of ammonium sulphate is that it is difficult to remove from the precipitated protein. Sodium sulfate is an alternative but it has to be used at 35–40°C for adequate solubility.

The use of organic solvents at lower temperature (less than –5°C) precipitates proteins by decreasing the dielectric constant of the solution. The organic solvents should be miscible in water to be effective. Acetone, ethanol, methanol, and isopropanol are commonly employed organic solvents.

Heating also can promote the precipitation of proteins by denaturing them. It is often used to eliminate unwanted proteins in a solution. However, the selective denaturation without harming the desired protein products can be difficult and often risky.

10.5.2 Chromatography

Chromatographic processes always involve a mobile phase and a stationary phase. The mobile phase is the solution containing solutes to be separated and the eluent that carries the solution through the stationary phase. The stationary phase can be adsorbent, ion-exchange resin, porous solid, or gel, which are usually packed in a cylindrical column.

A solution composed of several solutes is injected at one end of the column and the eluent carries the solution through the stationary phase to the other end of the column. Each solute in the original solution moves at a rate proportional to its relative affinity for the stationary phase and comes out at the end of the column as a separated band. Depending on the type of adsorbent or the nature of the solute-adsorbent interaction, they are called adsorption, ion-exchange, affinity, or gel filtration chromatography.

The basic principles of adsorption, ion-exchange, and affinity resins have been explained in the previous section on adsorption. Chromatography is similar to adsorption because both involve the interaction between solute and solid matrix. However, they are different in a sense that chromatography is based on the different rate of movement of the solute in the column, while adsorption is based on the separation of one solute from other constituencies by being captured on the adsorbent.

Gel filtration chromatography uses gel such as cross-linked dextrans, polyacrylamide, and agarose as the stationary phase. The gel contains pores of defined sizes. The smaller molecules penetrate the pore structure to a greater extent and therefore have a longer retention time than the larger molecules.

Chromatography has been used to purify proteins and peptides from complex liquid solutions extensively on the laboratory scale. When chromatographic processes can be scaled up from laboratory scale to a larger scale, the diameter rather than the height of a column should be increased. In this way only the flow rate is increased without changing the retention time and other operating parameters. For large-scale chromatography, a stationary phase has to be selected which has the required mechanical strength and chemical stability properties.

For designing and operating a chromatograph, the yield and purity of the separated products are the two important parameters to be controlled, which can be estimated as follows: The yield of a solute i collected between two times t_1 and t_2 during chromatographic separation can be calculated as (Belter et al., p.188, 1988)

$$\text{yield} = \frac{\text{amount of solute } i \text{ eluted}}{\text{total amount of solute}} = \frac{\int_{t_1}^{t_2} y_i \, L_S dt}{\int_0^\infty y_i \, L_S dt} \tag{10.32}$$

where y_i is the concentration of a solute i and L_S is the solvent flow rate. The change of the solute concentration can be approximated by the Gaussian form.

$$y_i = y_{i_{max}} \exp\left[-\frac{(t/t_{max} - 1)^2}{2\sigma^2}\right] \tag{10.33}$$

where $y_{i_{max}}$ is the maximum concentration, t_{max} is the time for the maximum, and σ is the standard deviation of the peak.

Substitution of Eq. (10.33) into Eq. (10.32) and integration results in

$$\text{yield} = \frac{1}{2}\left\{\text{erf}\left[-\frac{t_1/t_{max} - 1}{\sqrt{2}\sigma}\right] - \text{erf}\left[-\frac{t_2/t_{max} - 1}{\sqrt{2}\sigma}\right]\right\} \tag{10.34}$$

The purity of a solute i collected between two times t_1 and t_2 during chromatographic separation can be calculated as

$$\text{purity} = \frac{\text{amount of solute } i \text{ eluted}}{\text{amount of impurity eluted}} = \frac{\int_{t_1}^{t_2} y_i \, L_s dt}{\sum_j \int_{t_1}^{t_2} y_j \, L_s dt} \tag{10.35}$$

10.5.3   Electrophoresis

When a mixture of solutes is placed in an electrical field, the positively charged species are attracted to the anode and the negatively charged ones to the cathode. The separation of charged species based on their specific migration rates in an electrical field is termed *electrophoresis*.

It is one of the most effective methods of protein separation and characterization. The chief advantages of this method are that it can be performed under very mild conditions and it has high resolving power, resulting in the clear separation of similarly charged protein molecules. However, in order to use this separation technique, the components of a mixture must have an ionic form, and each component must possess a different net charge.

When a charged particle q moves with a steady velocity u_E through a fluid under an electric field E_f, the electrostatic force on the particle is counter-balanced by the fluid drag on the particle. For globular proteins, the drag force can be approximated from Stokes law. Therefore, the balance is

so that

$$qE_f = 3\pi\mu d_p u_E \tag{10.36}$$

$$u_E = \frac{qE_f}{2\pi\mu d_p} \tag{10.37}$$

where the steady velocity of particle u_E is known as electrophoretic mobility, which is proportional to the electrical field strength and electric charge, but is inversely proportional to the viscosity of the liquid medium and the particle size.

However, the increase of the electric field strength increases not only the electrophoretic mobility, but also the mixing caused by the gas release from the electrodes and the free convection caused by electrical heating, which are two major problems in designing electrophoresis setup. To minimize these problems, electrodes are usually located in separate compartments and the electrophoresis is performed in a gel, known as gel electrophoresis. Among various gels available, polyacrylamide gels are most commonly used. They are thermostable, transparent, durable, relatively inert chemically, nomonic, and easily prepared. The disadvantage of gel chromatography is that setting up the gel is tedious and time consuming and large-scale operation is not possible.

10.5.4 Membrane Separation

As we discussed earlier for the solid-liquid separation technique, filtration separates particles by forcing the fluid through a filtering medium on which solids are deposited. The conventional filtration involves the separation of large particles ($d_p > 10$ μm) by using canvas, synthetic fabrics, or glass fiber as filter medium.

We can use the same filtration principle for the separation of small particles down to small size of the molecular level by using polymeric membranes. Depending upon the size range of the particles separated, membrane separation processes can be classified into three categories: microfiltration, ultrafiltration, and reverse osmosis, the major differences of which are summarized in Table 10.2.

Table 10.2 Pressure Driven Membrane Separation Processes
(Lonsdale, 1982)

ProcessSize	Cutoff	Molecular Wt. Cutoff	Pressure Drop (psi)	Material Retained
Micro-filtration	0.02 – 10μm		10	Suspended material including microorganisms
Ultra-filtration	10 – 200 Å	3 00 – 300k	10 – 100	Biologicals, colloids, macromoleculres
Reverse Osmosis	1 – 10Å	< 300	100 – 800	All suspended and dissolved material

Microfiltration refers to the separation of suspended material such as bacteria by using a membrane with pore sizes of 0.02 to 10 μm. If the pore size is further decreased so that the separation can be achieved at the molecular level, it is called *ultrafiltration*. The typical molecular weight cutoff for ultrafiltration is in the range of 300 to 300,000, which is in the size range of about 10 to 200Å. *Reverse osmosis* is a process to separate virtually all suspended and dissolved material from their solution by applying high pressure (100 to 800 psi) to reverse the osmotic flow of water across a semipermeable membrane. The molecular weight cutoff is less than 300 (1 to 10Å). Protein products are in the range of the molecular cutoff for ultrafiltration.

Membrane filters are usually made by casting a polymer solution on a surface and then gelling the liquid film slowly by exposing it to humid air. The size of the pores in the membranes can be varied by altering the composition of the casting solution or the gelation condition. Another common technique is to irradiate a thin polymeric film in a field of α-particles and then chemically etch the film to produce well-defined pores.

The solvent flux across the membrane J is proportional to the applied force, which is equal to the applied pressure reduced by the osmotic pressure $\Delta\Pi$ as (Belter et al., p. 255, 1988),

$$J = L_p(\Delta p - (\sigma\Delta\Pi)) \tag{10.38}$$

where L_p is the membrane permeability and σ is the reflection coefficient ($a = 1$ if the membrane rejects all solute and $\sigma = 0$ if the solute passes through membrane freely). The solvent flux is the volume of solvent per unit membrane area per time, which is the same as the solvent velocity. If the solution is dilute the osmotic pressure is

$$\Delta\Pi = R'\, TC^* \tag{10.39}$$

where C^* is the solute concentration at the membrane surface.

One of the major problems with the membrane separation technique is concentration polarization, which is the accumulation of solute molecules or particles on the membrane surface. Concentration polarization reduces the flow through the membrane and may cause membrane fouling. It can be minimized by using cross-flow type filter and maintaining a high liquid flow parallel to the membrane surface by recirculating the liquid through thin channels. There are four basic membrane modules commonly employed, plate-and-frame, shell-and-tube, spiral-wound, and hollow-fiber membrane. All of these modules are cross-flow type. Concentration polarization can be further reduced by prefiltering the solution, by reducing the flow rate per unit membrane surface area, or by backwashmg periodically.

The flux of solute from the bulk of the solution to the membrane surface is equal to the solute concentration times solvent flux CJ. At steady state, it will be countered by the molecular diffusion of the solute away from the membrane surface as

$$CJ = -D\frac{dC}{dz} \tag{10.40}$$

which can be solved with the boundary conditions,

$$C = C^* \qquad\qquad \text{at } z = 0 \tag{10.41}$$
$$C = C_b \qquad\qquad \text{at } z = \delta$$

to yield

$$J = \frac{D}{\delta}\,\ln\frac{C^*}{C_b} \tag{10.42}$$

where δ is the thickness of the laminar sublayer near the membrane surface within which concentration polarization is assumed to be confined. The concentration ratio, C^*/C_b is known as the polarization modulus.

Example 10.4

Figure 10.13 shows a typical batch ultrafiltration setup. As the solution is pumped through the filter unit, the permeate is collected and the retentate is recycled. The volume of the solution reduces with time and the solute concentration increases. Develop a correlation for the time required to reduce the solution volume from V_0 to V. Assume that the concentration polarization is negligible. Also assume that the membrane totally rejects the solute.

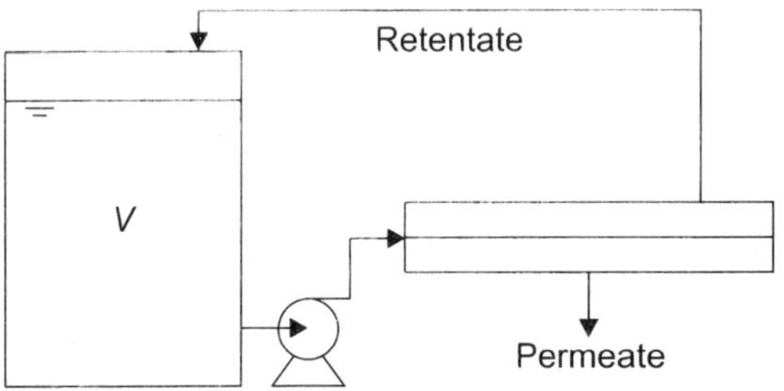

Fig. 10.13   Typical batch ultrafiltration setup.

Solution:

The decrease of the solution volume is equal to the membrane area A times the solvent flux across the membrane J.

$$\frac{dV}{dt} = -AJ \tag{10.43}$$

Substitution of Eq. (10.38) into Eq. (10.43) yields

$$\frac{dV}{dt} = AL_p\,(\Delta p - \sigma\Delta\Pi) \tag{10.44}$$

Since the solute is completely rejected by the membrane, $\sigma = 1$. Therefore, substituting Eq. (10.39) into Eq. (10.44) and rearranging gives

$$\frac{dV}{dt} = -AL_p\Delta p \left(1 - \frac{R'TC^*}{\Delta p}\right) \tag{10.45}$$

If the concentration polarization is negligible, the solute concentration on the membrane surface C^* is equal to the bulk concentration C_b. Since the membrane rejects all solute, the total solute $(m' = C_bV)$ is constant. Therefore,

$$\frac{dV}{dt} = -AL_p\Delta p \left(1 - \frac{R'TC^*}{\Delta p}\right) \tag{10.46}$$

The integration of the Eq. (10.46) with the initial condition, $V = V_0$ at $t = 0$, and the rearrangement for t yields

$$t = \frac{1}{AL_p\Delta p}\left[(V_0 - V) + \left(\frac{R'Tm'}{\Delta p}\right)\ln\left(\frac{V_0 - R'Tm'/\Delta p}{V - R'Tm'/\Delta p}\right)\right] \tag{10.47}$$

10.6   NOMENCLATURE

A	area of filtering surface, m^2
a	acceleration, m/s^2
C	solute concentration, kg/m^3
C_b	solute concentration at the bulk solution, kg/m^3
C^*	solute concentration at the membrane surface, kg/m^3
D	molecular diffusivity, m^2/s
d_p	particle diameter, m
E	mass (or mass flow rate) of extract phase, kg (or kg/s) E_F electric field strength, NC^{-1}
F	mass (or mass flow rate) of feed stream, kg (or kg/s)
J	solvent flux, the volume of solvent per unit membrane area per time, m/s

k	D'Arcy's filter cake permeability, m^2
Δp	pressure drop across the filter and the cake or pressure drop across the membrane, Pa
K	distribution coefficient, dimensionless
k_L	constant for the Langmuir isotherm, dimensionless
L_p	permeability, m/Ns
L_S	mass (or mass flow rate) of diluent or solvent, kg (or kg/s)
m, n	constants for the Freundlich equation,
m'	total amount of solute, kg
q	electric charge of a particle, C
R_C	resistance coefficient for filter cake, 1/m
R_M	resistance coefficient for filter medium, 1/m
R	mass (or mass flow rate) of raffinate stream, kg (or kg/s)
R'	gas constant, gas constant, J/kmol K
r	radial distance from the center of a centrifuge to a particle, m
S	mass (or mass flow rate) of solvent, kg (or kg/s)
S_S	mass (or mass flow rate) of adsorbent, kg (or kg/s)
s	compressibility coefficient, dimensionless
T	temperature
t	time, s
u_E	electrophoretic mobility, m/s
V	solution volume, m^3
V_f	volume of filtrate collected, m^3
vt	terminal velocity, m/s
X	mass fraction of solute in solution in solute-free basis, dimensionless
x	mass fraction of solute in solution (in raffinate phase for extraction), dimensionless
Y	mass of solute per mass of adsorbent, dimensionless
y	mass fraction of solute in extract phase (for extraction) or in adsorbent (for adsorption), dimensionless
α	specific cake resistance, m/kg
α_0	specific cake resistance for incompressible cake, m/kg

β	selectivity of extraction, dimensionless
γ	recoverability of solute, dimensionless
δ	thickness of the laminar sublayer, m
μ	fluid viscosity, kg/m s
$\Delta\Pi$	osmotic pressure, N/m^2
ρ	density, kg/m^3
ρ_c	mass of cake solids per unit volume of filtrate, kg/m^3
ρ_s	density of solid particle, kg/m^3
σ	reflection coefficient, dimensionless
ω	angular velocity, s^{-1}

10.7 PROBLEMS

10.1 An antibiotic, cycloheximide, is to be extracted from the clarified fermentation beer by using methylene chloride as solvent. The distribution coefficient K is 23. The initial concentration of cycloheximide in the feed is 150 mg/L. The recovered solvent containing 5 mg/L of cycloheximide is being used with the flow rate 1 m^3 /hr. The required recovery of the antibiotic is 98 percent.

   a. If you use four countercurrent stages, how much feed can you process per hour (F)?

   b. If you use four crosscurrent stages with equal solvent flow rate (0.25 m^3/hr), how much feed can you process per hour (F)?

10.2 Aspartic acid needs to be isolated from a fermentation broth. The initial concentration of aspartic acid in the solution is 1.0×10^{-3} g/mL. We need to recover 98 percent of the aspartic acid. The amount of aspartic solution is 1 m^3. The isotherm for the adsorption of aspartic acid into an anion exchanger (Duolite A162) is given as follows (Cowan et al., 1987)

$X \times 10^3$ (g solute/g water)	0.02	0.03	0.05	0.07	0.17	0.55	1.1	1.5	2.0
Y^* (g solute/g adsorbent)	0.05	0.07	0.12	0.11	0.19	0.21	0.20	0.23	0.24

   a. Which isotherm, Freundlich or Langmuir, fits the data better? Determine the experimental constants for the isotherm.

   b. If we use single-stage contact filtration, how much Duolite A162 is needed?

   c. If we use two-stage crosscurrent filtration with an equal amount of adsorbent for each stage, what is the total amount of absorbent?

    d. If we use two-stage countercurrent filtration, how much total absorbent is needed?

10.3 Monoclonal antibodies (MoAb) will be separated from the supernatant of a mammalian culture by using DEAE-Sephacel as the adsorbent. The supernatant contains 100 μg/mL of MoAb. Adsorption follows the Langmuir isotherm with Y'_{max} =116 mg of MoAb per mL of adsorbent (settled volume) and K'_L = 0.5 × 10^{-3} mg of MoAb per mL of supernatant (Desai et al, 1987). The amount of supernatant is 1 L. If you use 2 mL of DEAE-Sephacel for a single-stage, what percent of the MoAb will be recovered?

10.8 REFERENCES

Ambler, C. M., "Centrifugation," in Handbook of Separation Techniques for Chemical Engineers, ed. P. A. Schweitzer. New York, NY: McGraw-Hill Book Co., 1979, pp. 4.55–4.84.

Belter, P. A., E. L. Cussler, and W. Hu, *Bioseparations: Downstream Processing for Biotechnology.* New York, NY: John Wiley & Sons, 1988.

Cain, C. W., "Filtration Theory," in *Handbook of Separation Techniques for Chemical Engineers,* ed. P. A. Schweitzer. New York, NY: McGraw-Hill Book Co., 1979, pp. 4.3--4.8.

Cliffe, K., "Downstream Processing," in *Biotechnology for Engineers,* ed. A. Scragg. Chichester, England: Ellis Horwood Ltd., 1988. pp. 302–321.

Cowan, G. H., I. S. Gosling, and W. P. Sweetenham, "Modelling for Scale-Up and Optimisation of Packed-bed Columns in Adsorption and Chromatography," in *Separations for Biotechnology.*, eds. M. S. Verrall and M. J. Hudson. Chichester, England: Ellis Horwood Ltd., 1987. pp. 152–175.

Desai, M. A., J. G. Huddleston, A. Lyddiatt, J. Rudge, and A. B. Stevens, "Biochemical and Physical Characterisation of a Composite Solid Phase Developed for Large Scale Biochemical Adsorption," in *Separations for Biotechnology,* eds. M. S. Verrall and M. J. Hudson. Chichester, England: Ellis Horwood Ltd., 1987, pp. 200–209.

Dorr-Oliver Inc., *Filtration Leaf Test Procedures.* Stamford, CT: Dorr-Oliver Inc., 1972.

Keshavarz, E., M. Hoare, and P. Dunnill, "Biochemical Engineering Aspects of Cell Disruption," in *Separations for Biotechnology,* eds. M. S. Verrall and M. J. Hudson. Chichester, England: Ellis Horwood Ltd., 1987, pp. 62–79.

Lonsdale, H. K., "The Growth of Membrane Technology," *J. Membrane Sci.* **10** (1982):81–181.

Perry, R. H. and C. H. Chilton, *Chemical Engineers' Handbook,* (5th ed.), pp. 19-57–19-85. New York, NY: McGraw-Hill Book Co., 1973.

Schmidt-Kastner, G. and C. F. Gölker, "Downstream Processing in Biotechnology," in Basic Biotechnology, eds. J. Bu'lock and B. Kristiansen. London, England: Academic Press, 1987, pp. 173–196.

Treybal, R. E., *Mass-Transfer Operations* (3rd ed.), p. 128, pp. 479–488. New York, NY: McGraw-Hill Book Co., 1980.

SUGGESTED READING

Belter, P. A., E. L. Cussler, and W. Hu, *Biosepamtions: Downstream Processing for Biotechnology.* New York, NY: John Wiley & Sons, 1988.

Verrall, M. S. and M. J. Hudson, Eds., *Separations for Biotechnology.* Chichester, England: Ellis Horwood Ltd., 1987.